Individual-based Modeling
and Ecology

PRINCETON SERIES IN THEORETICAL AND COMPUTATIONAL BIOLOGY

Series Editor, Simon A. Levin

Individual-based Modeling and Ecology, by Volker Grimm and Steven F. Railsback
Mathematics in Population Biology, by Horst R. Thieme

GH
3o2
.G75
2005

Individual-based Modeling and Ecology

Volker Grimm

Steven F. Railsback

PRINCETON UNIVERSITY PRESS

PRINCETON AND OXFORD

LIBRARY OF CONGRESS CATALOGING-IN-PUBLICATION DATA

Grimm, Volker, 1958-

    Individual-based modeling and ecology / Volker Grimm ; Steven F. Railsback.

      p. cm. — (Princeton series in theoretical and computational biology)

    Includes bibliographical references and index.

    ISBN 0-691-09665-1 (hardcover : alk. paper) — ISBN 0-691-09666-X (pbk. : alk. paper)

    1. Population biology—Mathematical models.   2. Biotic communities—Mathematical models.   I. Railsback, Steven F. II. Title. III. Series.

QH352.G75 2005

577′.01′5118—dc22   2004022435

British Library Cataloging-in-Publication Data is available

This book has been composed in Times and Stone Sans

Printed on acid-free paper. ∞

pup.princeton.edu

Printed in the United States of America

10 9 8 7 6 5 4 3 2

Grau, teurer Freund, ist alle Theorie,
und Grün des Lebens goldner Baum.

(Gray, dear friend, is all theory,
and green life's golden tree.)

—MEPHISTO TO THE STUDENT, IN *FAUST*
BY JOHANN WOLFGANG VON GOETHE

Contents

Preface

Any new approach to science has to go through its infancy: like a child learning to walk, the first steps of a new approach are exploratory, insecure, and not directed by a clear idea of where the steps might lead—yet these steps are watched with great interest and excitement. Individual-based ecological modeling has been in its infancy for the past two decades. The rapidly growing interest in individual-based models (IBMs) certainly is encouraged by the enormous increases in computing power that now make it practical to simulate large numbers of individuals in virtual populations. However, individual-based modeling has also been fueled by another kind of power that has grown rapidly in recent years: the desire of ecologists to understand natural complexity and how it emerges from the variability and adaptability of individual organisms.

Early advocates of IBMs claimed that a shift in focus from populations to individuals would lead to new fundamental insights and have the potential to unify ecological theory. Indeed, numerous IBMs have demonstrated the potential significance of individual characteristics to population dynamics and ecosystem processes. Even in its infancy, individual-based modeling has changed our understanding of ecological systems. However, we have also learned that a shift in focus from populations to individuals does not *automatically* lead to better and more general ecological theory or to more effective strategies for solving applied problems. Freedom from the constraints of analytical modeling has come at a price: it allows IBMs to be more complex than the analytical models of classical ecology, and therefore harder to develop, understand, and communicate. Moreover, IBMs have made little use of reusable building blocks. Instead, many have been built anew from scratch and are often based on ad hoc assumptions not clearly linked to any theoretical framework. Many modeling projects have not addressed general theoretical questions at all. And many IBM-based projects have been beset with methodological and computational problems.

Scientists sometimes tend to rush to a new approach that promises to solve previously intractable problems, and then revert to familiar techniques as the unanticipated difficulties of the new approach are uncovered. Individual-based modeling may be ripe for a backlash because of the problems encountered to date. However, we believe very strongly that these problems can be overcome, often by adapting existing techniques from other fields of science. Instead of abandoning individual-based modeling because it has been less productive than expected, our goal is to show how this approach can be used productively.

Our primary objective in this book is to provide guidelines for making individual-based modeling more coherent and effective. We provide strategies and methods for optimizing model complexity ("pattern-oriented modeling") and coping with the problems that arise from the complexity of IBMs, and we propose a general, theory-based research program for individual-based modeling. The basic notion underlying this research program is that IBMs allow us to pursue a genuinely new and different way of doing ecology. We call this new approach "individual-based ecology." Individual-based ecology aims at developing theories of the adaptive behavior of individuals, but within the context of their population and environment. Moreover, individual-based ecology aims at understanding the mutual relationship between the adaptive behavior of individuals and system-level properties of populations, communities, and ecosystems. We view individual-based ecology as an approach for *understanding*, not simplifying, the complexity of nature.

Another of our objectives is to further integrate ecology with the general approach to science known as "Complex Adaptive Systems" (CAS). Individual-based ecology can be viewed as a subset of CAS, which attempts to develop general understanding of systems driven by interacting, adaptive agents. Much of the important work on CAS has been conducted (also in the past decade or so) using completely artificial systems, with a focus on identifying general principles instead of on mimicking nature. There is much that ecologists can learn from CAS, yet individual-based ecology is likely to make great contributions to CAS by forcing its theory to confront nature's reality.

This book was written for people interested in individual-based modeling and ecology from many perspectives: students (both undergraduate and graduate) and instructors; researchers using or considering individual-based approaches; empirical ecologists who want to understand how their work could support individual-based analysis; natural resource managers interested in using IBMs to address management problems; and reviewers of research proposals and scientific articles that include IBMs.

Users of agent-based models in other sciences should also find this book valuable. Although we focus on ecology, many of our concepts and techniques are readily applied to the general problem of understanding and modeling how a system's dynamics emerge from the characteristics of its individuals.

This book is a combination of monograph and textbook. We designed it primarily as a reference work. To us, it does not yet seem possible to write a pure textbook on individual-based modeling. In contrast to classical theoretical ecology, which is based on calculus and other established mathematical techniques, the procedures and tools for individual-based modeling are still too experimental to be presented in textbook fashion. The lack of established procedures and tools is also the reason for the heterogeneous character of the book. Textbooks on classical theoretical ecology can stay at the level of theory and strategy, but to get individual-based ecology off the ground we must cover the whole range of scientific activity, offering not only a new approach to theory and a new conceptual framework but also providing the details typically found in the "engine room" where the real work is done. Even though we cannot yet provide step-by-step guidance for building and using IBMs, in this book we gladly get our hands dirty with all aspects of applying IBMs to real-world problems.

Part 1 addresses the modeling process and pattern-oriented modeling. These chapters should be useful to all modelers because the modeling strategies they describe are not restricted to IBMs. The modeling approaches introduced in Part 1 are used throughout the rest of the book but can also stand alone as a guide to starting an ecological modeling project. In part 2 we start focusing on IBMs and individual-based ecology. We address fundamental issues such as What is theory in individual-based ecology? and How do we think about and describe IBMs? Then, in chapter 6 we delve into over thirty example IBMs to illustrate what has already been learned from individual-based approaches. Part 3 is the "engine room," where we provide guidance on the day-to-day work of building and using IBMs: formulating the details, developing software, doing the analysis, and even publishing results. All of these tasks are different when we use IBMs, and individual-based ecology cannot mature until we learn to do them well. In part 4 we return to the more strategic level. Chapter 11 discusses the relation of IBMs and classical analytical models and how the strengths of the two approaches can be combined. Finally, in chapter 12 we provide our outlook on the potential, and limitations, of individual-based modeling and ecology and where we hope this approach will lead us. Along the way we use a number of terms that are collected in the glossary.

Finally, a few words on what this book is *not* about and what it does *not* include are necessary. First, we do not pose individual-based ecology as a replacement for traditional approaches to ecology and ecological modeling. Instead, we present individual-based approaches as a new tool that ecologists can use to tackle new kinds of problems, a supplement to how we currently do ecology. Ecologists whose interests are only in addressing population-level questions with analytical models may find chapters 2, 3, and 11 useful but should otherwise not expect to find much of interest in this book.

Second, we originally hoped that this book would also collect and review specific methods for modeling many individual-level processes, such as feeding, mortality, and competition. Such methods are described throughout the book but only to provide examples. We quickly realized that a comprehensive collection and review would be a very large project and would rapidly be out of date. Instead, we will look for other ways that the community of individual-based ecologists can collaboratively collect and share theory and techniques.

Last, we chose not to address IBMs and research on evolutionary ecology: we do not consider any models of how traits or populations evolve. One reason for this choice is that there already is considerable literature on this topic and incorporating it would make the book much larger and less focused. But a more important reason is our personal interest in solving everyday problems of real systems, using models that are readily testable. Although the problems of evolutionary ecology are fascinating and need to be addressed with individual-based methods, we see them as less urgent than learning to understand the ecosystems we can observe and protect right now.

Acknowledgments

This book is in many ways the product of three organizations that have fostered the development of individual-based modeling. First is Volker Grimm's home organization, the Department of Ecological Modelling (ÖSA) at the UFZ Center for Environmental Research Leipzig-Halle, headed by Christian Wissel, the *spiritus rector* of ecological modeling in Germany. The ÖSA is the medium in which many of the ideas presented in this book were incubated. Second, Steve Railsback is one of many scientists whose involvement with IBMs was initially supported by the North American electric power industry through the "Compensation Mechanisms in Fish Populations" (CompMech) program. The CompMech program was organized by the Electric Power Research Institute, with Jack Mattice and Doug Dixon as program managers; and much of the pioneering work on IBMs was conducted under CompMech funding at Oak Ridge National Laboratory and a number of universities. Third is the Swarm Development Group (SDG; www.swarm.org), a nonprofit organization dedicated to the advancement of agent-based modeling as a tool for understanding complex systems. Having its origins in Chris Langton's Swarm project at the Santa Fe Institute, SDG now maintains software for agent-based modeling and supports a global community of scientists and software developers that use agent- and individual-based approaches. Many of the ideas presented in this book arose from discussions among members of the SDG community and from the Swarm software itself.

Volker Grimm thanks his two mentors Christian Wissel, who made him think about what modeling is, and Janusz Uchmański, who made him think individual-based. The skillful modeling work of Florian Jeltsch and Thorsten Wiegand gave rise to many of the ideas summarized in this book as "pattern-oriented modeling." Particularly acknowledged is also the collaboration with Uta Berger, Norbert Dorndorf, Lorenz Fahse, Hanno Hildenbrandt, Christian Neuert, Christine Rademacher, Hans Thulke, and Tomasz Wyszomirski.

Steve Railsback acknowledges all he has learned, during his foray of the past decade into individual-based modeling, from collaborators including Jim Anderson, Jarl Giske, Tamara Grand, Geir Huse, Rollie Lamberson, Glen Ropella, Kenny Rose, and his many friends at the ÖSA. In particular, Steve's belief in the power of interdisciplinary collaboration arises from his work with two remarkably talented and productive people: ecologist Bret Harvey and programmer Steve Jackson. Steve also thanks the UFZ for providing a research visitor's grant in 2002.

A number of experienced software professionals contributed to chapter 8, both by patiently teaching us the ideas and techniques related in the chapter and by reviewing drafts. We especially acknowledge the contributions of Marcus Daniels, Hanno Hildenbrandt, Steve Jackson, Phil Railsback, Rick Riolo, and Glen Ropella.

Chapter 10 grew out of a panel discussion of publication issues in agent-based simulation, at SwarmFest 2001 (one of the annual Swarm users' conferences). Contributors to this discussion include Gary An, Jim Anderson, Don DeAngelis, Doug Donalson, and Roger Nisbet.

We sincerely thank Uta Berger and Don DeAngelis, who took the time to comment on many chapters and whose encouragement throughout this project was invaluable.

We also thank the following colleagues for providing review comments, expertise, figures, or unpublished materials: Silke Bauer, Ginger Booth, Jim Bown, David Cope, Doug Donalson, Winnie Eckardt, Lorenz Fahse, Gerd Gigerenzer, Jarl Giske, John Goss-Custard, Tamara Grand, Bret Harvey, Charlotte Hemelrijk, Hanno Hildenbrandt, Geir Huse, Andreas Huth, Jane Jepsen, Frederick Knowlton, Stephanie Kramer-Schadt, Philippe Laval, Casey Lu, Michael Müller, Tamara Münkemüller, Chris Mullon, Mary Orland, Michael Potthoff, Christine Rademacher, Björn Reineking, Oswald Schmitz, Espen Strand, Jörg Tews, Karin Ulbrich, Egbert van Nes, Ute Visser, Gideon Wasserberg, Thorsten Wiegand, Eckart Winkler, and the students of Rollie Lamberson's Mathematics 580 course at Humboldt State University.

Finally, we gratefully thank those whose inspiration and patience made this project possible: our parents, Volker's wife Louise and daughter Edda, and Steve's wife Margaret.

PART 1
Modeling

Chapter One

Introduction

The essence of the individual-based approach is the derivation of the properties of ecological systems from the properties of the individuals constituting these systems.

—Adam Łomnicki, 1992

1.1 WHY INDIVIDUAL-BASED MODELING AND ECOLOGY?

Modeling attempts to capture the essence of a system well enough to address specific questions about the system. If the systems we deal with in ecology are populations, communities, and ecosystems, then why should ecological models be based on individuals? One obvious reason is that individuals are the building blocks of ecological systems. The properties and behavior of individuals determine the properties of the systems they compose. But this reason is not sufficient by itself. In physics, the properties of atoms and the way they interact with each other determine the properties of matter, yet most physics questions can be addressed without referring explicitly to atoms.

What is different in ecology? The answer is that in ecology, the individuals are not atoms but living organisms. Individual organisms have properties an atom does not have. Individuals grow and develop, changing in many ways over their life cycle. Individuals reproduce and die, typically persisting for much less time than the systems to which they belong. Because individuals need resources, they modify their environment. Individuals differ from each other, even within the same species and age, so each interacts with its environment in unique ways. Most important, individuals are *adaptive*: all that an individual does—grow, develop, acquire resources, reproduce, interact—depends on its internal and external environments. Individual organisms are adaptive because, in contrast to atoms, organisms have an objective, which is the great master plan of life: they must seek *fitness*, that is, attempt to pass their genes on to future generations. As products of evolution, individuals have traits allowing them to adapt to changes in themselves and their environment in ways that increase fitness.

Fitness-seeking adaptation occurs at the individual level, not (as far as we know) at higher levels. For example, individuals do not adapt their behavior with the objective of maximizing the persistence of their population. But, as ecologists, we are interested in such population-level properties as persistence, resilience, and patterns of abundance over space and time. None of these properties is just the sum of the properties of individuals. Instead, population-level properties *emerge* from the interactions of adaptive individuals with each other and with their environment. Each individual not only adapts to its physical and biotic environment but also makes up part of the biotic environment of other individuals. This circular causality created by adaptive behavior gives rise to emergent properties.

If individuals were not adaptive, or were all the same, or always did the same thing, ecological systems would be much simpler and easier to model. However, such systems would probably never persist for much longer than the lifetime of individuals, much less be resilient or develop distinctive patterns in space and time. Consider, for example, a population in which individuals are all the same, have the same rate of resource intake, and all reproduce at the same time. The logical consequence of this scenario (Uchmański and Grimm 1996) is that, of course, the population will grow exponentially until all resources are consumed and then cease to exist. Or consider a fish school as an example system with emergent properties (Huth and Wissel 1992, 1994; Camazine et al. 2001; section 6.2). The school's properties emerge from how individual fish move with respect to neighboring individuals. If the fish suddenly stopped adjusting to the movement of their neighbors, the school would immediately lose its coherence and cease to exist as a system.

Now, if ecologists are interested in system properties, and these properties emerge from adaptive behavior of individuals, then it becomes clear that understanding the relationship between emergent system properties and adaptive traits of individuals is fundamental to ecology (Levin 1999). Understanding this relationship is the very theme underlying our entire book: how we can use individual-based models (IBMs) to determine the interrelationships between individual traits and system dynamics?

But can we really understand the emergence of system-level properties? Ecological systems are, after all, complex. Even a population of conspecifics is complex because it consists of a multitude of autonomous, adaptive individuals. Communities and ecosystems are even more complex. If an IBM is complex enough to capture the essence of a natural system, is the IBM not as hard to understand as the real system? The answer is no—*if* we have an appropriate research program. But before we outline this program, which we refer to as "individual-based ecology," let us first look at three successful uses of individual-based models. Although these examples (which are described in more detail in chapter 6) address completely different systems and questions, they have common elements that play an important role in individual-based ecology.

1.2   LINKING INDIVIDUAL TRAITS AND SYSTEM COMPLEXITY: THREE EXAMPLES

1.2.1   The Green Woodhoopoe Model

The green woodhoopoe (*Phoeniculus purpureus*) is a socially breeding bird of Africa (du Plessis 1992). The social groups live in territories where only the alpha couple reproduces. The subdominant birds, the "helpers," have two ways to achieve alpha status. Either they wait until they move up to the top of the group's social hierarchy, which may take years, or they undertake scouting forays beyond the borders of their territories to find free territories. Scouting forays are risky because predation, mainly due to raptors, is considerably higher while on a foray. Now the question is: how does a helper decide whether to undertake a scouting foray? We cannot ask the birds how they decide, of course, and we do not have enough data on individual birds and their decisions to answer these questions empirically.

What we have, however, is a field study that compiled a group size distribution over more than ten years (du Plessis 1992). We can thus develop an IBM and test alternative theories of helper decisions by how well the theories reproduce the observed group size distribution (Neuert et al. 1995). These theories represent the internal model used by the birds themselves for seeking fitness. It turned out that a heuristic theory of the helpers' decisions, which takes into account age and social rank, caused the IBM to reproduce the group size distribution at the population level quite well (figure 1.1), whereas alternative theories assuming nonadaptive decisions (e.g. random decisions) did not. Then, after a sufficiently realistic theory of individual behavior was identified, questions addressing the population level could be asked—for example, on the significance of scouting distance to the spatial coherence of the population. It turned out that even very small basic propensities to undertake long-ranging scouting forays allow a continuous spatial distribution to emerge, whereas if the helpers only search for free alpha positions in neighboring territories the population falls apart (section 6.3.1; figure 6.5).

1.2.2   The Beech Forest Model

Without humans, large areas of Middle Europe would be covered by forests dominated by beech (*Fagus silvatica*). Foresters and conservation biologists are therefore keen to establish forests reserves that restore the spatiotemporal dynamics of natural beech forests and to modify silviculture to at least partly restore natural structures. But how large should such protected forests be? And how long would it take to reestablish natural spatiotemporal dynamics? What forces drive these dynamics? What would be practical indicators of naturalness in forest reserves and managed forests? Because of the large spatial

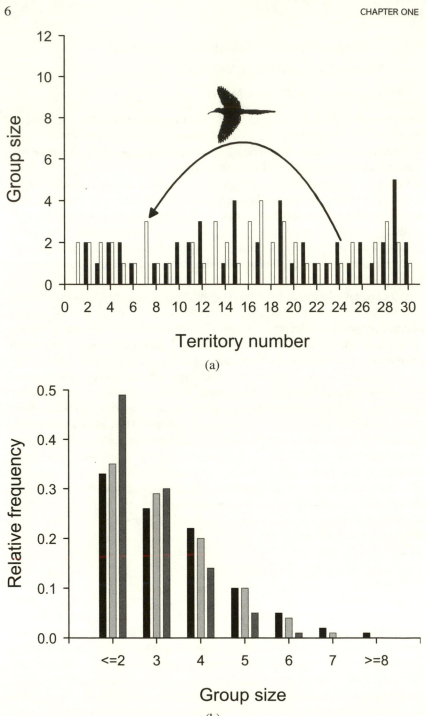

(a)

(b)

and temporal scales involved, modeling is the only way to answer these questions. But how can we find a model structure that is simple enough to be practical while having the resolution to capture essential structures and processes?

What we can do to find the right resolution of the model is use patterns observed at the system level. For example, old-growth beech forests show a mosaic pattern of stands in different developmental phases (Remmert 1991; Wissel 1992a). The model must therefore be spatially explicit with resolution fine enough for the mosaic pattern to emerge. Another pattern is the characteristic vertical structures of the developmental stages (Leibundgut 1993; Korpel 1995). For example, the "optimal stage" is characterized by a closed canopy layer and almost no understory. The model thus has to have a vertical spatial dimension so that vertical structures can emerge (figure 1.2). Within this framework the behavior of individual trees can be described by empirical rules because foresters know quite well how individual growth and mortality depend on the local environment of a tree. Likewise, empirical information is available to define rules for the interaction of individuals in neighboring spatial units.

The model BEFORE (Neuert 1999; Neuert et al. 2001; Rademacher et al. 2001, 2004; section 6.8.3), which was constructed in this way, reproduced the mosaic and vertical patterns. It was so rich in structure and mechanism that it also produced independent predictions regarding aspects of the forest not considered at all during model development and testing. These predictions were about the age structure of the canopy, spatial aspects of this age structure, and the spatial distribution of very old and large trees. All these predictions were in good agreement with observations, considerably increasing the model's credibility. The use of multiple patterns to design the model obviously led to a model that was *structurally realistic*. This realism allowed the addition of model rules to track woody debris, which was not an original objective of the model. Again, the amount and spatial distribution of coarse woody debris in the model forest were in good agreement with observations in natural forest and old forest reserves (Rademacher & Winter 2003). Moreover, by analyzing hypothetical scenarios where, for example, no windfall occurred, it could be shown that storms and windfall have both desynchronizing (at larger scales) and synchronizing (at the local scale) effects on the spatiotemporal dynamics of beech forests. The model

---

Figure 1.1 Individual decisions and population-level phenomena in the woodhoopoe IBM of Neuert et al. (1995). (a) Group size (*black*: male; *white*: female) in thirty linearly arranged territories. Subdominants decide whether to undertake long-distance scouting forays to find vacant alpha positions. (b) Observed group size distribution (*black*), and predicted distributions for the reference model (scouting decisions based on age and social rank; *light gray*); and a model with scouting decisions independent of age and status (*dark gray*). (After Neuert et al. 1995.)

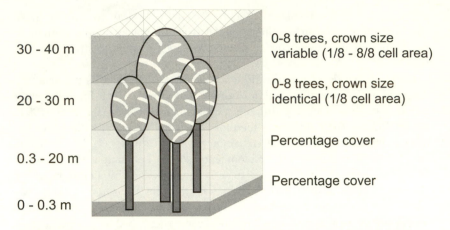

30 - 40 m   0-8 trees, crown size
                 variable (1/8 - 8/8 cell area)

                 0-8 trees, crown size
20 - 30 m   identical (1/8 cell area)

                 Percentage cover
0.3 - 20 m
                 Percentage cover
0 - 0.3 m

Figure 1.2 Vertical structure of the beech forest model BEFORE (Neuert 1999; Rademacher et al. 2001). (Modified from Rademacher et al. 2001.)

thus can be used for answering both applied (conservation, silviculture) and theoretical questions.

1.2.3   The Stream Trout Model

Models have been used to assess the effects of alternative river flow regimes on fish populations at hundreds of dams and water diversions. However, the approach most commonly used for this application, habitat selection modeling, has important limitations (Garshelis 2000; Railsback, Stauffer, and Harvey 2003). IBMs of stream fish have been developed as an alternative to habitat selection modeling (e.g., Van Winkle et al. 1998). These IBMs attempt to capture the important processes determining survival, growth, and reproduction of individual fish, and how these processes are affected by river flow. The trout literature, for example, shows that mortality risks and growth are nonlinear functions of habitat variables (depth, velocity, turbidity, etc.) and fish state (especially size), and that competition among trout resembles a size-based dominance hierarchy. River fish rapidly adapt to changes in habitat and competitive conditions by moving to different habitat, so modeling this adaptive behavior realistically is essential to understanding flow effects.

However, existing foraging theory could not explain the ability of trout to make good trade-offs between growth and risk in selecting habitat under a wide range of conditions. A new theory was developed from the assumption that fish select habitat to maximize the most basic element of fitness, the probability of surviving over a future period (Railsback et al. 1999). This survival probability considers both food intake and predation risk: if food intake is insufficient, the individual will starve over the future period, but if it feeds without regard for risk, it will likely be eaten. The new theory was tested by demonstrating that it could reproduce, in a trout IBM, a wide range of habitat selection patterns observed in real trout populations (Railsback and Harvey 2002).

Once its theory for how trout select habitat was tested, the IBM's ability to reproduce and explain population-level complexities was analyzed (Railsback et al. 2002). The IBM was found to reproduce system-level patterns observed in real trout including self-thinning relationships, "critical periods" of intense density-dependent mortality among juveniles, density-dependence in juvenile size, and effects of habitat complexity on population age structure. Further, the IBM suggested alternatives to the conventional theory behind these patterns (section 6.4.2).

In an example management application, the trout IBM was used to predict the population-level consequences of stream turbidity (Harvey and Railsback 2004). Individual-level laboratory studies have shown that turbidity (cloudiness of the water) reduces both food intake and predation risk. Whereas the population-level consequences of these two offsetting individual-level effects would be very difficult to evaluate empirically, they can be easily predicted using the IBM: over a wide range of parameter values, the negative effects of turbidity on growth outweighed the positive effects on risk.

1.3 INDIVIDUAL-BASED ECOLOGY

The preceding examples address different systems and problems, and the models differ considerably in structure and complexity. What they have in common, however, is the general method of formulating theories about the adaptive behavior of individuals and testing the theories by seeing how well they reproduce, in an IBM, patterns observed at the system level. The main focus may be more on the adaptive behavior of individuals, as in the woodhoopoe and stream trout examples, or on system-level properties, as in the beech forest example, but the general method of developing and using IBMs is the same.

This general method of using IBMs is a distinctly different way of thinking about ecology. We therefore have taken the risk of coining a new term, *individual-based ecology* (IBE), for the approach to studying and modeling ecological systems that this book is about. Classical theoretical ecology, which still has a profound effect on the practice of ecology, usually ignores individuals and their adaptive behavior. In contrast, in IBE higher organizational levels (populations, communities, ecosystems) are viewed as complex systems with properties that arise from the traits and interactions of their lower-level components. Instead of thinking about populations that have birth and death rates that depend only on population size, with IBE we think of systems of individuals whose growth, reproduction, and death is the outcome of adaptive behavior.

Instead of going in the field and only observing population density in various kinds of habitat, with IBE we also study the *processes* by which survival and growth of individuals are affected by habitat (and by other individuals) and how the individuals adapt.

The following are important characteristics of IBE. Many of these have more similarity to interdisciplinary complexity science (e.g., Auyang 1998; Axelrod 1997; Holland 1995, 1998) than to traditional ecology:

1. Systems are understood and modeled as collections of unique individuals. System properties and dynamics arise from the interactions of individuals with their environment and with each other.
2. Individual-based modeling is a primary tool for IBE because it allows us to study the relationship between adaptive behavior and emergent properties.
3. IBE is based on theory. These theories are models of *individual* behavior that are useful for understanding *system* dynamics. Theories are developed from both empirical and theoretical ecology and evaluated using a hypothesis-testing approach. The standard for accepting theories is how well they reproduce observations of real individuals and systems.
4. Observed patterns are a primary kind of information used to test theories and design models and studies. These patterns may be system-level patterns or patterns of individual behavior that arise from the individuals' interactions with the environment and other individuals.
5. Instead of being framed in the concepts of differential calculus, models are framed by complexity concepts such as emergence, adaptation, and fitness.
6. Models are implemented and solved using computer simulation. Software engineering, not differential calculus, is the primary skill needed to implement and "solve" models.
7. Field and laboratory studies are crucial for developing IBE theory. These studies suggest models of individual behavior and identify the patterns used to organize models and test theory.

We do not, of course, propose that IBE replace existing branches of ecology such as behavioral ecology or classical population ecology. We do not claim that IBE is the new "right" way to do ecology and that other approaches should be abandoned. Instead, IBE is a way to apply a variety of concepts, most of them already fundamental to ecology and other sciences, to kinds of problems that cannot be addressed by approaches that look only at individuals or only at populations. IBE is simply a new addition to the toolbox that ecologists can use to solve particular problems.

The IBE research program we develop in this book is based on but differs from earlier statements of the role of IBMs in ecology (e.g., Huston, DeAngelis, and Post 1988). These differences reflect the experience gained during the past twenty years or so, which has demonstrated both the potential and the specific problems of the individual-based approach. To understand how IBE deals with these problems, it is important to understand the problems and the reasons why they were not detected earlier. Therefore, in the following sections we give an overview of the development of the IBM approach, including the research programs outlined by the pioneers of IBM. We explain why it is important to clearly distinguish IBMs from the other modeling approaches which also consider individuals. Then, we briefly summarize the current status of individual-based modeling and list the most important challenges of the approach. Addressing these challenges is another major focus of this book.

1.4 EARLY IBMS AND THEIR RESEARCH PROGRAMS

Modeling the behavior of individuals and testing whether this behavior leads to realistic system-level properties is a natural idea. Therefore, IBMs were developed occasionally, and independently of each other, as soon as adequate computers were available (e.g., Newnham 1964; Kaiser 1974; Thompson et al. 1974; Myers 1976). Two early models were very influential and contributed significantly to the establishment of IBMs: the JABOWA forest model (Botkin et al. 1972) and the fish cohort growth model by DeAngelis, Cox, and Coutant (1980). The purpose of JABOWA was to model succession in mixed-species forests and thereby predict species composition. JABOWA was based on the notion that the interactions that drive forest dynamics are local. JABOWA gave rise to a full pedigree of related models (Liu and Ashton 1995, fig. 1; Shugart 1984; Botkin 1993) and probably is one of the most successful ecological simulation models ever developed. Reasons for the success of JABOWA include that it can be parameterized rather easily and its results are easily tested. (See section 6.7.5 for more details on JABOWA and other forest IBMs.)

The fish cohort model of DeAngelis, Cox, and Coutant (1980) is a similar success story. The model was able to predict accurately the outcome of laboratory experiments in which minute changes in the initial size distribution of the population led to completely different distributions at the end of the growth period. The reasons for this sensitivity to initial conditions were positive feedback mechanisms including asymmetric competition and cannibalism. As did JABOWA, the fish cohort model of DeAngelis et al. gave rise to a full family of fish cohort models (DeAngelis et al. 1990; Van Winkle, Rose, and Chambers 1993).

Interestingly, neither of these two influential models was presented as part of a larger program to develop individual-based modeling as an approach to ecology. Rather, the individual-based approach was chosen for pragmatic reasons: it would simply not have been possible to tackle these problems with classical approaches that ignore individual differences and local interactions. This pragmatic motivation of JABOWA and the fish cohort model is in contrast to the work of two other pioneers of the IBM approach, H. Kaiser and A. Łomnicki, whose attitude may be referred to as "paradigmatic" (Grimm 1999). They explicitly discussed the limitations of the classical ecological modeling paradigm and speculated about a new individual-based paradigm that could lead to fundamentally new insights.

Kaiser (1979) first constructed classical models to explain certain phenoma, for example, that the number of male dragonflies searching for mates along the shoreline of a lake was almost independent of the number of males foraging in the neighborhood of the lake. Kaiser then identified a number of limitations of these classical models: it was not possible "to trace the systems properties back to the behaviour of the individual animals"; the models contained parameters, for example, the arrival rate of male dragonflies at the shoreline, which have no direct biological meaning because "the dragonfly males have no means of observing the arrival rate"; and the parameters of the model were fit to one set of observations that reflected one certain environment, and there was no way to extend the model to situations beyond the original one. Kaiser concluded that the classical models did not offer much explanation of the processes determining population dynamics. In contrast, the IBMs developed by Kaiser used simple behavioral rules or physiological mechanisms for which empirical parameters were available. The *populations* had certain properties because *individuals* behaved in a certain way. This characteristic allowed the models to be extended—cautiously—to situations that were not observed in the field, such as longer or shorter shorelines and other temperatures.

The other paradigmatic pioneer, A. Łomnicki (1978, 1988), focused on the problem of why some individuals should leave a habitat of optimal quality and disperse to suboptimal habitat. Classical population models could not answer these questions because in classical models individuals are all the same. Within the framework of classical theory, the only solution to the problem of dispersal to suboptimal habitat was group selection: individuals behave suboptimally for the benefit of the population. Classical theory, Łomnicki argued, thus contradicts one of the most fundamental assumptions of evolutionary theory: that individuals (or their genes) are the units of natural selection, not groups of individuals. The only way to solve this dilemma is to construct models that include differences among individuals. As a central mechanism of population regulation, Łomnicki assumed that resources are unequally partitioned and that this inequality increases when resources become scarce (section 6.5.1). Ironically, the model used by Łomnicki to demonstrate regulation by unequal resource partitioning does not simulate individuals as discrete entities, but consists of two coupled difference equations. Although Łomnicki's attitude is strongly paradigmatic, claiming that classical theory leads ecology into a "blind alley," he still used the classical modeling approach.

Neither the work of Kaiser nor that of Łomnicki had a strong impact on the early development of the IBM approach. Kaiser received little attention because he published mainly in German. Łomnicki stuck to using analytical models, and his exclusive focus on resource partitioning and population regulation was too narrow to influence a larger array of modelers and ecologists.

The visionary article by Huston, DeAngelis, and Post (1988), "New computer models unify ecological theory," is widely regarded as having established the use of IBMs as a self-conscious discipline. Interestingly, this article does not discuss the paradigmatic notions of Kaiser and Łomnicki; Kaiser is ignored completely, and Łomnicki is only mentioned briefly. Instead, the article starts with the statement that "individual-based models allow ecological modelers to investigate types of questions that have been difficult or impossible to address using the [classical] state-variable approach" (p. 682). These questions include the significance of individual variability and local interactions among individuals. Huston et al. saw the main potential of IBMs as their ability to "integrate many different levels in the traditional hierarchy of ecological processes" (p. 682) because all ecological phenomena can eventually be traced back to the physiology, autecology, and behavior of individuals.

Today it is impressive to note how clearly all these pioneers saw both the pragmatic and paradigmatic potential of IBMs (see also the insightful early review of Hogeweg and Hesper 1990). On the other hand, these real pioneers cannot be blamed for not having foreseen all the challenges and limitations of the IBM approach so that these problems could have been recognized and tackled earlier. The first of these problems is to distinguish IBMs explicitly from other types of models.

1.5   WHAT MAKES A MODEL AN IBM?

Kaiser (1979) and Huston, DeAngelis, and Post (1988) defined IBMs as models that describe individuals as discrete and autonomous entities, but they did not precisely distinguish IBMs from classical models. The first and frequently cited volume about IBMs, entitled "Individual-based models and approaches in ecology" (DeAngelis and Gross 1992), also does not clearly delineate what an IBM is. The models considered in this volume range from IBMs as defined by Kaiser and Huston et al., to analytical models dealing with distributions of individual properties instead of discrete entities, to cellular automata that do not necessarily describe individuals at all. By the middle of the 1990s, the term "individual-based" had become so fuzzy that it became increasingly difficult to tell if IBMs really had the potential to unify ecological theory and to overcome the limitations of classical modeling approaches. Therefore, Uchmański and Grimm (1996) proposed four criteria that distinguish what we consider IBMs in this book, those reflecting the research programs of the IBM pioneers, from other more or less "individual-oriented" models that acknowledge the individual level in some way but still adhere mainly to the classical modeling paradigm.

The four criteria are: (1) the degree to which the complexity of the individual's life cycle is reflected in the model; (2) whether or not the dynamics of resources used by individuals are explicitly represented; (3) whether real or integer numbers are used to represent the size of a population; and (4) the extent to which variability among individuals of the same age is considered.

The degree to which the life cycle is reflected in a model (criterion 1) is important because individuals of most species change significantly in the course of their life: they need more and, often, different resources while they are growing; in different states of their development they interact with different biotic and abiotic elements of their environment; and individuals can adapt life history characteristics as they grow and develop, for example, growing or reproducing more slowly when resources are scarce or competition is high. IBMs thus have to consider growth and development in some way; otherwise they neglect essential aspects of the "ecology of individuals" (Uchmański and Grimm 1996).

The second criterion refers to resources exploited by individuals. Models that simply assume a constant carrying capacity for resources cannot be fully individual-based because they ignore the important, and often local, feedback between individuals and resources. Moreover, carrying capacity is typically a population-level concept, often used to describe density dependence in a population's growth rate. Such population-level concepts have little meaning at the individual level: individuals usually cannot know the overall density of their population but instead are affected by their local resources.

The third criterion is obvious: individuals are discrete so population size necessarily is an integer. However, sometimes classical models are made "individual-based" merely by rounding the real-number results to integer numbers. But the model's population dynamics are still fine-tuned using real numbers, whereas in real populations where individuals usually interact locally and only with a limited number of other individuals, this fine-tuning does not exist. Truly individual-based models are built using the mathematics of discrete events, not rates.

The fourth criterion distinguishes models using age, size, or stage distributions from IBMs. In distribution models, differences among individuals belonging to the same group (e.g., age-class) are ignored. In reality, however, even individuals of the same age, or size, may develop along different pathways so that after some time the variation among individuals within a class is comparable with the variation among class averages (Pfister and Stevens 2003). Neglecting this degree of freedom in population structure could mean ignoring important mechanisms determining population dynamics.

This classification scheme has been described as interfering "with the insightful process of comparing models at different levels of detail" (Bolker et al. 1997), but this was not the intent of its authors (Uchmański and Grimm 1997). Their objective was not to distinguish models as true and false, or useful and useless, but to provide classification criteria necessary to answer the question of whether IBMs can lead to a fundamentally new view of ecological systems and processes (Uchmański and Grimm 1996). This question cannot be answered unless IBMs are clearly delineated from other kinds of models, which we here refer to as "individual-oriented."

Of course, numerous models do not fulfill all of these four criteria but nevertheless provide important theoretical insights. Matrix models describing age- or stage-structured populations are powerful for determining the intrinsic rate of increase and the stable age or stage structure of exponentially growing populations (Caswell 2001). More sophisticated distribution models successfully describe laboratory populations of planktonic species (Dieckmann and Metz 1986), or patterns in fish communities (Claessen, de Roos, and Persson 2000). Models of predator-prey systems that describe individuals as discrete units having local interactions but no life cycles or variability can demonstrate the stabilizing effect of local interactions and the emergence of striking spatial patterns (de Roos, McCauley, and Wilson 1991; figure 1.3; see also Donalson and Nisbet 1999; section 6.6.1). All these models consider individuals to some extent but still refer to the framework of classical models and theory. They ask: what do we gain—compared with using classical, highly aggregated models—if we include, for example, the discreteness and local interactions of individuals (Durrett and Levin 1994)? But none of these "individual-oriented" models allows us to fully "trace the systems properties back to the behaviour of the individual animals" (Kaiser 1979, p. 116).

Individual-oriented models are, like classical models, indispensible, useful, and sometimes fascinating tools, but they should indeed be "separated" (Bolker et al. 1997) from IBMs. This separation is necessary if we are to compare the classical framework, which describes ecological systems as relatively simple and characterized by system-level state variables, to the view that ecological processes and systems emerge from the traits of adaptive individuals.

1.6   STATUS AND CHALLENGES OF THE
INDIVIDUAL-BASED APPROACH

The individual-based approach is now firmly established in ecology. Hundreds of publications have been based on IBMs, prompting Grimm (1999) to review fifty IBMs of animal populations published in the decade after the paper of Huston, DeAngelis, and Post (1988). Earlier reviews of IBMs (DeAngelis et al. 1990; DeAngelis, Rose, and Huston 1994; Hogeweg and Hesper 1990) provide useful summaries of existing IBMs, but Grimm focused on the degree to which the vision that IBMs "unify ecological theory" (Huston, DeAngelis, and Post 1988) has been fulfilled. The conclusion of this review was rather sobering: although every model served its purposes and was thus

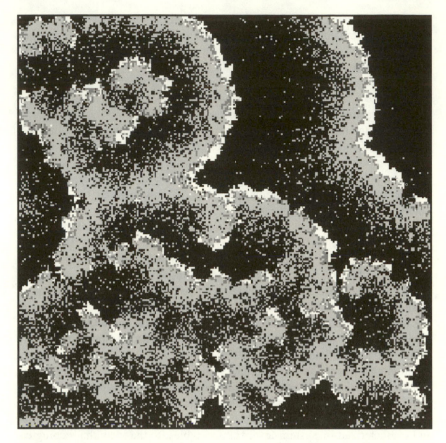

Figure 1.3 Spatial patterns produced by the model by de Roos, McCauley, and Wilson (1991) of an artificial predator-prey system. The model world consists of 256×256 grid cells, which can be in the states empty (*black*), prey (*white*), predator (*gray*), or predator handling prey (*dark gray*). Predator individuals are just "points" that jump to neighbor cells and eat, if present, prey "points." (Figure produced using a program written by H. Hildenbrandt.)

useful, ecology as a whole seemed to have learned less from the individual-based approach than originally expected. The main reason for this conclusion was that few IBMs addressed general issues of theoretical population ecology like persistence, resilience, or regulation. Similarly, new theoretical issues like emergence (chapter 5) or self-organization were rarely discussed; IBM applications seemed driven more by pragmatic motivations then by paradigmatic ones. Grimm (1999) also concluded that most IBMs were: (1) developed for specific species with no attempt to generalize results; (2) rather complex, but lacking specific techniques to deal with this complexity; and (3) too elaborate to be described completely in a single paper, making communication of the model to the scientific community incomplete (note that both JABOWA and

the fish cohort model of DeAngelis, Cox, and Coutant 1980, mentioned earlier as particularly influential IBMs, were each completely described in one
report).

The pioneers' vision that IBMs would induce a paradigm shift and unify
ecological theory has clearly not come true automatically. The promise of the
individual-based approach still exists (as we will try to show in this book), but
the experience gained in one and a half decades of individual-based modeling shows that two closely linked problems have been underestimated. First
is the complexity of IBMs, which "imposes a heavy cost compared with
the other model types" (DeAngelis et al. 1990, p. 585) in understanding,
testability, data requirements, and generality (Murdoch et al. 1992). Second
is the lack of a theoretical and conceptual framework for individual-based
modeling, leading to the widespread use of ad hoc assumptions and preventing a more coherent development of the approach (Hogeweg and Hesper
1990).

Because IBMs deal with many entities, spatial scales, heterogeneities, and
stochastic events, they are necessarily more complex than classical, analytically
tractable models. Many IBMs are complex as indicated by such conventional
measures as the number of variables, parameters, or rules in the model. However, even IBMs that are relatively simple by conventional measures can be
complex in new ways, such as in the number of unique individuals; the number,
type, and order of interactions among individuals; and the number of ways a
model population can reach any particular state. This complexity, along with
the lack of an overall theoretical framework for individual-based modeling, has
resulted in the following challenges to the productive use of IBMs. (Do not
panic! This book shows how to meet these challenges.)

Development. Developing IBMs is a challenge because much more of the
complexity of the real world is acknowledged and not ignored a priori. Designing the model structure and resolution is a more time-consuming and complex
task than when developing classical models, which are constrained to a coarser
representation of reality.

Analysis and Understanding. The more complex a model, the more difficult
it is to analyze and understand. Many theorists and modelers therefore assume
that an increase in complexity inevitably reduces a model's potential to provide understanding. Critics of IBMs argue that complex models are as hard to
understand as the real world and therefore of little use.

Communication. Classical models are easy to communicate because they
are formulated in the common language of mathematics. IBMs, on the other
hand, have essential characteristics that cannot be described by equations and
parameters. As yet, we lack a common, concise language for communicating
IBMs. Moreover, many IBMs are too big to describe completely in publications. Therefore, IBMs are often not fully open and available to the scientific
community, which is probably the most serious threat to the credibility of the

whole approach (Lorek and Sonnenschein 1999; Grimm 1999, 2002; Ford 2000).

Data Requirements. The more kinds of entities, scales, and events a model represents, the more parameters are needed. Sufficiently precise parameter values are, however, notoriously difficult to obtain in ecology. IBMs have therefore been criticized as being too "data hungry"—especially IBMs designed for specific, applied problems. For example, the usefulness of spatially explicit population models, many of which are individual-based, has been questioned because adequate parameter values are never available (Beissinger and Westphal 1998).

Uncertainty and Error Propagation. Data available to parameterize IBMs are likely to be uncertain. Thus, keeping the number of parameters low seems wiser because many uncertain parameters might combine to produce extremely high uncertainty in system-level results. This error propagation potential appears capable of rendering IBMs totally useless for solving applied problems and of limiting the testability of IBMs, yet this aspect has received very little investigation.

Generality. Classical models using population size as a state variable are considered most general because they ignore almost every aspect of real species and populations. As more details are included (e.g., adding age or stage structure, space, habitat dynamics, dispersal), models become less general. Each detail added to a model makes it more specific to a particular population. How, then, can IBMs be general or produce theory when they are highly detailed? It has even been argued that using IBMs necessarily means relinquishing the "holy grail" of general ecological theory (Judson 1994).

Lack of Standards. Classical theoretical ecology has a suite of standard models with known properties. These standard models serve as building blocks for all kinds of structured classical models. It is no longer necessary to explain or even to justify the use of these standards. If, for example, a model addressing synchronization of local population dynamics in different patches assumes that the local dynamics are described by the so-called Ricker equation, the assumption is familiar and noncontroversial. Using this standard assumption makes analysis, communication, and comparison to other similarly structured models easier. In contrast, most IBMs have been built from scratch, using ad hoc assumptions not guided by general concepts. The lack of standard, widely accepted building blocks makes individual-based modeling both inefficient and controversial. This lack also makes it difficult to compare models and develop theory. If two IBMs produce different results, it is much more difficult to explain and learn from the differences when the IBMs have different structures and use many nonstandard assumptions.

Many of these same challenges have limited the progress of individual-based (or "agent-based," the term used in fields other than ecology) approaches in other fields of science. A period of initial excitement and exploration (e.g., Waldrop 1992; Arthur 1994; Axelrod 1984, 1997) has not been followed by as much rapid progress as some undoubtedly expected. Our observations have been that agent-based modeling has not become a widespread, paradigm-altering tool even among scientists focused on complexity (although there are important and exciting exceptions to this generalization). Building and learning from the models, and producing results of general theoretical interest, have proved difficult.

1.7 CONCLUSIONS AND OUTLOOK

The individual-based approach is no longer new, being established as a distinct approach for more than ten years now and having intrigued ecologists for over twenty years. IBMs can address types of questions that cannot be addressed with classical models. From many IBMs of real and hypothetical systems, we have learned much about the ecological significance of local interactions, individual variability, and so on. However, the most notable result of our experience with IBMs so far may simply be an understanding of the approach's many challenges and fundamental differences from classical ecology. The potential of IBMs remains high yet largely unfulfilled. To see this potential realized in the future, it is now time to formulate strategies for coping with the challenges we have listed in the preceding section.

This book presents our research program for IBE, much of which concerns strategies for coping with the problems that have limited IBMs so far. These strategies, outlined here, are adapted from existing theory and practice in ecological and simulation modeling, analysis of complex systems, and software engineering.

Pattern-oriented Modeling. The term "IBM" contains not only the word "individual" but also the word "model." So far, methodological work on IBMs has focused too much on individuals and their significance and not enough on modeling. Perhaps the most decisive modeling issue is how to find the optimal level of complexity for an IBM. Using multiple patterns at different levels of ecological process ("pattern-oriented modeling") helps optimize model complexity, parameterize models, and make models testable and general.

Theory. In IBE, "theory" mainly concerns how to represent individual-level behavior in a way useful for explaining system-level processes. These theories could also be referred to as "models" or "assumptions," but referring to "theory" underlines the research program of IBE: to develop a general theoretical framework for describing individual behavior. The rationale of this program is that generality should be easier to achieve at the individual level than at aggregate levels because all individuals follow, as pointed out earlier, the same master plan: seeking fitness. Individuals must continually decide—in the literal or the more metaphorical sense—what to do next, and these decisions are based on the individual's internal models of the world. It seems reasonable to believe that individuals of many types have similar internal models and traits that are based on fitness seeking; and complexity science teaches us that individuals with identical adaptive traits but their own unique states, experiences, and environments can produce an infinite variety of system dynamics. Coherent and predictive theories of these traits will provide an important key to understanding ecological phenomena in general.

Design Concepts. Designing every element of a model requires decisions about variables, parameters, functional relationships, and the like; and if these decisions are not to be ad hoc, they must be based on a consistent set of concepts. Unfortunately, differential equations do not provide a useful conceptual framework for IBMs. Instead, a general conceptual framework for designing IBMs can be borrowed from the new discipline of Complex Adaptive Systems (CAS; Waldrop 1992; Holland 1995, 1998). Such concepts as emergence, adaptation, and prediction can provide an explicit basis for design decisions and reduce the need for ad hoc modeling decisions. These concepts also provide a common terminology for designing and describing IBMs.

Software Design and Implementation. Software development is inevitably a major part of an IBE project, and project success requires software that is well designed and thoroughly tested. Computer models are the primary tools of IBE and, as in any other science, the rate and nature of progress are highly dependent on the quality of the tools. Successful conduct of IBE often requires software expertise beyond the meager training ecologists now typically receive.

Simulation Experiments. We can only understand and learn from simulation models such as IBMs if we design and execute controlled simulation experiments. Thus, the art of analyzing IBMs is in designing experiments whose outcome can (at least partly) be predicted, and falsified, and to combine such experiments in a way that we get a comprehensive understanding of the key structures and processes of ecological systems. The ability of this experimental approach to produce new and general insights has been demonstrated in a number of studies.

Communication. The complexity of IBMs and newness of IBE makes scientific communication more important yet more challenging. Both models and software need full documentation, and often separate publications are required to describe an IBM and then its research or management applications. A model, or any scientific idea, is successful if it is memorized in total or part by peers who then use it in future work. Improving this "memetic fitness" (Blackmore 1999) of IBMs is critical to the success of IBMs and IBE.

Where will we be in another decade or two? We envision IBE being conducted by interdisciplinary teams having expertise in simulation modeling and complex systems science, software engineering, and the biology and ecology of the organisms and systems being studied. As in other kinds of ecology, toolboxes of standard IBE modeling practices, theory, software, and analysis methods will gradually be developed and refined as more models are designed and tested and more theory is developed. These toolboxes will allow us to build models rapidly and conduct analyses of many ecosystem dynamics and complexities that we currently cannot explain. IBE and more traditional approaches will continue to contribute to each other in many ways (e.g., see chapter 11). However, what will continue to set IBE apart is its goal not to simplify ecological complexity but to *understand* complexity and how it emerges from the adaptive traits of individuals.

Chapter Two

A Primer to Modeling

Modeling is presented as a discipline that draws (in the first instance) on the perception of the detective rather than the expertise of the mathematician.

—Anthony Starfield and Andrew Bleloch, 1986

2.1 INTRODUCTION

Individual-based modeling is, above all, modeling. If we want to make individual-based modeling effective and coherent, we must understand what modeling really is and how it works. Therefore, in this chapter we introduce general guidelines for developing models, referring readers to other authors (especially Starfield, Smith, and Bleloch 1990; Starfield and Bleloch 1986; and Haefner 1996) for more detailed introduction to the principles of modeling. These guidelines also set the stage for the remainder of the book: subsequent chapters address the modeling tasks introduced here.

Intuitively, we know a model is some sort of simplified representation of a real system. But why do we build models, and what do models have in common? The answer is fundamental and independent of the context in which we build models: the purpose of modeling is to solve problems or answer questions, and the common feature of models is that they are developed under constraints (Starfield, Smith, and Bleloch 1990). A model may address a scientific problem, a management problem, or just a decision in everyday life: any attempt to solve such problems is constrained by scarcity of information and time. We can never take into account all the elements of the real world that influence a problem. We cannot know everything about a problem, and if we *did* know everything, we would not be able to process the flood of information. For solving real-world problems, simplified models are the only alternative to blind trial and error, which usually is not a good approach.

A simple real-world problem is choosing a checkout queue in the supermarket, with the objective of minimizing the time spent waiting to check out. Very often we choose the shortest queue because we apply a very simple model: the length of a queue predicts the time spent waiting in it. We make—usually subconsciously—numerous simplifying assumptions in our model, such as that all customers require the same handling time and that all checkers are equally efficient. Certainly, the model is oversimplified. We could have observed the employees for a while, trying to identify the most efficient one; we could have checked the shopping carts of the customers in all queues to predict the handling time of each. But gathering such additional information would require time, which conflicts with our objective of minimizing the time spent checking out. Moreover, even if we invest time in gathering information, we still can never be sure that we choose the fastest queue. How, for example, could we know which customers will be very slow in filling out a check? How could we predict which checkout stands will open or close, or which checkers will go on break, as we wait?

This example demonstrates that modeling is problem solving under constraints. We know that the model we use is not perfect, but—facing the constraints of information and time—we also know that even a simple model will probably lead to a better solution to the problem than not using any model at all, for example, by simply entering the first queue encountered. And we know that a more complex model will not necessarily provide a better solution to the problem of minimizing the waiting time.

A completely different model of the queue system would be used, though, if we had a different objective, such as minimizing the waiting time of *all* customers. Models addressing this problem differ fundamentally from the single-customer model in that they cannot ignore the variability in customer handling time and employee efficiency: this variability is just what increases the average waiting time of all customers. Multicustomer models predict that the use of one queue for all customers, common at banks and airport check-in counters (and now used, quite famously, at one particularly crowded New York City supermarket), significantly reduces total waiting time compared to separate queues for each employee.

The difference in the structure of the two queue models is due to their different purposes. The lesson from this difference is that a model must not be viewed as merely a representation of a system but as a "purposeful representation" (Starfield, Smith, and Bleloch 1990). A model's structure depends on the model's purpose because the purpose helps decide which aspects of real system are essential to model and which aspects can be ignored or described only coarsely. Now, what does all this mean for individual-based modeling? We can draw three main lessons from the checkout queue example:

1. Mere "realism" is a poor guideline for modeling. Modeling must be guided by a problem or question about a real system, not just by the system itself. The problem to be solved provides a filter that should be passed only by those elements of the real system considered essential to understanding the problem. Without a clearly stated problem, we have no such way to filter what should versus what should not be included in a model, often with fatal consequences. Models become unnecessarily complex and are never finished because there are always more details to add. In fact, Mollison (1986) observed that modelers following "naive realism" as a guideline can easily be identified by their promise that their model will be "finished soon." IBMs certainly can be more "realistic" than the highly aggregated models of classical population ecology but not in the naive sense of just including more details of the real world. IBMs are more realistic when we believe that individual behavior is an essential process affecting the problems we want to solve.

2. Constraints are essential to modeling. It is a widely held myth that a model cannot be developed before we have sufficient data and a comprehensive understanding of the system (Starfield 1997). The opposite is true: our knowledge and understanding are always incomplete and this, exactly, is the reason to develop models. Models are useful because we want to solve problems *despite* our lack of knowledge and understanding. If we knew and understood everything, why bother with models and theory? Admittedly, constraints on information, understanding, or just time are painful when we want to solve a problem and we naturally complain about these constraints. But, as Starfield, Smith, and Bleloch (1990) point out, constraints reward clear thinking, forcing us to hypothesize about factors essential to the problem. One of the main problems with IBMs is that they are less constrained by technical limitations than classical models. IBMs can include many more factors than analytical models. Thus, to improve individual-based modeling, the important role of constraints has to be acknowledged and new, powerful constraints must be identified. This will be a major theme in the remainder of this book, especially chapter 3.

3. Modeling is "hardwired" into our brains. Modeling is not a specific method used only by specialists calling themselves "modelers"; we are all modeling all the time, for each decision we make, because each decision is constrained by a lack of information and time. Most of the time we are not aware that we are modeling, but instead we use powerful modeling *heuristics* to solve problems. These heuristics are mental problem-solving resources that are often more effective than logical reasoning for dealing with complex systems.

2.2   HEURISTICS FOR MODELING

Starfield, Smith, and Bleloch (1990) define a heuristic as a "plausible way or reasonable approach that has often (but not always) proved to be useful" (p. 21), or simply as "rules of thumb." If we want to make scientific modeling, including individual-based modeling, efficient, it is important to be aware of these powerful heuristics so that we learn to use them consciously. While developing an IBM, the following heuristics should be used as a checklist; modelers can review the list to make sure they use the most helpful heuristics. If used in this way, the heuristics will help exorcise the ghost of naive realism that lurks behind many IBMs (Grimm 1999), because the heuristics reinforce the fundamental notion of models as purposeful, not complete, representations of a system.

These heuristics should be checked at all hierarchical levels of a modeling project, not only for the entire model but also for particular parts of a model—for example, submodels representing the abiotic environment and its dynamics, behavioral traits of individuals, or physiological processes. The heuristics we describe here were formulated by Starfield, Smith, and Bleloch (1990) and are only the most important and general heuristics. Modelers can profit from their own experience by adding their own heuristics to this list.

1. Rephrase the problem to be solved with the model. This is obviously important if someone else's problem is to be solved, for example, a management problem. Rephrasing the problem helps make sure the modeler understands it. In science we typically address problems we define ourselves, but rephrasing the problem is still valuable because it forces us to be explicit: what *exactly* is the question we want to ask? Good science requires good questions, and good questions are clear and explicit. If the question is not clear, we probably are aiming too high, trying to understand too many things at the same time. Especially when we are dealing with complex systems, as in ecology, rephrasing a problem in a productive way is not trivial; in fact, it can be decisive. For example, "explaining dynamics of a beech forest" would be too broad of a problem, providing no guidance on the model's structure. The model could describe individual trees, or plant-animal interactions, or primary production, or whatever. Rephrasing the problem as "explaining spatiotemporal dynamics" is a step in the right direction because now we know that the model has to include spatial structures in some way.

2. Draw a simple diagram of the system to be modeled. This technique is useful because most of us are very poor in drawing. To make sure that the objects we draw can be identified, we focus on drawing essential aspects. We might even produce caricatures (C. W. Clark and Mangel 2000) that exaggerate essential aspects. Thus, while drawing simple diagrams of the system's objects and processes, we make use of powerful filters hardwired into our brains. These visual filters can be more effective than verbal arguments, perhaps because human perception and imagination are mainly visual. The vague term "simple diagrams" means exactly this: the diagrams require no formal structure, just draw and have confidence in your brain! Simple diagrams are also ideal for communication, especially in early stages of model development. Modeling requires communication, and simple diagrams are efficient for explaining how we understand the problem and what elements of the modeled system we consider essential.

3. Imagine that you are inside the system. This heuristic is of particular importance for IBMs. The perspective from inside the system keeps us from imposing our external perspective on the system's objects. For example, as external observers we may know the size or density of a population, but this information usually is not available to the individuals in the population. While imagining that we are an object inside the system, we can ask: What is going on around me? What affects me, and what do I affect?

4. Try to identify essential variables. The model we are developing will represent a real system, so the question arises, Which variables are essential for representing the system? There are, for example, questions about populations for which it is sufficient just to know the number of individuals, so that population size (or density) is the only essential state variable. However, there are many questions, including the ones we focus on in this book, where additional variables are essential. Such variables often include the location, size, age, sex, and social rank of individuals.

5. Identify simplifying assumptions. The purpose of modeling is to find a simplified representation of the real system that can be used for solving a problem. Simplification includes using only a small number of variables to represent the system. Every aspect of the real system that is not represented by a variable is assumed to be constant and uniform. We know that this assumption is not true literally, but we assume that the simplification does not interfere with solving the problem. If first drafts of a model turn out to be too complex to be useful, we have to simplify further, for example, by aggregating variables.

6. Use "salami tactics." This technique is one of the most powerful modeling heuristics: if we have no idea how to find the answer to the problem in one big step, we can *approach the problem in many small steps*. It is particularly valuable when the problem we address is the dynamics of essential variables. Usually, we have no idea how to deduce the dynamics over long times, but often we can rather easily predict what happens over the next small time interval. This prediction can be made from simple bookkeeping of the most important processes that affect the essential variables. Because we are predicting only a short time interval, we often can use the approximation that changes are linear. In the case of discontinuous changes caused by some event (e.g., a disturbance), we can specify the probability that the event will occur and how it will affect the variables. Salami tactics are routinely used in simulation models: time is "sliced" into steps short enough to predict changes between time steps. In spatial models, space can also be cut into small pieces such as grid cells, so spatial processes can be represented as acting among adjacent cells. The "salami tactics" heuristic is also powerful when applied to the process of modeling itself, especially when modeling complex systems. Instead of trying in one step to formulate a complete model containing all essential system characteristics, it is usually better to start with a strongly and deliberately simplified "null model" (Haefner 1996). Then, the model's complexity can be increased step by step.

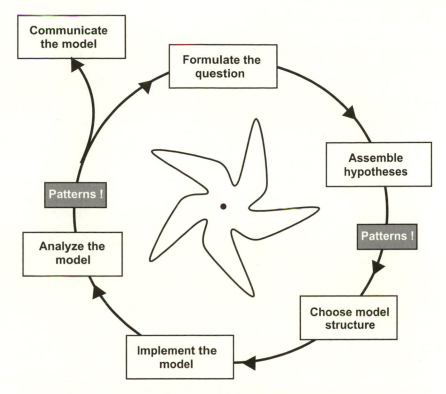

Figure 2.1 The six tasks of the modeling cycle. Considering patterns is especially important when choosing model structure and analyzing the model.

2.3   THE MODELING CYCLE

The heuristics described in the preceding section are powerful but not sufficient by themselves as a general guide to modeling because they do not include the full cycle of tasks performed in developing and using models. Modeling is an iterative process (Haefner 1996; Thulke et al. 1999) in which several tasks are performed repeatedly. We now describe these tasks and the typical "modeling cycle" (figure 2.1). Some of the tasks are nearly identical to the heuristics described above, and most tasks are discussed in more detail in later chapters.

2.3.1   Task 1: Formulate the Question

This task corresponds to the "rephrase the problem" heuristic. Modeling requires deciding which aspects of the real system to represent and at what resolution. Without clearly formulating the question or problem the model is to address, we could not make these decisions. With the question clear, we can consider every known element and process of the real system and decide whether we believe them to be essential for addressing the question or problem.

2.3.2 Task 2: Assemble Hypotheses for Essential Processes and Structures

Every answer to this question—whether an element or process is essential for addressing the modeling question—is a hypothesis that could be true or false. Modeling means exactly this: to build a model with working hypotheses and then to test whether these hypotheses are useful and sufficient for explaining and predicting observed phenomena. But where do these hypotheses come from? They reflect our first, preliminary understanding of the system, which in fact is a first "conceptual model" (DeAngelis and Mooij 2003). Without a conceptual model, consisting of hypotheses about what is important, we could not start the modeling cycle. This means in particular that if we have no idea how a certain system works, we cannot develop a model.

The hypotheses of the conceptual model are formulated verbally (and often graphically) and are based mainly on two sources: theory and experience. Theory provides a framework through which we perceive a system. If, for example, our theoretical background is ecosystem theory, we will view ecosystems as a system of compartments containing nutrients and energy, with fluxes of nutrient and energy driving the system's dynamics. Population ecologists will focus on a few population rates and on census time series. In this book, the focus is simultaneously on the adaptive behavior of individuals and on characteristic properties, or patterns, of the system.

Experience, the other source of hypotheses about the system, is shaped by theory or by the way we use the system. Theory constrains the field data we collect and the experiments we perform; empirical research is thus never theory-free (Fagerström 1987). Therefore it is important to also consider the empirical knowledge of those who simply use the system (e.g., natural resource managers) or who just know them well (e.g., naturalists). Every naturalist or natural resource manager knows much more than can be expressed in hard data. Often, this qualitative knowledge is latent and will only be expressed if the right questions are asked by the modeler. Empirical knowledge can easily be expressed in "if-then" rules. For example, forest managers who have observed, a hundred times or more, how a canopy gap in a beech forest closes over time, can formulate empirical rules for this process: either the neighboring canopy trees will spread to close the gap or one of the younger trees of the lower canopy will grow into the gap. It may not be possible to predict which of these two processes will occur in any particular gap, but experienced managers can estimate probabilities of the two alternative outcomes.

Assembling the first conceptual model of the system will require some time, especially when the system is complex, and often it will be necessary to cycle through tasks 1 and 2 several times before we can proceed to task 3. We might initially have formulated the question of the modeling project in a way that does not easily lead to useful hypotheses about the essentials. Or, while formulating our working hypotheses, we might realize that we can again rephrase the question because formulating the working hypotheses forced us to think more deeply and clearly about the question.

2.3.3 Task 3: Choose Scales, State Variables, Processes, and Parameters

The next task is to make the working hypotheses quantitative by translating them into a specific model structure and into equations and rules describing the dynamic behavior of the model's entities. To do this, first of all we have to select the variables necessary to describe the state of the system (thereby defining the model's structure); the essential processes that cause changes of the state variables; and the parameters that are used in our models of the essential processes (i.e., equations and rules) and which quantify when, how much, and how fast the variables change (defining the model's dynamics).

The basic structure of IBMs is a collection of discrete individuals, because we consider the discreteness and adaptive behavior of individuals essential. We therefore have to select variables describing the state of individuals, parameters describing individual behavior, and variables and parameters describing the individuals' environment. Of course, we would have to use thousands of variables if we want to describe an individual and its environment completely. But complete description is not our purpose. Instead, we ask ourselves, Which characteristics of an individual are really essential to the question we are trying to answer? Location, age, size, and sex are essential for many questions but not for all. Other examples of important variables are the energy an individual has stored, the individual's social rank, the number of mating attempts it has made so far, the distance it has traveled.

Most often, the initial list of variables will look intimidating because it is too long. Each additional state variable makes a model harder to develop, parameterize, implement, analyze, and understand—so we should fight hard to limit the list of variables. Empirical biologists who know the diversity of their systems well often find it especially hard to boil all this diversity down to a handful of variables. A good heuristic is, at this point, to shorten the list of variables *to the threshold of pain, or a bit farther*. This heuristic is useful because "pain" means that we—for the time being—consider all remaining variables absolutely essential.

In addition to working with lists of variables, it often makes sense to draw simple graphical representations of the model's elements, for example, simplified

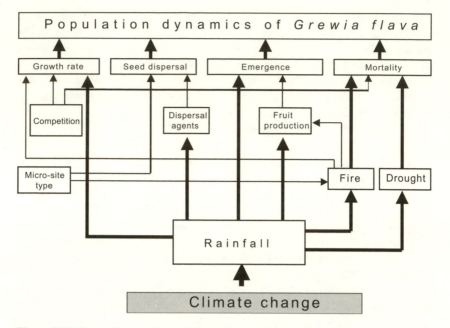

Figure 2.2 Influence diagram of a spatially explicit population model of *Grewia flava*, a woody plant species in the southern Kalahari Desert. This diagram was developed and used to decide on model structure (state variables) and the processes to be included in the model. *Bold arrows* indicate processes through which population parameters and variables are affected by annual rainfall. (Modified after Tews 2004.)

Forrester diagrams (see, e.g., Haefner 1996) or "influence diagrams" (figure 2.2; Jeltsch et al. 1996; Brang et al. 2002) where boxes delineate structural elements or processes and arrows indicate influence: "Element A has an influence on element B." Influence diagrams are also useful for aggregating processes to keep initial model versions simple and manageable.

Variables represent the structure of the model system, whereas parameters, which are used in equations and rules, represent processes. Parameters are constants that quantify the relationships between variables. Making this quantification constant usually involves simplifying assumptions. For example, when changing dollars to euros as a traveller, we use the simple model $N(\$) = aN(€)$, with a being the exchange rate (Starfield and Bleloch 1986). Assuming that a is constant is a simplifying assumption. In fact, a is not constant and depends on very complex processes that are hard to predict. But for the purpose of a traveler—to assess the price of something in a foreign currency—the simplifying assumption is reasonable. (The same assumption is *not* reasonable for the purposes of professional currency traders.) Parameters determine the resolution by which we describe processes. Later on in the modeling cycle, we may decide to change this level of resolution, either by aggregating several model processes into one parameter, or by replacing a constant parameter by a submodel, for example, equations and rules that dynamically produce values of the parameter.

The selection of variables, parameters, and the equations and rules they are used in is inseparably bound to the selection of the spatial and temporal scales of our model. "Scale" has two aspects: the *grain*, that is, the smallest slice of time or space we are going to consider, and *extent*, that is, the total time or area to be covered by the model. (Note that in ecology "large scale" often is used to refer to large extents, whereas originally, in geography [i.e., for maps], "large scale" referred to a large resolution or a small grain; Silbernagel 1997.)

One of the major recent advances in ecology is understanding how the scales we select affect results of field studies and models (Levin 1992). Most spatial models in ecology are grid-based, so that their spatial extent is the size of the entire grid, and their spatial grain is the size of a grid cell. The choice of the extent depends on the spatial processes and structures to be modeled, for example, long-distance dispersal events or a mosaic of habitat patches of different quality. The extent should also be large enough to avoid significant edge effects, unless these effects are important to the problem to be solved. The grain should be defined by the distance below which we believe spatial effects can be ignored. For example, the location of animals within a territory can often be ignored so that the grain, that is, the size of the grid cell, is determined by the average size of a territory (Jeltsch, Müller, et al. 1997; Thulke et al. 1999). On the other hand, if we consider the variation in the size of territories essential, we have to choose a grain considerably smaller than the mean size of territories (figure 2.3). For plants and mobile, nonterritorial animals, environmental variability is a major consideration in selecting the spatial grain: over what distances do environmental conditions change significantly? Additional guidance and examples for selecting spatial scales are provided by Laymon and Reid (1986), Bissonette (1997), Mazerolle and Villard (1999), Storch (2002), and Trani (2002).

Similar considerations are used to determine grain and extent of the temporal scales of a model. The grain, or time step, is the time span over which we ignore details of temporal variation; instead, we consider only the net change in variables over the entire time step. Some IBMs that describe physiological states and behavioral motivations of animals use time steps of fifteen (Wolff 1994) or even only five minutes (Reuter and Breckling 1999). Other typical and more or less "natural" time steps are days, seasons, or years. For slowly developing systems, larger time steps may be sufficient: the BEFORE model of beech forest structure uses fifteen years (Neuert 1999; Rademacher et al. 2004; section 6.8.3).

At the end of task 3 we will have—preliminarily—decided on the spatial and temporal extent and grain of the model, its variables and parameters, and the equations and rules to describe the processes identified in task 2. Before we

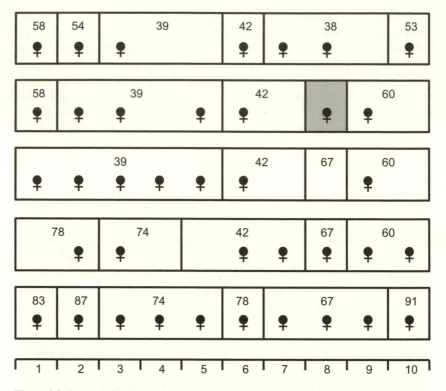

Figure 2.3 Size and distribution of male territories of the wall lizard (*Podarcis muralis*) in five consecutive years (top to bottom) (Hildenbrandt et al. 1995; Bender et al. 1996). Numbers denote the identity of male territorials, ♀ symbols indicate the location of occupied female territories within a male's territory. As individuals die, neighbors or a nonterritorial lizard (e.g., individual 67 in year 3) takes over the free space. At the very bottom, the grid underlying the territories is shown. Territories may vary in size between one and five grid cells. A gray grid cell is not occupied by a male. (After Hildenbrandt et al. 1995.)

can proceed to the next task, we might have to go back to task 2 or even task 1. While considering variables, parameters, scales, equations, and rules, we may change our mind about the hypothesized processes and structures, or about the formulation of the problem in the first place. Nevertheless, at some stage we proceed.

2.3.4 Task 4: Implement the Model

Starfield, Smith, and Bleloch (1990) quote a heuristic coined by the engineer Billy V. Koen: "At some point in the project, freeze the design." To freeze the design does not mean to freeze it forever, but any modeling project will reach a stage where the model's design cannot be improved until the model is implemented in computer code. Only the implemented model, an "animated" entity (Lotka 1925) with a "life" of its own, can show us the logical consequences of the model's formulation. Once we implement a model and start analyzing its results, the modeling process can truly become cyclic: we analyze the consequences of our model assumptions, develop new assumptions, implement them, generate new consequences, etc.

It is important to enter this modeling cycle soon, especially when we are dealing with complex systems. Facing the complexity of an ecological system, we might feel more and more confused, insecure, and lost while designing a model. As a consequence, we may hesitate and rethink the design again and again. Neophytes in modeling often get trapped in this way, but they must realize that the very reason to develop a simulation model is that we are *not*, by any means, able to understand the problem merely by reasoning because it is too complicated. Therefore, an important heuristic for breaking this psychological barrier is to *start with a ridiculously oversimplified model*, called a "null model" (Haefner 1996). The null model might be stripped of virtually all the complexity that confuses us: we might make all individuals the same, make the environment homogeneous and constant, let the individuals be stupid. The purpose of the null model is simply to get started, to enter the modeling cycle. The null model is much easier to analyze than the full model. The null model may be even so simple that we can predict its results: "Of course! What else could have happened with such stupid individuals?" The null model is just a trick to get the modeling cycle started and therefore does not interfere with the heuristic to freeze the design once we cannot improve the design anymore without implementing the model. It is always a good idea first to implement and analyze a model that is much simpler than we expect the final model to be. This is salami tactics: don't make steps that are too big while developing a model!

Implementing the model means designing and writing its computer software. Before we can do this, we have to decide on the order in which model processes occur, that is, the "schedule." Flow charts are useful for developing and visualizing the sequence in which model processes are executed (Starfield, Smith, and Bleloch 1990; Haefner 1996). However, the order of events in a simulation model is not necessarily specified entirely by the modeler; events can also be scheduled by the model entities themselves ("event-based" simulations; section 5.10).

Many simple ecological models can be implemented in software by novice programmers (see examples in Starfield, Smith, and Bleloch 1990). However, even simple IBMs involve special software challenges. Especially important is that the software must allow the modeler to observe and conduct experiments on all parts of an IBM: the software must not only implement the model but also provide a virtual laboratory for experiments on the model. We must be able to observe individual behavior, patterns over space and time, and other such results that are unique to IBMs. A second special consideration is that software errors are especially difficult, yet important, to detect in IBMs. In chapter 8 we present detailed recommendations for implementing IBMs (see also Haefner 1996; Ropella, Railsback, and Jackson 2002). Here, we simply advise modelers that the quality and efficiency of the software is of utmost importance to the quality and efficiency of the modeling project. Without a competent implementation, models cannot be analyzed and revised effectively and the modeling cycle will grind to a halt.

The last thing to do in task 4 is to specify initial values for all variables and specify values for all parameters. Parameterization is an important issue, especially for complex models such as IBMs, and is discussed in more detail in chapter 9.

2.3.5   Task 5: Analyze, Test, and Revise the Model

Nonmodelers often believe that designing a model is the most difficult part of modeling. However, some sort of model can always be formulated and implemented rather quickly. The real challenge is building a model that produces meaningful results. Even experienced modelers need at least ten times longer to analyze, test, and revise a model than they need to design and implement the first version.

As with designing models, there are many useful heuristics for analyzing models in general, and simulation models in particular. In chapter 9 we discuss analysis of simulation models extensively. Here we only discuss the most general heuristic for analyzing models: *decide what currency is used to rank different versions of the model*. Analyzing a model means comparing and ranking different versions, so we need a basis for comparison, a currency that tells us if we improved the model or not. Of course, the basic currency is how useful the model is for answering the question for which it was developed. The model will be useful only if it captures the essentials of the system that are needed to address the question. But how can we be sure that it captures these essentials? Somehow we have to learn how much we can trust the model. Testing models means exactly this: assessing the confidence with which we can apply model results to our real system and problem. The only way to assess this confidence is to evaluate the extent to which the model behaves in the same way, and has the same properties, as the real system. "Pattern-oriented modeling," explained in the next chapter, is a formal way of making this evaluation.

The purpose of analyzing and testing a model is to improve it. Improving a model may include simplifying it by excluding elements that turn out not to be essential, adjusting its resolution when elements turn out to be described too coarsely (or too finely), modifying its representation of processes or its structures, making it better at reproducing observations, making it easier to understand, and making it more predictive. Therefore, analyzing the model can lead to cycling back through tasks 4, 3, and 2—and even to task 1 if the analysis improves our understanding so much that we can again rephrase the original question in a more productive way.

However, we cannot stay in this modeling cycle forever. What we need is a "stopping rule" (Haefner 1996) defining the point at which we consider the model good enough. Haefner (1996) recommends that the stopping rule be specified at the start of the project because this rule has an influence on model design. For example, it makes a big difference for the model's design if the purpose of the model is to predict general trends rather than specific variables within a standard deviation of not more than, say, 5 percent. However, stopping rules, like all other elements of the modeling process, may change during model development. In the real world, probably the most important stopping rule is the constraints of time and resources: most modeling projects stop just because the funding is depleted. But even this seemingly trivial stopping rule should influence model design: when we know that we have resources for only two researchers for two years, we should reject model designs that are obviously too complex to be dealt with in only 2 × 2 person-years.

2.3.6   Task 6: Communicate the Model and Its Results

When we finally have a model in which we have sufficient confidence and which provides answers to our original question or problem, we are still not finished. This book is about scientific modeling, which means that we have to communicate both model and results to the scientific community, or to the managers who are going to use our models. Observations, experiments, findings, insights—all these become scientific only when communicated in a way that allows others to reproduce the observations and experiments independently and to achieve the same insights. The same holds for models. However, IBMs are hard to communicate because they cannot be described unambiguously with a few equations and parameters. Therefore, all the hard work of building and analyzing an IBM makes no contribution to science until we find a way to communicate our model completely and unambiguously. We must plan from the start of a project to devote sufficient resources to the task of communicating the model. In practice this may imply that we have to make our model even simpler—that we stop the cycle of model design and analysis even earlier so we have time for communication; and that we use tools for implementing and analyzing the model that facilitate communication, such as software platforms that are widely used and provide visual output. Communicating models and their results is no less important than developing and analyzing models. We therefore devote chapter 10 to this issue.

2.4   SUMMARY AND DISCUSSION

This chapter provides guidance on the entire modeling process. The guidance has three main elements: the fundamental understanding that a model is a *purposeful* representation, a list of powerful yet versatile model design heuristics, and a description of the general modeling cycle as six tasks. Following these general guidelines can make individual-based modeling more efficient and coherent than it typically has been.

The historic inefficiency and incoherence of many IBMs is not to be blamed on the IBMs' developers. These problems reflect an important gap in how we educate ecologists, conservation biologists, natural resource managers, and biologists in general. Few universities provide modeling courses at a level as basic as this chapter. Most textbooks on theoretical ecology focus on models but fail to address the *process* of modeling; and confine themselves to analytical models, not simulation models. To our knowledge, the books of Starfield, Smith, and Bleloch (1990), Starfield and Bleloch (1986), and Haefner (1996) are the only textbooks in biology, ecology, and natural resource management that really introduce the modeling process and which partly or completely (Haefner 1996) focus on simulation models. Computer simulation is an extremely powerful new tool for scientists, including biologists (Casti 1998). The modeling and software skills needed to do simulation well could be just as important as calculus and statistics, yet are rarely taught to biologists.

The most important general guidelines presented in this chapter are that:

- Modeling is problem-solving under constraints.
- A model is a purposeful representation.
- Modeling includes several tasks that must be cycled through repeatedly.
- Models should be simplified to the threshold of pain, at least initially.
- Parameterizing, analyzing, and communicating models is at least as important as designing the models, especially for simulation models.

But do these guidelines really apply to all kinds of models? Scientists may refer to completely different things when they talk of models, and many different categories of models have been labeled: tactical and strategic models (Holling 1966; May 1973); descriptive, complex systems, and conceptual models (Wissel 1989); predictive, explanatory, and prescriptive models (Casti 1998); minimal and synthetic models (Roughgarden et al. 1996), to name but a few of the categories identified in the literature (see also section 11.2). The differences among all these model types are due to the different purposes models can serve but are not a reflection of differences in the basic modeling process.

The main purposes of models in science are description, explanation (also referred to as understanding), and prediction. The criteria for telling good models from bad depend on the model type, which is determined by model purpose.

Models that provide good explanations may fail completely in prediction (Darwin's theory of natural selection being an example; Casti 1998); models almost perfect in prediction may fail in explanation (e.g., the Ptolemaic model of planetary movement; Casti 1998); and models can be good in both explanation and prediction, as is Newton's famous $F = ma$. However, there is no fundamental difference in the modeling process behind all these different models: a specific question was raised about a real system, then working hypotheses about essential processes were formulated, essential variables and parameters were specified, and—if possible—the model was implemented and analyzed. Likewise, classical ecological models and IBMs are different not because they originate from different modeling processes but because they are based on fundamentally different hypotheses of what elements of the real system need to be represented in a model.

Now, if everyone developing IBMs follows the general guidelines provided in this chapter, will IBMs become as effective and coherent as they could be? Not necessarily. The reason is that computer models are less constrained technically than mental or mathematical models. Even if we try hard to limit the complexity of an IBM, its design can still be complex compared with those of classical models. What we need are additional guidelines for coping with the complexity of bottom-up simulation models and IBMs. In the next chapter we introduce a strategy—pattern-oriented modeling—that supplements the guidelines in this chapter for organizing IBM projects and designing IBMs. The implementation, analysis, testing, and communication phases of modeling are also more difficult for simulation models and IBMs. Figure 2.1 is an overview of the general modeling cycle described in this chapter and also provides a road map of the following chapters that focus on each part of this cycle.

Chapter Three

Pattern-oriented Modeling

A change without a pattern is beyond science.

—Boris Zeide, 1991

3.1 INTRODUCTION

Models of complex systems should be neither too simple nor too complex if they are to be useful. Complexity in a model causes many difficulties, yet models that are too simple cannot explain much. To find the right level of complexity, we rely on the principle of parsimony, or "Occam's razor": if we have two models that both explain a phenomenon, we should stick to the simpler one. However, we must be careful to use Occam's razor appropriately: the simpler model should be chosen only if it really explains the phenomenon! With IBMs the phenomenon we want to explain is the mutual relationship between the behavior of individuals and ecological systems. If a model is too simple, for example, by ignoring individuals or their variability, we may not to be able to explain this mutual relationship. Albert Einstein famously restated the principle of parsimony as "things should be made as simple as possible, but not simpler." To modelers, Einstein's statement is a compelling summary of what we try to do.

The trade-off between the difficulties caused by complicated models and the need to explain complex systems means that with bottom-up models such as IBMs the relationship between a model's payoff and complexity no longer is monotonic and negative (as assumed for classical models) but hump-shaped (Figure 3.1). There is a zone of intermediate complexity where the payoff is high. This zone might be called the "Medawar zone," because Peter Medawar described a similar relationship between the difficulty of a scientific problem and its payoff (Loehle 1990). But how do we design models so that they end up in the Medawar zone? How can we find the "right" level of detail? How do we decide which aspects of a real ecological system should be included in a model and which should be ignored?

Before we can find the complexity that maximizes an IBM's payoff, we must ask what determines the payoff of a model. When we use IBMs for ecology, we are trying to learn something about the real world, not just about the properties of virtual worlds in the computer. Therefore, appropriate criteria for deciding

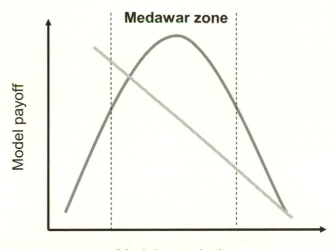

Figure 3.1 A model's payoff—how much we can learn from it—versus its complexity. For IBMs (*humped curve*), a zone of intermediate complexity—the "Medawar zone" (named after Peter Medawar; Loehle 1990)—exists where payoff is maximized. For analytical models of classical theoretical ecology (*straight line*), payoff is high only for very simple models and then declines continuously with complexity. This unequivocally negative effect of complexity is often—erroneously—assumed for IBMs and other bottom-up simulation models.

on a model's structure are those which allow us to *test* the model—to understand the IBM and to compare it with observed phenomena of the real world. An IBM's payoff is thus determined to a large degree by its testability (Grimm 1994). To develop IBMs that end up in the Medawar zone, we should therefore think about how we can make the models testable in the most productive way.

The way to make IBMs testable in a productive way is obvious yet underused in ecological modeling: models should be designed so that their properties and dynamics can be compared with patterns observed in real systems. If an IBM is too simple, realistic patterns will not emerge from it; if the model is too complex, we will not be able to understand *how* patterns emerge from it. Patterns thus provide the criteria for optimizing a model's complexity. Modeling that uses real patterns for designing, testing, and parameterizing models is referred to as pattern-oriented modeling (POM; Grimm 1994; Grimm et al. 1996; Grimm and Berger 2003; Wiegand et al. 2003). POM is useful for any kind of modeling, not just individual-based.

The fundamental idea of this chapter, that the structure of models be determined by the need to test the models against patterns, does not conflict with the basic guidance of chapter 2 that the structure of a model is determined by its purpose. Whatever scientific or management problem we want to solve with a model, the credibility and, therefore, the testability of the model will be important. Moreover, in science the patterns themselves are often the problems to be solved: the purpose of a model or theory is often to explain specific patterns.

In fact, science abounds with examples of patterns that determined the structure of models, not just analytical or simulation models but also our general understanding and theories of how important systems work. It is very worthwhile to scan introductory textbooks and historical accounts of scientific discovery for patterns and how they have been detected and used in physics, chemistry, biochemistry, genetics, geosciences, and similar fields. For example, biologists are familiar with how the double-helix structure of DNA was discovered by Watson and Crick; the identification and explanation of patterns (e.g., in specific x-ray photographs, and in the general behavior of nucleic acids and crystals) were critical to this discovery (Watson 1968). The more aware we are of how patterns determine the structure of succesful theories and models, the more likely it is that we detect and use patterns successfully ourselves.

In the following sections we discuss the significance of patterns for modeling and describe the four main tasks of POM within the modeling cycle describe in chapter 2. The remainder of the book, in particular chapter 6, contains numerous examples of POM and the use of patterns.

3.2   WHY PATTERNS, AND WHAT ARE PATTERNS?

A pattern is any display of order above random variation. The presence of a pattern indicates the presence of specific mechanisms that cause the pattern. Patterns that are defining characteristics of a system are likely to be indicators of *essential* underlying processes and structures; nonessential properties are unlikely to leave clearly identifiable traces in the system. Patterns thus provide information on the essential properties of a system, but this information is available only in a "coded" form. The purpose of POM is to "decode" this information (Wiegand et al. 2003).

The attempt to identify patterns and then to decode them to reveal the essential properties of a system is the basic research program of any science and of natural sciences in particular. Physics and other natural sciences provide numerous examples of patterns being the key to understanding the essence of systems: classical mechanics (Kepler's laws), quantum mechanics (atomic spectra), cosmology (the red shift), molecular genetics (Chargaff's rule), and paleontology (mass extinctions and the iridium layer at the Cretaceous boundary).

In ecology, however, this basic pattern-oriented research program seems to be less acknowledged (but see, e.g., Watt 1947; Levin 1992; Turchin 2003). Even when patterns have been addressed using ecological models, the pattern-oriented approach has usually not been explicitly used as a modeling strategy, so its full potential has not been realized. The well-known population cycles of snowshoe hares and lynx in Canada provide an instructive example. These cycles are certainly a pattern, but it is relatively easy to reproduce cycles with many kinds of mechanisms (Czárán 1998). Therefore, a number of different models could explain the cycle pattern—this pattern by itself provided no ability to falsify any of the alternative models, so we could not infer which model best explained the cycles. Recently, however, another pattern in the hare-lynx cycles was also considered: the period length of the cycles is almost constant, whereas the amplitude of the peaks varies chaotically (Blasius, Huppert, and Stone 1999). This second pattern could only be reproduced by a model with a specific structure—a food chain of vegetation, hare, and lynx—and a previously ignored mechanism: in times of low hare abundance the lynx may switch to other prey (presumably squirrels). Considering *both* patterns eliminated many of the competing models and greatly reduced the uncertainty in our understanding of the hare-lynx cycles.

This example is particularly revealing because of the frequent complaint that there are so few clear patterns in ecology. Two or more seemingly weak patterns (constant period *and* chaotic amplitude) may reveal more of the essence of a system than one single strong pattern because it is usually harder to reproduce multiple patterns simultaneously than to reproduce just one pattern. Multiple patterns often arise from different essential properties of a system, so addressing several patterns is a more powerful way to identify models that capture the full essence of a system (and to reject models that do not; see "multicriteria assessment" of models; J. H. Reynolds and Ford 1999; Ford 2000).

Unfortunately, empirical ecologists often use experimental designs that do not address patterns and therefore often have relatively little value for modeling. Field and laboratory experiments commonly use factorial designs that elegantly answer a few specific questions, but rarely provide the broad patterns of system response that help build and test models (individual-based or otherwise) that elicit underlying processes. Suter (1996) provides an example from ecological risk assessment: field studies often use replicate samples to determine whether some ecological indicator is statistically different at a contaminated site, compared with an uncontaminated site. Studies that instead look for patterns in how ecological indicators vary along gradients in contaminant concentration are much more likely to be useful for developing and testing models (and for supporting management decisions).

3.3   THE TASKS OF PATTERN-ORIENTED MODELING

The term "pattern-oriented modeling" is admittedly redundant because modeling should be pattern-oriented anyway. But many classical ecological models are not pattern-oriented but more "free-style": simple, unspecific, and not easily tested (Grimm 1994). Such free-style models are presented in many ecology textbooks and thus narrow the ecologists' view of what constitutes a useful model. The term "pattern-oriented" was introduced to remind us of the basic research program of science: ascertaining the essential properties of a system by decoding its patterns. The term POM is similar in purpose to "evidence-based medicine," a term introduced in medicine to remind doctors of the obvious, that their therapy should be based on evidence instead of tradition, gut feeling, or whatever (Gigerenzer 2002).

POM as described here is not genuinely new. Many modelers apply this method intuitively (e.g., Jeltsch 1992; Jeltsch et al. 1992; Jeltsch and Wissel 1994; Wood 1994; Ratz 1995; Johst and Brandl 1997; Lewellen and Vessey 1998; Blasius, Hupport, and Stone 1999; Casagrandi and Gatto 1999; Doak and Morris 1999; Bjørnstad et al. 1999; Briggs et al. 2000; Claessen, de Roos, and Persson 2000; Elliot et al. 2000; Turchin et al. 2000; Ellner et al. 2001; Fromentin et al. 2001). There have also been attempts to describe the usage of patterns as a general strategy (e.g., Kendall et al. 1999; Turchin 2003), but most of this work is concerned with selecting the most appropriate analytical model to reproduce a population time series. In contrast, POM as it is presented here encompasses modeling in general and in particular bottom-up models such as IBMs. DeAngelis and Mooij (2003) independently developed a notion of modeling that is very similar to POM, which they refer to as "mechanistically rich" modeling.

What is new about POM, as we describe it here, is making the use of patterns explicit and integral to a general protocol of ecological modeling. The main feature of POM is the use of *multiple patterns* to develop and test models and thereby identify essential elements of ecological systems. POM is not different from the general modeling cycle described in chapter 2, but augments this cycle. Figure 2.1 indicates where the following four tasks, specific to POM, fit into the modeling cycle.

3.3.1   Identify Multiple Patterns

The first POM task is to identify patterns to use in structuring and testing a model. Of primary importance are patterns that the model is designed to explain. When a model is designed to explain specific patterns, then the whole modeling project is easier because it has a specific target to constrain and guide model design and testing (Thulke et al. 1999). However, many models (including many IBMs) are designed to address a problem that is not captured in a well-known pattern. Instead, for example, we might design an IBM to predict a population's response to some alteration that has never been observed: how would a beech forest's age structure change if different harvest practices are used? How would a river fish population respond to major changes in the magnitude and variability of flow?

Identifying the patterns against which a model will be tested is also a very important part of this task, and these patterns can be just as important to a model's success as identifying the problem or pattern a model addresses. The confidence we develop in a model depends on how we test it against these patterns. Models used to predict system response to unobserved conditions must especially be tested thoroughly to develop confidence in the mechanisms underlying system responses. Further, the diversity of patterns used to test an IBM determines what structures and processes must be in the model. Patterns that all emerge from the same structures and processes might be successfully reproduced by a simple IBM, but a variety of patterns emerging from many structures and processes can only be reproduced by a more complex but structurally realistic model.

In selecting the patterns used to structure and then to test and parameterize an IBM, it is important not to focus only on "strong" patterns. Strong patterns are strikingly different from random variation and therefore seem to be strong indicators of underlying processes. However, as the hare-lynx population cycle example shows, a single strong pattern may be not sufficient to eliminate competing explanations and identify the most appropriate model structure. A combination of several seemingly "weak" patterns can be more powerful for finding good model designs (Railsback 2001b; Railsback and Harvey 2002): each pattern may eliminate a competing model design so the combination of multiple patterns leaves only one clear way to proceed. The power of multiple kinds of weak information is widely known. For example, it is virtually impossible to identify a person if we know only their age, but as we add information such as their sex, profession, and birthplace, we rapidly eliminate alternatives and focus on the right person. In the subsequent tasks of POM, we use each pattern as a test that can reject false hypotheses for how some part of an IBM should be represented. A variety of simple or weak patterns can provide a rapid and easy way to filter out unuseful approaches and to identify useful model designs.

Not only system-level patterns, but also patterns at the bottom level—individual behavior, local dynamics—are both powerful and necessary in POM. An IBM may reproduce system-level patterns well, but if the behavior of the individuals is unrealistic then the IBM is not yet useful. This situation has in fact been quite common in IBMs. A good example is a very simple IBM of salmon migration down a large river-reservoir system that one of us helped develop. With calibration, this IBM could reproduce observed population-level patterns of migration timing; however, examining the individuals found that calibrated swimming speeds were far faster than real salmon can swim. Only the examination of individual-level patterns revealed that the IBM's structure was inadequate. Like system-level patterns, individual-level patterns used to structure and test an IBM must not be hard-wired into the model but must emerge from the interaction of the individuals with each other and their environment (section 5.2).

A caveat with identifying multiple patterns is that the patterns need to be compatible with each other in the processes they emerge from and the scales at which they occur. Patterns caused by processes not important to a model's objectives will not be useful for testing the model (or, worse, can distract modelers from the task at hand and tempt them to add unnecessary complexity to the model). Likewise, patterns occurring at spatial or temporal resolutions incompatible with a model will not be useful—we cannot test an IBM that uses daily time steps and 100-meter grid cells by how well it reproduces patterns of individual interactions that occur minute to minute over a few meters.

The following discussions of how patterns are used shows how selecting patterns, especially for testing an IBM, tends to be an iterative process, with new patterns being chosen as model design and testing proceeds.

3.3.2   Use Patterns to Design an IBM

In POM, we use patterns along with (or as) the problem the model addresses, to guide design of the model's structure, resolution, and processes. This is the general idea of POM: we decide to use a pattern to guide model design because we believe the pattern contains information about essential structures and processes. To reveal this essential information we provide model structure, resolution, and processes that make it possible for the observed pattern to emerge. The model must include the state variables through which the patterns are expressed and the processes that cause the patterns to emerge, and must use spatial and temporal scales over which the patterns are detected.

Patterns can guide model structure by showing us what kinds of object or entity need to be in an IBM and what state variables are needed. For example, in designing the BEFORE beech forest IBM (section 6.8.3), the modelers identified patterns in vertical structure (single- vs. multilayered canopies) as characteristic of real forests. This pattern suggested that the model should not only represent the horizontal structure of natural beech forests, as in earlier models (Wissel 1992b), but also the vertical structure (Neuert 1999; Rademacher et al. 2004). Therefore, the modelers designed a simple vertical structure with trees grouped into four different height classes. This design does not hardwire the vertical structure of a beech forest into the model but allows the modelers to test the IBM by whether typical vertical structures emerge or not. Other examples of how patterns determine and constrain model structure include:

- Using spatial patterns requires the model to be spatially explicit (e.g., Levin 1992).
- Temporal patterns (including constancy) in population age structure require the model to represent individual age and an age-structured population.

- Patterns in how life history strategies vary among biotic or abiotic environments require the IBM to represent the individuals' life cycle (Uchmański and Grimm 1996).
- In benthic marine systems, patterns in larval settlement with elevation require representation of topography (Grimm, Günther, et al. 1999).
- Patterns in how habitat selection depends on growth potential and mortality risk require that the IBM represent how both growth and mortality risk vary over space and time (Railsback and Harvey 2002).

Patterns guide the selection of a model's spatial and temporal resolution because the model's resolution must be compatible with the resolutions over which the pattern is detected and over which the processes causing the pattern act. For example, the IBM built to explain fish schooling patterns (section 6.2.2) needs very fine spatial and temporal grains because the processes that cause schooling happen very rapidly and over distances of a few centimeters. The spatial extent of this IBM must be large enough to include the whole fish school and the distance it moves while the schooling patterns of interest emerge. The temporal extent must also be long enough for the schooling patterns to emerge fully. In strong contrast, the beech forest model (section 6.8.3) was designed to explain patterns of forest dynamics that take place over long time periods and large distances, so the model was designed with a large spatial extent, long time steps (fifteen years), and long temporal extent (with model runs of a thousand years and more). However, these same patterns are caused by processes (interactions among individual trees) that occur over small distances. These patterns determined that the IBM needed a relatively fine spatial grain of a few meters.

Once an IBM's general structure and resolution have been determined, patterns are then used to determine what processes need to be represented. Of course we cannot know, simply by looking at patterns, what processes caused them to arise; yet patterns can provide strong clues that help us start deciding what processes need to be represented in an IBM and how. We can examine the observed patterns that we want our IBM to explain or be tested against and use our judgment (and as much knowledge of the real system as we can accumulate!) to make a good guess at the processes causing the patterns. We can, for example, determine what environmental variables or individual interactions appear responsible for the patterns. What individual- or system-level responses characterize the pattern? How do the environmental variables or interactions at the bottom of the pattern appear to be linked to the responses that characterize the pattern?

The herring migration case study (section 6.2.3) provides a good example of how a pattern can point to underlying processes. In this case, the IBM was designed to explain an observed pattern with two components. First, the observed behavior was a sudden change in where a herring school overwinters— usually the school overwinters in the same location year after year, but in occasional years the location changes. The second component of the pattern was the observation that these changes in overwintering location occur in years when the experienced, adult migrants made up an exceptionally low fraction of the whole school. Thinking about how the observed change in an individual's biological environment—fewer experienced migrants as neighbors in the school—might be a cause for the system-level response of a new migration destination immediately suggested to the modelers (who were already familiar with the theory that schooling fish simply follow their neighbors; Huth and Wissel 1992) a process causing the pattern.

Telling us what we do *not* need to include in an IBM is a crucial benefit of POM. IBMs probably suffer more often from too many variables and processes than from too few, and patterns provide criteria for leaving things out. If variables or processes are not necessary to reproduce the patterns of interest, they can be left out. Even better, when we use explaining patterns as a filter for putting things *into* an IBM, then the model will automatically be as simple as it can be without being too simple (Wiegand et al. 2003). In contrast, using "realism" as a guide for model structure provides no criteria for what can be left out.

POM is valuable for suggesting model structures and processes, but simply suggesting hypotheses for a model's design is not enough. We want to test these hypotheses, and we need to test *how* we represent processes in an IBM; so we move on to the next task.

3.3.3   Use Patterns in Model Analysis and Testing

As we discussed in chapter 2, model testing and analysis requires a currency that we can use to compare alternative model versions. What measure of an IBM's "goodness" can we use to compare versions and identify the best? POM provides such a currency: we can evaluate alternative models by how well they reproduce the patterns we have observed in real systems.

Pattern-oriented testing and analysis of an IBM can proceed by identifying alternative versions of the IBM and then seeing how well each version reproduces the patterns selected in task 1 for this purpose. Alternative versions may differ by using different representations of an important process, for example, different rules for how the individuals make some key decision (see the next chapter on "IBE theory"). Or alternative versions may differ only in using different parameter values or in having fundamentally different structures.

At this point the benefits of using multiple patterns for testing are very apparent. We should be very happy if each of the alternative models reproduces some of the patterns but only one model reproduces *all* of the patterns: then we can be confident that we not only have identified a useful model but have also found a powerful way to falsify hypothesized models. However, if none of the model versions can reproduce all the patterns well (and we remain confident in the realism and appropriateness of the patterns), then we must conclude that the IBM is not capturing the system's essential characteristics that are captured by the patterns. If several alternative models reproduce all the patterns, we can (if we want) search for new patterns that have the power to falsify one of the models; this search may require additional study of the system being modeled.

Once we find a model version that adequately reproduces all the observed patterns, we can analyze the model in more detail. This analysis (the subject of chapter 9) addresses the following kinds of questions. Which processes and structures of the model are responsible for the patterns? Can the model be simplified while still reproducing the patterns (sections 9.4.4, 11.4.2)? And, of course, we must still tackle the original problem that the model was intended to solve, perhaps using the same pattern-oriented approach. These kinds of analysis can be informal or can be conducted as inferential hypothesis testing with the patterns (and, sometimes, statistical analyses such as those described by Hilborn and Mangel 1997) used to exclude inadequate models.

The use of patterns to test and analyze an IBM has implication for the model's software. The software, right from the beginning, needs to provide efficient tools for observing and comparing patterns. Because human perception is mainly visual, visual model output—graphical interfaces—often facilitates pattern comparison (Grimm 2002; Mullon et al. 2003; chapters 9 and 10).

3.3.4 Use Patterns for Parameterization

Finding appropriate parameter values is almost always an issue in ecological modeling. Model output can depend both quantitatively and qualitatively on parameter values, but usually we have precise values for few parameters. For other parameters we may be able to specify biologically meaningful ranges, while some parameter values may be completely unknown. If values of too many parameters are too uncertain, the model's output may be too uncertain to allow the model to be tested or to answer the original question it was intended for.

With POM, patterns can be used to determine parameter values indirectly (Wiegand et al. 2003; Wiegand, Revilla, and Knauer 2004; Wiegand, Knauer, et al. 2004). This is similar to the traditional method of calibration where a parameter is tuned until model output shows some desired behavior. The new aspect of calibration with POM is that not only one but several parameters can be simultaneously determined by calibration, not despite but because of the model's complexity. The traditional view on model complexity is that it has only disadvantages, but if the model was constructed following the POM approach, it is likely to be structurally realistic so that its richness in structure and mechanisms can be exploited to estimate parameter values. If a model is capable of reproducing multiple patterns simultaneously, then parameter values can be estimated by excluding values that produce model output not matching the observed patterns. For example, Hanski (1994, 1999; Wiegand et al. 2003) uses this kind of *indirect parameterization* to find values of parameters of real metapopulations. His simple but structurally realistic metapopulation model (the Incidence Function Model) includes the position and size of habitat patches and a simple relationship between patch size and local extinction risk. Parameters for this relation are estimated by fitting model output to empirical presence-absence data from the real network of patches. Similarly, Wiegand et al. (1998) used specific patterns in the census time series of brown bears (including information about family structure) to narrow down the uncertainty in a model's demographic parameters. (This pattern-oriented parameterization is well known in other disciplines as "inverse modeling"; e.g., Burnham and Anderson 1998.) We discuss pattern-oriented, indirect parameterization more thoroughly in chapter 9.

3.4 DISCUSSION

In this chapter we describe what pattern-oriented ecological modeling is, why it is useful, and how it can be used within the modeling cycle. Patterns can guide how we design a model's structure and resolution, provide a currency for testing and comparing versions of a model, and allow parameters to be determined indirectly. Still, this discussion does not prove that using patterns helps us find the range of model complexity where the model's payoff is greatest. To that end, we present in chapter 6 several studies where the pattern-oriented approach led to particularly successful IBMs. A great part of this success was that the pattern-oriented IBMs turned out to be "structurally realistic," which means they succeeded in capturing the essence of an ecological system so well that they could produce testable, independent, predictions of system properties that were not even considered during the IBM's development and testing. When such independent predictions can be made and validated, we have even greater confidence that the model captures the system properties underlying the patterns we built the model to explain (section 9.9).

Of course, POM has its limitations and problems. First of all, patterns must be chosen carefully. Of primary concern is making sure the patterns chosen to guide model design and testing are themselves realistic. The human mind is inclined to perceive patterns all the time, even if they do not exist, and in field biology and natural history myths can be common. It is not unusual for patterns (e.g., "Species *A* is more fond of habitat type *X* than is species *B*") to be widely believed and sometimes reported in the literature even though poorly supported by evidence. Especially common are patterns that may emerge from site-specific conditions and then be misinterpreted as universal. (IBMs can be great tools for exposing such misinterpretations.) DeAngelis and Mooij (2003) discuss an example in which a flawed census time series was used as a pattern for parameterization. On the other hand, there may be patterns that are generally realistic and very useful despite occasional exceptions. The modeler must carefully scrutinize the literature supporting patterns and be wary of experimental design artifacts or site-specific conditions that reduce a pattern's value.

A second limitation is that we must never naively infer that the mechanisms producing a pattern in a model must be the mechanisms at work in the real system. The pattern produced by a model may be correct while the model's mechanisms are completely wrong. The most prominent example is the Ptolemaic model of planetary movement, which reproduced the planet's trajectories in the sky well while assuming they revolve around the earth (Casti 1998). In ecology, multiple models exist to explain such patterns as cyclic population dynamics, the species-area relationship of MacArthur and Wilson (1967), and the linearity and slope of the self-thinning trajectory of plant monocultures. Obviously, several different models for the same pattern cannot all be true—most of them must be partly, or even completely, wrong.

These two potential problems are additional reasons why it is valuable to (1) build an IBM around a specific, well-understood pattern or problem, and (2) at least start the pattern-oriented analysis of an IBM using a variety of simple, well-documented, and well-understood patterns. Care should be taken to search for many patterns and use them gradually, pattern by pattern, to increase confidence in the model. The highest degree of confidence is provided by successful independent predictions, which indicate that the model is structurally realistic. But even after successfully testing a model's independent predictions, we must remember that the model is still just a model and will never be able to represent all aspects of the real system. Modeling and models have a momentum of their own (Grimm 1994), and models that are initially successful pose the risk that modelers will stop distinguishing sufficiently between the real system and their model. (The ethologist Konrad Lorenz recommended that a good exercise for scientists is, every day before breakfast, to discard a pet hypothesis.)

A third potential limitation of POM, discussed extensively in succeeding chapters, is the question of *emergence*: do the patterns in the IBM really emerge from the model's structure or are they accidentally *imposed*: hardwired into the model by giving individuals behaviors that force the pattern to arise? A good way to test whether patterns have been imposed on an IBM is to replace rules for individual behavior with ones that are obviously absurd and see if the patterns are still produced. For example, the beech forest model BEFORE (section 6.8.3) could be tested by assuming a young tree dies whenever it is shaded at all, or never dies but waits forever to grow into a canopy opening. If the forest model still produced realistic patterns with these absurd rules implemented, then the patterns probably were somehow hardwired into the model structure.

POM is certainly not a panacea for ecological modeling in general or for bottom-up modeling in particular. It is also not a simple step-by-step recipe, but a general approach to be adapted to each problem and model. Nevertheless, POM clearly helps us avoid the pitfalls of both "free-style" models that lack testability and relevance to real systems, and overly detailed IBMs that are too complex to understand. Likewise, thinking about POM can help empirical ecologists develop study designs that, in combination with models, are more likely to improve our understanding of ecological mechanisms. Using POM is, after all, simply to follow the general research approach of all science: to search systematically for the mechanisms underlying the patterns we observe.

PART 2

Individual-based Ecology

Chapter Four

Theory in Individual-based Ecology

The most important challenge for ecologists remains to understand
the linkages between what is going on at the level of the physiology
and behavior of individual organisms and emergent properties
such as the productivity and resiliency of ecosystems.
                                                    —*Simon Levin, 1999*

4.1 INTRODUCTION

In chapter 1 we identified the complexity of IBMs as a major challenge to their
efficient and coherent use. Therefore, in chapters 2 and 3 we presented general
modeling guidelines and pattern-oriented modeling as ways to help modelers
end up in the "Medawar zone" of complexity that provides a high payoff (figure 3.1). However, these techniques are not sufficient to provide *coherence*
to the use of IBMs. By coherence we mean that different IBMs are consistent with each other in some important, general, ways. Without coherence,
it is difficult to compare the structure of different IBMs and to integrate the
insights produced by individual IBMs into a general body of knowledge. Consistency and coherence among models are—in any science—the role of theory.
Without coherence we will have no theory, and without theory IBMs will lack
coherence.

A key conclusion of Grimm's (1999) review of more than fifty early IBMs
was that the insights gained from individual IBMs could not easily be inte-
grated into a general understanding of how either ecosystems or IBMs work.
An important reason for this difficulty was discussed in chapter 2: the structure of a model is largely determined by its purpose. If different IBMs have
different purposes and no common underlying theme, they cannot be coherent.
Individual-based modeling has lacked a master plan or theoretical framework that we can use to relate different IBMs to each other, develop general
knowledge, and reduce the need to reinvent, in each new IBM, how we model
individuals.

Now that we can take advantage of the experience provided by pioneering IBMs and new approaches to theory in CAS (section 5.1), we are in a
much better position to establish an approach to theory for individual-based

ecology (IBE) and IBMs. The theoretical framework we propose in this book formulates *theories* of the *adaptive behaviors of individuals* and tests the theories by seeing how well they reproduce, in an IBM, *patterns observed at the system level*. These theories are in one sense just models or hypotheses, but calling them theories emphasizes that they are the basic, underlying elements of IBE: models of individual behavior that explain system dynamics. These models are "theories" because they, after maturation, can be reused in different IBMs to explain a broad range of phenomena. This concept of theory is widely used in other sciences: when engineers want to predict how a complex electronic circuit behaves, or how a large building responds to novel loads, they build models based on simple theory of how individual elements of these systems—transistors and resistors, steel beams—behave and interact. While we use the word "theories" for models of individual behavior, we use the term "IBE theory" for the whole body of theories in IBE (including theories about ecological systems produced via IBE) and the process by which we develop theories. (A further note on terminology: we use "trait" as a general term for a model of something individuals do in an IBM. A theory is a special kind of trait.)

IBE theory specifically addresses the challenge laid out by Simon Levin in this chapter's motto. In IBE we model system properties as emerging from the traits of individuals. Therefore, IBE theory must provide a means of explaining the links between individual traits and system behavior. However, we are not merely interested in seeing what system patterns emerge from some specific individual trait. Instead, we want to identify and understand the individual traits that give rise to the specific system behaviors we are interested in. Conventional ecological theory does not address these links: the classical theory of population ecology includes models built only at the population or community level, and the theory of behavioral ecology typically addresses only the individual level and not how individual behavior explains emergent system behavior. In IBE, therefore, we need to develop new theory that links the individual and higher levels. The lack of such theory has been a great obstacle both to understanding ecosystem complexity and to the practice of using IBMs.

In this chapter we propose a process for developing theory of how individual traits explain system behaviors. The goal of the process is to find theories of how individuals interact with each other and the environment that have been tested and shown to produce, in IBMs, useful representations of population-level behavior. These theories will eventually fill a toolbox that can be used to assemble IBMs for theoretical and applied studies. Near the end of this chapter we describe an example of IBE theory development using the individual-based, pattern-oriented analysis process we advocate. Various terms are defined in this chapter; these are also collected in the glossary.

4.2 BASIS FOR THEORY IN IBE

The approach to theory we present in this chapter is a direct adaptation of the conventional scientific method attributed to Francis Bacon: the cycle of proposing hypotheses, devising and conducting experiments with the power to distinguish the most valid hypotheses, and repeating the cycle with refined hypotheses and more powerful experiments. Platt (1964) argued that this cycle is critical for propelling science forward. This conventional scientific method has previously been adapted to the problem of understanding the relationship between individual traits and system behavior in complex systems. Auyang (1998) outlines a general approach, referred to as "synthetic microanalysis," for developing theory in such systems. Auyang's approach, adapted to our ecological context, includes devising theories for the traits of individuals, representing these theories in an IBM of the system, and testing the theories by how well the IBM then reproduces patterns observed in nature at the system level. This approach has several characteristics which clearly distinguish it from other approaches to ecological theory (Auyang 1998).

- Theory is neither holist (system-level) nor reductionist (individual-level). We do not assume that ecological systems can be understood from only the system level, but we also do not assume that a system is simply the sum of its individual parts. Systems have properties of completely different types than the properties of individuals, and theory must explain these system properties.
- Theory must therefore be multilevel, linking traits of individuals to properties of the system. We are not interested in understanding all aspects of individual behavior but instead are interested in developing models of individuals that explain important system properties.
- Observational and experimental science at both the individual and system level is the basis of theory development. Such empirical science is important both for discovering the phenomena driving the system and for testing theories.

The merits of a hypothesis-testing approach to ecological theory have been debated quite famously. In particular are objections to the notion of falsifying hypotheses against data: (1) our data are often highly uncertain or influenced by preconceived ideas of theory, and (2) all our theories are wrong at some level, so rejecting wrong theories leaves us with nothing (Chitty 1996; Fagerström 1987; Turchin 2003, chap. 1). Our approach accommodates these realities, first by using a wide range of evidence to evaluate alternative theories, not rigid tests against single data sets; and by evaluating theories not by whether they are "wrong" but by how useful they are for solving specific problems in specific contexts.

4.3 GOALS OF IBE THEORY

The process we outline here for developing IBE theory has the goal of producing theories with four critical qualities: testability, generality, integration among levels of ecology, and usefulness in applied ecology.

4.3.1 Testability

Certainly, "minimal" models (Roughgarden et al. 1996; section 11.2) are important aids for thinking about ecosystems: devising concepts and exploring logical possibilities. But at some stage ecologists want to learn something about the real world, not only about our ability to think about the world. Therefore, the ability to test hypothesized models with increasing levels of rigor is critical to making progress. For theory to be testable, hypothesized models must lead to specific predictions, and there must be practical means of testing the predictions with increasing levels of rigor. The ability to use an IBM as a virtual laboratory for testing theory is an important advantage of IBE. Whereas population-level ecological models have often proved difficult to test, IBMs can make many easily tested predictions at both the individual and population level (Murdoch et al. 1992; DeAngelis and Mooij 2003). The pattern-oriented process we propose in this chapter can be a practical and powerful way to compare the validity of alternative theories: it allows us to test models of individual traits, often even before the IBM used in testing is fully calibrated to any particular site or data set.

4.3.2 Generality

Intuitively, we think of "general" theory as being independent of specific contexts. Theory in physics is very general in this way, but physics deals with matter and forces, which are indeed independent of history and context. Seeking the same sort of generality in ecology has not proved to be very productive (Grimm 1999; Ghilarov 2001). Organisms are not atoms, and the "forces" of ecology are not fundamental properties of matter and space but emerge from interaction among individuals and their environment. In ecology, therefore, useful theories are likely to be context-specific and the search for generality must include the search for the limits of this generality. Any theory or statement about an ecological system is specific to how we define the system (Jax, Jones, and Pickett 1998) and how we view the system: at which scales, using which state variables, considering which types of disturbances, and so forth (Levin 1992; Grimm and Wissel 1997).

The quest for general theories at the population and community levels has often neglected the search for appropriate context. Many ecological theories at the system level are context-free—for example, they do not consider temporal or spatial resolution and variability. In contrast, IBE seems to force us to deal with context because IBE theory focuses on what individuals *do*: how they adapt to their situation, an inherently context-specific problem.

The need to address context in IBMs has led some to assume that IBMs are always specific instead of general, especially that they are highly dependent on context-specific data instead of being built from general theory. While many early IBMs have indeed been very specific, the IBE theory development process is designed to identify general models of individual behavior. The process helps us find theories that work in many contexts while also (just as importantly) delineating the contexts in which a theory does and does not apply. Instead of thinking of general ecological theory as context-free, we think of theory in IBE as being general when we know it works in many specific contexts.

The search for general theory of how ecological systems function may be more productive when we look at how system functions emerge from individual traits because, using IBMs, we can test theories in many different contexts. Another reason is that complex systems are often much easier to understand when we consider the individual level. For centuries, astronomers struggled to understand the apparently complex movement of celestial bodies; but as soon as Newton developed a general model of gravitational interaction between individual bodies it became possible to predict accurately the motion of complex celestial systems. (Newton's law of gravity is an excellent example of a theory that is oversimplified at the individual level yet extremely useful for modeling systems.) Even though organisms are not as simple as planets, many kinds of organisms do many similar things (feed, grow, move, reproduce, etc.) so we can expect—and verify—that some individual traits are useful in IBMs of different species and systems. But a more important reason to expect individual-level theory to be general is that at the individual level, we have a very firm foundation for modeling behavior: the theory of evolution. The theory that the genetic traits of organisms have evolved because they provide fitness is one of the most powerful and least controversial concepts in biology and (as we discuss in chapter 5) can be directly applied to IBE theory.

4.3.3 Integration across Levels

Traditionally, ecology has been conducted at discrete levels: behavioral ecology focuses on what individuals do, population and community ecology address higher levels of organization, and evolutionary ecology addresses longer time scales and genetic adaptation. Ecologists are generally aware that these levels are not really separate: population dynamics, for example, are affected both by individual behavior and community factors such as interspecific competition. Ecological processes are important drivers of evolution, and evolution can affect community dynamics. An important aspect of IBE and its theory is that it links the various levels of ecology, contributing to the unification of theory promised by Huston, DeAngelis, and Post (1988). In the theory development process we describe in this chapter, natural history becomes an important resource in developing and testing theory, and theories of individual behavior (the realm of behavioral ecology) are tested and refined specifically to explain phenomena observed in population, community, and evolutionary ecology.

4.3.4   Usefulness in Applied Ecology

We believe strongly that both ecological theory and environmental management benefit when theory development is closely tied to applied ecology. Obviously, theories that have been tested and shown valid can contribute to better management of the systems they address. On the other hand, when theory development proceeds without being closely tied to real-world applications, there is a risk of theories being developed for reasons (e.g., mathematical or conceptual elegance) other than providing models useful for understanding real ecosystems (Suter 1981). Another risk when theory development is not tied to management applications is the temptation to search for instances in nature where a theory appears valid instead of searching for theories that best explain specific problems. When we closely link theory to management applications, we force our theory to solve specific problems; finding theory that solves specific problems is a productive path toward finding theory that is general in the sense of solving many problems.

   IBE theory is intended to be as useful for applied ecology as for basic ecology. At the individual level, the distinction between applied and theoretical ecology is actually hard to make; organisms often exhibit the same adaptive behaviors in response to both natural and human-influenced situations. There are, of course, types of disturbance caused only by man—for example, pesticide pollution. However, the IBE theory development process can be used without modification to develop models of how man-made disturbances affect individuals and ecosystems.

4.4   THEORY STRUCTURE

In this section we describe the general structure of our concept of IBE theory. Major elements of a body of theory are axioms, theories, and testing (or proof); we define what these elements are in IBE.

4.4.1   Fundamental Axiom

An axiom is a central assumption underlying a body of theory. Our approach to theory for IBE has one fundamental axiom: that phenomena occurring at higher levels of observation (populations, communities) emerge from the traits of individuals and characteristics of the environment that together determine how individuals interact with each other and their environment. In other words, IBE theory is based on the assumption that the dynamics of an ecological system can be modeled usefully by (1) modeling the characteristics of the system's environment that affect its individuals, (2) modeling the traits of the system's individuals that affect how the individuals interact with each other and their environment, and (3) simulating these interactions in an IBM.

4.4.2 Theories

The terms "theory" and "hypothesis" have many different meanings in science. In some sciences, hypotheses are not accepted as theories until they have undergone tests considered sufficient to exclude all alternative hypotheses, or until their ability to make accurate independent predictions is well established. In ecology the word "theory" traditionally has been applied to models in general, without regard to how they have been tested. We take a middle approach and call individual traits "theories" after they have been tested and shown useful in some specific contexts.

We define theories in IBE to be:

> Models of individual behavior that are useful for explaining population level phenomena in specific contexts, with contexts being characterized by the biotic and abiotic environment, sometimes including the individual's own state.

This definition means that in IBE we are not interested in theory that explains individual behavior in detail, only in models of individual behaviors that are readily used in IBMs to explain population dynamics. Often, IBE theory is a highly simplified representation of the real behavior of individuals.

Our focus in this chapter is on theories that are not easily tested in the field or laboratory. Many individual traits that are crucial for IBMs can be formulated and tested using controlled experiments on real organisms. For example, many IBMs use models of how an individual's growth varies with energy intake and metabolism; such models can often be developed and parameterized in the laboratory with relative ease. Other behaviors, however, are complex responses to a wide range of stimuli and therefore are not readily modeled using only empirical experiments. Useful models of adaptive behavior, crucial to IBMs, are often difficult to develop and validate using only field or laboratory observations.

4.4.3 Testing

Testing is essential to theory development in any science, as it is the process by which we eliminate inadequate hypotheses and establish the credibility of useful ones. In complex sciences we do not hope to prove that a theory is "true" but instead look for evidence of how general and useful a theory is. The validity of a theory increases as it is shown better able than alternative theories to reproduce observed phenomena in tests of increasing rigor.

Two kinds of experiments can be important in testing theory in IBE: experiments on real organisms and the pattern-oriented IBM analyses that we discuss later. Scientists using both kinds of experiment can follow the scientific method using strong inference, designing and conducting experiments to distinguish among alternative theories and then refining and retesting the theories.

4.5   THEORY DEVELOPMENT CYCLE

Because they are models of what individuals do, theories in IBE can be developed and tested using a cycle almost identical to the cycle for developing pattern-oriented models that we describe in chapters 2 and 3. Here, we adapt the pattern-oriented modeling cycle specifically to develop and test IBE theory. This cycle has six phases, corresponding to the tasks of modeling and pattern-oriented modeling described in previous chapters. The theory development cycle compares alternative theories by seeing how well each theory, when implemented in an IBM, causes realistic patterns of individual and system behavior to emerge. (The cycle is illustrated with an example in section 4.6.)

4.5.1   Phase 1: Define the Trait and Ecological Contexts of Interest

The first step in any research should be to define the problem. In developing theory for IBE, the problem we address is to find useful models for a specific behavior of a specific kind of organism. Even though we may be trying to find "general" theories that are useful in many contexts, we must also define the contexts in which the theory will be tested and used.

4.5.2   Phase 2: Propose Alternative Theories

In this phase, the researcher identifies or devises potential theories for the individual-level trait of interest. The trait is a model of how individuals *decide*, literally or metaphorically, what to do in certain situations. This step is often an opportunity for the individual-based ecologist to take advantage of the existing natural history, autecology, and behavioral ecology of the organisms being studied. Proposed theories should be thoroughly grounded in the natural history and autecology of the organisms (Hengeveld and Walter 1999; Walter and Hengeveld 2000), and sometimes can fruitfully be based on existing theory of behavioral ecology (e.g., Sutherland 1996). However, our experience has been that existing empirical and theoretical science at the individual level does not always provide theories "ready to go" for IBE. Much of the existing ecological literature is descriptive, and many studies are too restricted in scope to test the generality of the potential theories they suggest. The individuals in many IBMs are confronted with a wide range of situations they must adapt to, whereas empirical research on behavior is usually restricted by practicalities to addressing only a few situations. Existing knowledge is essential for suggesting and screening potential IBE theories, but synthesis and adaptation of theoretical approaches (discussed extensively in chapter 5) are typically necessary.

Platt (1964) suggested that theory development is most rapid when *alternative* models are proposed and then tested using experiments designed to determine which model best predicts observations (see also Hilborn and Mangel 1997). Comparing alternative theories also helps researchers avoid the risks of becoming too attached to any one approach (Platt 1964). Alternative theories for specific individual traits can be generated from fundamentally different approaches to understanding the behavior of interest; for example, alternative traits could be developed from the fitness maximization and simple heuristics approaches to decision theory (section 7.5). On the other hand, it is also often necessary to compare alternative theories that are conceptually similar but different in their details.

4.5.3   Phase 3: Identify Test Patterns

In phase 5 proposed theories are tested by analyzing how well they explain a set of test patterns in individual-based simulations. These test patterns are identified in phase 3 because their selection can affect design and implementation of the IBM used for testing. Identifying the test patterns is a critical step because these patterns are the data set against which theories are tested. The validity of a theory is limited by the power of the tests it has been subjected to, and the choice of test patterns determines how much power the analyses have to distinguish among alternative theories and establish the range of contexts in which a theory is useful.

Test patterns are patterns of behavior that emerge in an IBM and depend on the individual trait for which theories are being tested. These patterns can occur at the individual level and at higher levels. Test patterns can range from very simple individual behaviors to extensive time series of population responses, or spatial patterns. Qualitative patterns can be very useful, at least early in the development of a theory, because a theory can be tested against qualitative patterns before the IBM used in testing (phase 4) is fully parameterized and calibrated.

4.5.4   Phase 4: Implement the Proposed Theories in an IBM

An IBM provides the virtual ecological system in which proposed theories are tested. When a theory is to be tested, it is implemented as an individual trait in an IBM. The IBM must also represent the biological and environmental processes to which the trait responds, and any other processes that are necessary to reproduce the test patterns.

As a test-bed for theory, IBMs have many advantages over natural systems, especially the abilities to test a theory under many conditions, to conduct completely controlled and reproducible experiments, and to make thorough and accurate observations without affecting results. A major disadvantage of this approach is, of course, that the extent to which results obtained from an IBM apply to real ecosystems depends on the design and implementation of the IBM. This is a chicken-or-the-egg problem: we implement theories in an IBM to test them, but the IBM must be built from theories. A flawed IBM could certainly produce misleading conclusions about the validity of a proposed theory. However, this problem is avoidable because IBMs can be customized to test one trait at a time. By judicious selection of what processes are modeled as emergent versus imposed (chapter 5), an IBM can be customized so that everything other than the theory being tested is "hardwired" and noncontroversial. A theory that has been successfully tested can of course then become part of an IBM used to test other theories.

4.5.5   Phase 5: Analyze the IBM to Test the Proposed Theories

A proposed theory for an individual trait is tested by determining whether it, when implemented in an IBM, causes the test patterns to emerge. For each test pattern, the IBM is used to simulate the conditions under which the pattern is expected to emerge, and observations are taken from the IBM to determine whether the pattern was indeed reproduced. (Analysis methods are discussed further in chapter 9.)

The analysis of the IBM must address more than just whether the proposed theory succeeded in reproducing the test patterns. As in any model-testing exercise, it is also important to consider and document the inferential power of the test: its ability to distinguish among alternative theories. Using the pattern-oriented approach, statistical measures of inferential power usually are not feasible (but see Hilborn and Mangel 1997). Instead, we can qualitatively evaluate the power of a theory test by addressing such questions as:

• How many test patterns were used, and how general or specific were they? It is very important to understand the concept, which may at first seem counterintuitive, that testing proposed theories against a wide variety of specific patterns is better able to distinguish the most general theory than is testing theories against general patterns. The more general a pattern is, the more likely it is that multiple traits are able to reproduce it. On the other hand, a single trait that can reproduce a wide variety of specific patterns (as illustrated in section 4.6) is most likely to be a useful and general theory.

- Did more than one proposed theory explain the test patterns? If so, then the test patterns did not provide sufficient inferential power to distinguish among theories.
- How robustly were test patterns reproduced in the IBM? If a test pattern is widely observed in nature but only reproduced in an IBM under a narrow range of parameter and input variable values, then the test results are weak.

4.5.6   Phase 6: Repeat the Cycle to Refine Theory and Tests

Repeating and refining the cycle of theory development and testing may be necessary to find satisfactorily general and useful theories. In the case that none of the proposed theories succeeds in reproducing the test patterns in the IBM, then the cycle must start over at phase 1. If more than one proposed theory passes the pattern-oriented IBM analysis, or if the analysis had low inferential power, then it can be desirable to start the cycle over at phase 3 by identifying additional test patterns that provide greater resolution to the experiment. Often when a proposed theory appears good in concept, it takes several cycles through the testing phases to determine the best detailed implementation of the concept. Identifying additional test patterns may require new field or laboratory studies specifically designed to distinguish between competing theories. Again, it is important to understand that the most general theories explaining an individual trait are most likely to be found by testing them against a wide range of specific patterns, not by testing against general patterns.

4.6   EXAMPLE: DEVELOPMENT OF HABITAT SELECTION THEORY FOR TROUT

To illustrate the process of IBE theory development, we describe development of theory for habitat selection in stream trout. This work used the stream trout IBM described in sections 1.2 and 6.4.2. One of the interesting lessons from this example is its illustration of a point made in section 4.3—that applied ecology is important to theory development. The need to simulate a specific individual behavior in a realistic management context forced the development of new, more useful theory for an important ecological problem.

   (A similar story is told by the papers of John Goss-Custard, Richard Stillman, and co-workers [Goss-Custard et al. 2001, 2002, 2003; Stillman et al. 2002, 2003; West et al. 2002], which address interference among overwintering shore-birds. The management problem at the start of their project in the 1980s was predicting the effect of habitat loss on the birds' winter mortality. Over the years, detailed and very well-tested models of individual foraging behavior have been developed. Now, this toolbox of established theory allows Goss-Custard et al. to adjust the model quickly to new species and habitats. The theory development cycle, and its relationship to specific management problems, was very similar to that of the trout-modeling project described here.)

4.6.1   Phase 1: Define the Trait of Interest and the Ecological Context

The trout IBM was designed for environmental impact assessment: comparing the effects of alternative stream flow and temperature regimes on abundance and production of trout populations. The ecological context included (1) spatial variation (at a resolution of several square meters) in the habitat variables driving growth rates and mortality risks; (2) temporal variation in stream flow and temperature, which affect growth rates and mortality risks; and (3) competition among trout for food and feeding habitat. Field studies indicate that moving to different habitat is the primary way trout adapt to changes in flow and temperature, so habitat selection was identified as a critical individual behavior that the IBM must represent realistically.

4.6.2   Phase 2: Select Alternative Theories for Habitat Selection

Previous IBMs for stream fish and the behavioral ecology literature were examined for useful, tested models of habitat selection within the trout model's ecological context (Railsback et al. 1999). An especially important aspect of the trout model's context is its assumption that habitat selection is driven by both growth potential and mortality risk. Two general approaches were examined. The first approach is that of deriving, from fitness maximization theory, a simplified rule for making trade-offs between growth and mortality risk. One such rule is that fitness is maximized by selecting habitat to minimize the ratio of mortality risk to growth rate (Gilliam and Fraser 1987); a similar rule was derived by Leonardsson (1991). This approach was rejected for the trout IBM because the assumptions needed to derive these simplified rules are highly incompatible with the IBM's ecological context. For example, these rules were intended to apply only to juveniles and typically assume that all locations offer positive growth rates whereas in the IBM (as in nature) there are many locations where fish would lose instead of gain weight. These rules also produce results that seem incompatible with a basic understanding of natural history— for example, by predicting that trout with already high energy reserves would take on substantially higher mortality risks to gain even more food.

The second theoretical approach to habitat selection examined is the dynamic state variable modeling approach (Mangel and Clark 1986; Houston and McNamara 1999; C. W. Clark and Mangel 2000). The state-based approach is a less-simplified model of fitness maximization, assuming animals select their habitat over time to maximize their expected number of offspring at some future time of reproduction. The expected number of offspring is a function of growth and accumulation of energy reserves, and also the expected probability of surviving until the future time. Survival to a future time depends on food intake as well as predation and other risks: food intake must be sufficient to avoid starvation. This approach has many conceptual advantages, including adaptability to many ecological contexts, freedom from some of the restrictive assumptions, and more closely representing the real problem (survival and reproduction) that animals are adapted to solve.

A significant problem with the state-based approach is its assumption that the growth potential and mortality risks in each habitat patch are static over time, which is incompatible with the IBM's context. To retain the benefits of the state-based approach while accommodating temporal variation in habitat conditions, a new theory was proposed: animals are assumed to make a very simple prediction of habitat conditions over a future time horizon and then select the habitat that maximizes their expected probability of surviving (and their growth to reproductive size, for juveniles) over the time horizon. This approach was called the "state-based, predictive" habitat selection theory; Railsback et al. 1999; Railsback and Harvey 2002; section 7.5.3).

To make the subsequent tests of the state-based, predictive habitat selection theory more rigorous and interesting, two additional traits were also tested. These traits assume trout select habitat to (1) maximize their immediate growth rate, and (2) maximize their immediate probability of survival.

4.6.3   Phase 3: Identify Test Patterns

Six patterns were selected to test the traits for trout habitat selection (Railsback and Harvey 2002). These patterns were selected from the published trout literature. Each is relatively simple (i.e., a "weak" pattern as discussed in chapter 3) and noncontroversial, yet is a specific response to a known change in conditions. Together the patterns include responses to a wide range of factors.

1. Hierarchical feeding. Trout of similar size exhibit preference for a single best feeding site, which is occupied by the most dominant individual. When this dominant individual is removed, the preferred site is then occupied by the next most dominant individual.
2. Response to high flow. When extremely high stream flows occur, trout move to slower water along the stream margin and then return to their previous location as flow recedes.
3. Response to interspecific competition. In the presence of another trout species with larger individuals, trout shift their habitat, usually to higher velocities.
4. Response to predatory fish. In the presence of large, predatory fish, juvenile trout use faster, shallower habitat.

5. Seasonal velocity preference. The average velocity used by trout increases with temperature among seasons.
6. Response to reduced food availability. When food availability is reduced, trout shift habitat to obtain higher food intake, and this shift happens before starvation is imminent.

It is important to realize that behaviors reproducing these patterns could each be hardwired into the model trout (which would be an example of *imposed* behavior; section 5.2), but doing so would not prove that the behaviors were a useful, general model of habitat selection. When we instead see if these behaviors *emerge* from a single theory for habitat selection, they provide a powerful test of the theory.

4.6.4   Phase 4: Implement the Proposed Habitat Selection Theories in an IBM

The three alternative habitat selection theories—the "state-based, predictive," "maximize growth," and "maximize survival" traits—were each implemented in the trout IBM. Simulations were designed to represent the conditions under which each of the six test patterns was expected to occur. These simulations could then be repeated, altering only the IBM's habitat selection trait.

4.6.5   Phase 5: Analyze the IBM to Test Theories

The trout IBM was analyzed to determine if the six test patterns were reproduced under the expected conditions, with the analyses repeated for each of the three alternative habitat selection traits. The results of the analyses (figure 4.1) show that the wide range of patterns was important in resolving the three traits. Patterns 1 and 3 were not reproduced by the "maximize survival" trait, excluding it as a useful theory for the trout IBM. The "maximize growth" and state-based traits both could reproduce the first three patterns, but pattern 4 showed that maximizing growth (or any other trait neglecting mortality risks) is not generally useful.

Patterns 5 and 6 proved to be the decisive tests, as they were not reproduced by *either* of the traits based on maximizing immediate growth or survival probability. Patterns 5 and 6 can be reproduced only by traits that consider how an animal's state is expected to change over a future period. When temperature (and hence metabolism) is increased, or food availability reduced, the immediate state of the individual is not changed but the individual's predicted *future* starvation risk is increased. Consequently, expected fitness is increased by changing habitat to obtain more food even if mortality risks are higher. The analyses indicated that the state-based, predictive trait is a useful theory for habitat selection in the IBM; the analyses also showed that conventional

Pattern	Maximize growth	Maximize survival	State-based, predictive
Hierarchical feeding	+		+
Response to high flow	+	+	+
Response to inter-specific competition	+		+
Response to predatory fish		+	+
Seasonal velocity preference			+
Response to reduced food availability			+

Figure 4.1 Summary of the pattern-oriented test of habitat selection theory for the trout IBM. Only the state-based, predictive trait for habitat selection caused all six patterns to be reproduced in the IBM.

foraging models based only on *immediate* food intake and mortality risks are unlikely to be general and useful in realistic contexts.

4.6.6 Phase 6: Repeat to Refine Theories and Tests

In subsequent development of the trout IBM, the habitat selection theory was revised to include choice of activity in conjunction with habitat selection (Railsback et al., in press). Trout routinely switch between two activities: feeding and hiding. Different habitat is used for the two activities, and under different conditions trout exhibit different preferences for feeding during daytime versus nighttime. The IBM was revised to represent selection of feeding and hiding activity along with habitat selection because both of these activities may be important in determining how flow and temperature affect trout. (For example, habitat for hiding instead of feeding may limit populations in some situations.) This change in context required revision of the previously developed theory.

The state-based, predictive habitat selection theory was modified to include selection of activity along with habitat. Under the revised theory, trout are assumed to examine all four possible activity combinations (daytime feeding and nighttime feeding, daytime feeding and nighttime hiding, etc.) for each potential habitat patch and then choose the best combination of activity pattern and habitat.

Testing this revised theory in the IBM immediately identified a weakness: adult trout ceased to grow, instead feeding only enough to avoid starvation and spending as much time as possible in hiding to avoid predation risk. This model prediction contradicted observations from the study site, where food is abundant and adult trout grow to large size. (This problem did not arise previously because the IBM previously assumed trout always feed during the daytime, in which case feeding enough to avoid starvation also produced realistic growth rates.)

The theory was again modified, this time by including a function representing how the fitness of adult individuals continues to increase as their size increases. This modification was certainly inspired by the need to make IBM results reproduce observed adult growth (a key observed pattern), but it also has a sound foundation in evolutionary theory. This function summarizes the many benefits to fitness of larger size, which include better ability to compete for food and mates, higher gonad mass, and better ability to protect eggs. The revised habitat and activity selection theory was implemented in the IBM and, with minor calibration, produced reasonable growth and survival rates. A set of eight patterns was identified for testing the revised theory. These patterns are ways that the frequency of diurnal versus nocturnal feeding have been observed to change in response to environmental or competitive factors. The test patterns were taken primarily from controlled experiments conducted independently by Metcalfe et al. (1999). The revised theory of habitat and activity selection successfully resulted in the eight test patterns being reproduced in the trout IBM.

This whole theory development exercise took advantage of a great deal of existing experimental ecology but identified important new topics for empirical study. The conclusion that behaviors such as selection of habitat and activity can be modeled as a state-based, predictive process calls for empirical testing and explanation. The theory could be refined via controlled experiments on real trout, perhaps in conjunction with simulation. Such experiments could, for example, test alternative theories about the ability of fish to predict future conditions and about the value of growth in decisions such as habitat selection.

4.7   SUMMARY AND DISCUSSION

The problem of linking behaviors of populations and ecosystems to the traits of individual organisms and the characteristics of the individuals' environment is now accepted as critical to ecology. However, ecology (like other complex sciences) currently not only lacks an established body of theory for how system-level properties arise from individual traits; we also lack a process for developing such theory. This chapter goes to the heart of the most basic problem we address in this book: *how do we learn how system-level properties of ecosystems arise from the traits of individuals?*

Fortunately, other scientists working on CAS have developed approaches to theory that are readily adapted to IBE. Their approaches that we adopt have an advantage the importance of which is difficult to overstate: they are based on the conventional scientific method in which new theories are proposed, tested against alternative theories in controlled experiments, and revised and retested using experiments of increasing rigor. The use of IBMs and pattern-oriented analysis provide the ability to follow this theory development cycle with reasonable effort and cost.

We propose an approach to the theory of IBE in which (1) the fundamental axiom is that system-level properties arise from the traits that determine how individuals interact with each other and their environment, and from characteristics of the individuals' environment; (2) theories are models of individual behavior that are useful for predicting system-level behaviors; and (3) theories are "proved" by developing experimental evidence of how general and useful they are. This approach to theory is designed to provide four important qualities. First is *testability*: the pattern-oriented analysis approach makes it practical to test theories thoroughly. Next is *generality*: this approach promotes the development of theories of individual traits that apply under wide ranges of conditions. More important, it provides a way to document how general theories are known to be; theories cannot honestly be considered general until they have been tested under a wide range of conditions. This theory development process can document the range of situations and contexts where a theory has been found useful and unuseful. The theory process also provides *integration* across ecological levels, as it can use natural history and behavioral ecology in producing models of individuals that are useful for understanding population- and community-level behaviors. Finally, the process is designed to produce theories with *applicability* to real-world ecological management problems. In fact, we argue (and illustrate in the example of developing habitat selection theory for a trout model) that application to real management issues advances theory development by forcing our theory to confront reality.

So far, the ecologists who have used IBMs have developed little reusable theory of the kind we describe in this chapter (Grimm 1999). The same can be said about any of the sciences addressing complex systems of adaptive individuals (e.g., economics, microbiology, sociology); science as a whole is just learning how to study such systems. One purpose of the example IBMs we present in chapter 6 is to identify models of individual traits that have been explored and found useful in some contexts.

As IBE progresses, our toolbox of established theory will grow. As more scientists take up the challenge of understanding the links between individual traits and system behaviors, they will identify theories and establish the contexts in which those theories are useful. Even more useful will be identifying general classes of theory from which models for specific species and contexts can be developed rapidly. The "state-based, predictive" approach mentioned in section 4.6 is one example, and it was adapted from another class of theory, dynamic state variable modeling. The general idea of the state-based, predictive approach—that individuals make decisions to maximize some realistic estimate of their future survival and reproductive potential—can be applied to many traits in addition to habitat selection. Other general classes of theory no doubt are applicable to many traits and organisms. In fact, much of chapter 5 is dedicated to general concepts that can guide and facilitate development of IBE theory. To us it seems very likely that the problem of IBE theory will, once tackled seriously by ecologists, prove more tractable than our experience with IBMs so far indicates. A few general classes of theory may allow us to identify useful theory for many specific traits and contexts with relative ease, and we may be able to assemble IBMs for many management applications completely from existing theory.

Why are we optimistic that ecologists can succeed in producing useful and general theory for IBMs? There are specific reasons why the approach to theory we propose here seems likely to be productive. First, it gives up on the notion of context-free generality and instead focuses on the more realistic goal of finding different theories that are useful in different, specific contexts. Second, instead of trying to impose system-level theory, we try to understand ecosystems as they really are: emergent properties of adaptive individuals. Models of how individual components behave and interact have been very successful at predicting the behavior of complex physical systems. As simplified representations of such systems, IBMs help us find useful theory for emergent system properties without getting lost in too much complexity. Finally, and perhaps most important, the pattern-oriented analysis process makes IBE theory relatively easy to test. The ability that IBMs provide to test and contrast alternative theories under wide ranges of conditions makes it practical to propel ourselves forward via the scientific method and strong inference, a power that is lacking in many other approaches to ecological theory.

Chapter Five

A Conceptual Framework for Designing Individual-based Models

To describe the complexity approach, we begin by pointing out six features of an economy . . . that together present difficulties for the traditional mathematics used in economics: Dispersed Interaction . . . No Global Controller . . . Cross-cutting Hierarchical Organization . . . Continual Adaptation . . . Perpetual Novelty . . . Out-of-Equilibrium Dynamics . . .

   Because of the difficulties outlined above, the mathematical tools economists customarily use, which exploit linearity, fixed points, and systems of differential equations, cannot provide a deep understanding
   —W. Brian Arthur, Steven Durlauf, and David A. Lane, 1997

5.1   INTRODUCTION

The modeling guidelines of chapters 2 and 3 and the process of theory development described in chapter 4 address the strategic level of individual-based modeling: they provide efficient strategies for designing IBMs to address specific problems and for developing theory of how individual traits determine system dynamics. Now, however, we turn to the process of actually designing an IBM: how do we represent those processes that we determined, at the strategic level, must be in the IBM?

   At this point, modelers in most fields turn to an established conceptual framework: well-known and widely used classical concepts that provide a way to think about the problem, frame a model, and derive results. Classical ecological modeling has the benefit of the familiar and noncontroversial modeling framework of differential calculus. By providing a framework, terminology, and notation that all scientists are familiar with, differential calculus is also a very efficient language for communicating models—a few equations and sentences about solution methods can describe most classical models thoroughly.

   Individual-based modeling has lacked this kind of established conceptual framework, with very important and detrimental consequences. First, without an established conceptual basis, IBMs are inherently more controversial than differential equation models. IBMs are more difficult to communicate and seem highly ad hoc—if a model cannot be described using established terminology and concepts, it tends to sound as if it were made up arbitrarily. In fact, their seemingly ad hoc nature has been a common criticism of IBMs (e.g., Hogeweg and Hesper 1990). Second, many IBMs have not been as good as they could have been because modelers did not understand what issues needed to be considered in designing the IBM, much less what approaches were most appropriate for each issue. A sound conceptual framework is the most fundamental tool for model-building: it keeps us from starting over with each new model and provides a basis for discovering general techniques and theory.

Fortunately, general concepts that characterize systems of adaptive individuals have been developed within a new approach to science referred to as Complex Adaptive Systems (CAS). Scientists working in CAS attempt to understand the dynamics of systems of adaptive individuals, often using individual-based (or "agent-based") computer simulation as a tool. (Note that "Adaptive" in CAS refers to the entities making up the system, not the system itself, because there is no clear concept of fitness at the system level; see Sommer 1996.) IBE can be considered a subset of CAS, although much of the founding work in CAS has been in completely abstract systems and in sciences other than ecology.

A general understanding of CAS can be very beneficial for ecologists, both as a new and refreshing way to view the world and as background for IBE. In addition to the pioneering work of economist Brian Arthur, the highly accessible books of Axelrod (1997), Holland (1995, 1998), Kauffman (1995), and Waldrop (1992) and the somewhat less accessible but rewarding work of Auyang (1998) all help develop an understanding of CAS and its general concepts. In the past decade, the literature of CAS has mushroomed, with the appearance of many books and journals (e.g., *Artificial Life*, *Complexity*, *Emergence*, and *Journal of Artificial Societies and Social Simulation*). The concepts presented in this chapter were extracted from the most basic elements of CAS and applied to ecology (Railsback 2001a). We do not review the extensive CAS literature comprehensively but instead seek a few general concepts for thinking about and modeling systems of adaptive individuals. We certainly expect this initial attempt at a conceptual framework for IBMs to evolve as both CAS and IBE evolve and crystallize.

Ecology and CAS are closely related; after all, ecologists have been studying complex systems for many decades. Many of the concepts at the heart of CAS are also at the heart of ecology and evolution. Concepts such as adaptation, fitness, interaction, and sensing are fundamental to how we think about organisms and populations. The relevance of some of these concepts to IBMs and IBE has in fact been strongly anticipated by ecologists without reference to CAS (e.g., Tyler and Rose 1994; Giske, Huse, and Fiksen 1998). The particular value of CAS to ecology is its determination to address explicitly how individuals and populations affect each other, even if doing so requires completely new conceptual approaches.

One objective of this chapter is to introduce a list of standard modeling concepts that can serve as a framework for building and communicating IBMs, much as differential calculus provides a standard set of model design decisions for classical models. But a further objective is to show, through examples and design guidance for IBMs, *how to think about IBMs* using these concepts. As a conceptual framework, the concepts we present should help with many modeling tasks, including the following.

- Designing IBMs. Thinking about how to address each of these concepts will help modelers identify and make important design decisions in a more informed way.
- Developing theory. The modeling concepts provide the framework within which we develop theory in IBE (chapter 4). In particular, the concepts of emergence, adaptive traits, and fitness are critical for devising theory of how individual traits explain system behaviors.
- Describing IBMs. The concepts provide terminology for essential characteristics of IBMs that are not easily described by equations. They should reduce the effort and improve the clarity of communicating an IBM to others (chapter 10).
- Classifying and evaluating IBMs. Reviewers of proposed or completed work based on IBMs can use the modeling concepts as an aid. Not all IBMs need consider all of these concepts explicitly, but the concepts can help make reviews thorough and specific.

Each of the next ten sections discusses one concept. We describe the concept and why it is important, and alternative ways to address it. We then provide guidance on specific model design decisions related to the concept: how do we put the concept to work in designing an IBM? At the end of the chapter we provide a conceptual design checklist for IBMs: a list of important model design issues related to the ten concepts. The checklist is intended to be a concise road map for thinking about and documenting the essential characteristics of an IBM.

5.2   EMERGENCE

One of the most decisive steps in designing an IBM is determining which behaviors of the system *emerge* from *adaptive traits* of the individuals.

In some ecological and CAS literature, the terms *emergence* and *adaptive* refer to evolutionary processes: new genotypes emerge as a result of genetic adaptation and natural selection. We have chosen (for reasons explained in the preface) not to address evolutionary or genetic processes in this book. However, these terms are very applicable to processes within the life-span of individuals.

Organisms continually adapt their behavior and state in response to internal and external conditions, and we ecologists are interested in the population dynamics that emerge from the adaptive behavior of individuals.

In our terminology for IBMs, a *trait* is an algorithm (or model, or theory; cf. chapter 4), specified by the modeler, for some behavior of the individuals. An *adaptive trait* (defined more completely in section 5.3.1) is a trait involving a decision or response to changes in the individual's environment or its internal state, made with the intent of improving the individual's eventual fitness. An individual's *behavior* is what the individual actually does during simulations, an outcome of the individual's traits and its experiences. This terminology is loosely based on the metaphor of a real organism's genetic traits being the "rules" it is born with, and its behavior being the result of applying those rules in whatever situation the organism finds itself in.

5.2.1 Emergent and Imposed System Behaviors

Scientists working on CAS enjoy debating the exact meaning of emergence, but for IBMs only a general definition is necessary. Emergence occurs at the system level: system behaviors may emerge from the traits of individuals as the individuals interact with each other. In many IBMs, individuals interact with their environment as well as with each other, so system behaviors can emerge in part from the simulated environment: different system behaviors can emerge from the same individuals when they are in different environments. Emergent system behaviors are not limited only to dynamics of system state variables (e.g., fluctuations in abundance). Emergent behaviors can also include patterns of individual behavior that arise within the system: things that individuals do in the system that they do not do alone (e.g., forming flocks or schools; see section 6.2); such patterns are properties of the entire system.

Not all system behaviors in an IBM are emergent. Three criteria are often used to distinguish system-level properties that are emergent:

- Emergent properties are not simply the sum of the properties of the individuals.
- Emergent properties are of a different type than the properties of the individuals (e.g., the spatial distribution of individuals is a system property of a type that none of the system's individuals has).
- Emergent properties cannot easily be predicted by looking only at the individuals.

That emergent properties are not easily predictable from individual traits does *not* mean that emergent behavior is mysterious or impossible to understand, or is always spectacular and unexpected. In fact, one of the primary goals of IBE is to understand how even the most basic, common properties of ecological systems emerge from the traits of individuals. Understanding emergence is a declared aim of IBE: "For a simulation model to be of any use, both obvious and nonobvious patterns must be explained and not brushed under the carpet of emergence as this amounts to an admission of failure" (Di Paolo, Noble, and Bullock 2000, p. 504).

Research in CAS has shown that it can be surprisingly easy to produce interesting and realistic emergent behavior in IBMs by including simple representations of the underlying mechanisms (e.g., Axelrod 1997; Holland 1998). Camazine et al. (2001) document many fascinating examples of emergent (or "self-organized") structures and behavior in biological systems. The book of Camazine et al. is especially valuable because it documents research processes that can determine how complex structures emerge from individual behavior. (Camazine et al. include IBMs in the category of "Monte Carlo models" in discussing techniques for understanding emergence.)

In contrast to emergent system behaviors are *imposed* behaviors—system properties that are directly predictable from, or tightly constrained by, individual traits. The difference between emergent and imposed behaviors and their significance to what we can learn from an IBM are illustrated by the following two examples.

Example 1: Individual Mortality Risk. Consider one IBM in which the mortality risk for individuals—their probability of dying during one time step—is constant: all individuals always have the same risk. Also consider a second IBM in which mortality risk varies and depends on individual behavior, such as whether individuals decide to forage or hide, or what habitat they choose to occupy. In the first IBM, the population's mortality rate (average number of deaths per time step) is *imposed*: it is readily predicted from the constant mortality risk of individuals. In the second IBM, however, the population mortality rate *emerges* from the traits individuals use to make decisions that affect risk, and from the habitat itself (what risks are present in which kinds of habitat); the mortality rate is not directly predictable from the model's parameters and rules.

Example 2: Upstream Migration of Salmon. When adult salmon are ready to spawn, they migrate upstream from the ocean, usually to spawn in the stream where they were born. Some salmon, however, "stray" into streams other than their natal stream. How could we design an IBM that represents this migration and the possibility of straying into nonnatal streams? One approach is to *impose* upstream migration behavior by giving each salmon traits that tell it which way to turn at each river junction so that it returns to the natal stream (figure 5.1). Straying could be imposed by giving each salmon a probability of randomly taking a wrong turn; adjusting these probabilities could force the model to reproduce observed rates of straying.

Alternatively, upstream migration behavior could be modeled as emerging from the mechanisms salmon use to navigate. Research on salmon has identified the primary mechanism they use to navigate upstream: if the fish are in water that contains the odor of their natal stream, they swim upstream; if they cannot smell

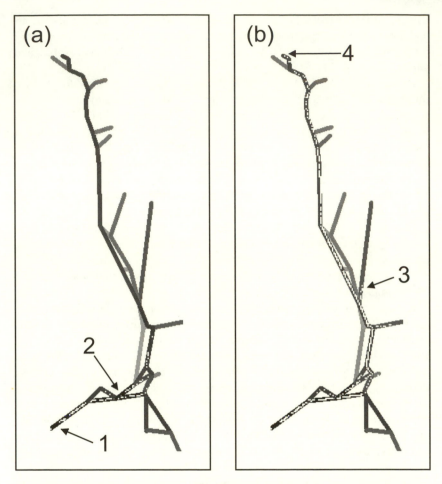

Figure 5.1 Emergent upstream migration of adult salmon in an unpublished IBM of the Sacramento River basin, California. Model salmon navigate by sensing whether they are downstream of the river segment where they were born. Early in the migration (a), the model salmon (*white dots*) enter the basin from the ocean via the Golden Gate at San Francisco (1). After several weeks (b), the salmon approach their natal segment, the upper Sacramento River below Keswick Reservoir (4). Realistic patterns that emerge include salmon following divergent pathways through the complex channels of the Sacramento–San Joaquin Delta (2) and a tendency to stray into other rivers such as the Feather (3).

their natal river, the salmon swim downstream (figure 5.1). This mechanism can be implemented in an IBM using two simple rules:

1. Periodically (e.g., once per day) each salmon senses whether it is in a river containing flow from its natal stream. If so, the salmon moves upstream; if not, it moves downstream.

2. When a salmon reaches a junction, it decides stochastically which river to move into, with the probability of moving into a river proportional to the river's flow rate.

The upstream migration patterns of the salmon then emerge from this trait for navigation and the simulated river network.

5.2.2   Comparison of Emergent versus Imposed Behaviors

When we decide to model some particular behavior of an IBM as emerging from adaptive individual traits, we are essentially choosing a *mechanistic* representation of the behavior. Instead of simply forcing the behavior to be exhibited, we model the underlying, individual-level mechanisms that give rise to the behavior. Like mechanistic modeling approaches in general, modeling an IBM behavior as emergent has the advantages of being more explanatory and general and the disadvantages of being more complex.

The primary problem in modeling system behaviors as emergent is, of course, finding individual traits that cause the system behaviors to emerge. This problem is exactly the one that chapter 4 addresses: finding theory for how system behaviors emerge from individual traits. So if we choose to model a system behavior as being emergent, we must find or develop the necessary theory— often, a major project. The emergent approach sometimes also requires more complex traits: individuals must make complex calculations or evaluate a large number of decision alternatives. Consequently, the challenge of testing, analyzing, and understanding an IBM (chapter 9) typically increases sharply with the number of emergent behaviors.

When we do model behaviors as emergent, we typically learn much more about the system. Modeling emergence requires us to develop an understanding of the real mechanisms driving individual behavior and the system. When we succeed in producing simple traits that capture these mechanisms, the IBM no longer simply reproduces observed behaviors but explains how the system works. Returning to the preceding example of modeling mortality, clearly only the second IBM would be useful for understanding mortality rates, for example, how they depend on habitat, population density, or behavior. A second point, also important but less obvious, is that we must also use the second IBM if we want to understand processes that depend on *how individuals adapt to mortality risks*. We cannot study how individuals behave to avoid risk using an IBM in which mortality risk does not vary.

Modeling behaviors as emergent can also make IBMs surprisingly general. A successful mechanistic model of some process should have the advantage of being generally applicable under a wide range of conditions, not just under the conditions used to estimate parameters (Kaiser 1979; DeAngelis and Mooij 2003). In the salmon migration example, the emergent behavior alternative is a much more general model: it is not necessary to program each salmon with a map to follow back to its natal spawning stream. Instead, fish find their way back to their natal stream with no information other than the identity of the stream; the stream network's flows or connections can be modified and migration resimulated with no changes in salmon traits. In fact, mechanistic individual traits often produce realistic emergent behaviors that were not even thought about as the traits were designed; examples of IBMs that produced such unexpected but realistic behaviors are in sections 6.2.2 and 6.8.3.

The alternative to emergence—imposing a system behavior—resembles an *empirical* approach: instead of representing the mechanisms driving the system, we simply force it to reproduce behaviors observed in real systems. When a mechanistic understanding of the system behavior is not of interest, this approach can be a simple, easy way to obtain the desired outcomes.

When we choose to impose some system behavior by giving individuals nonadaptive traits that force them to reproduce observed behaviors, the primary model design concern becomes whether such traits are appropriate under all the conditions that could arise in the IBM. This concern is essentially the extrapolation issue common to all empirical models: does the model work under unforeseen conditions? This risk can be serious because behaviors exhibited only occasionally, and therefore likely to be ignored in empirical modeling, can sometimes be very important. For example, IBMs of territorial animals often assume that individuals *always* maintain a territory. This assumption precludes the animals from using nonterritorial behaviors that can occasionally be highly adaptive: for example, animals may abandon their territories to seek refuge during high-risk events. If an IBM includes such events, it could predict that most animals die during an event that real animals would avoid by simply abandoning their territories and moving to refuge habitat. If observed behaviors are imposed in an IBM, great care must be taken to assure that the imposed behaviors are appropriate under all conditions that could arise in simulations.

Certainly, all processes of an IBM must be imposed at some level—otherwise system behavior would have to be modeled as emerging from the fundamental properties of subatomic particles (somewhat beyond the scope of this book). When we model emergence in an IBM, we represent how system behavior depends on individual behavior and how individual behavior depends on lower-level processes; these lower-level processes must at some point be imposed using empirical relations.

5.2.3 Design Guidance for Emergence

One of the most fundamental IBM design decisions is selecting the model outcomes that emerge from adaptive individual traits versus the outcomes that are imposed. The highest-level outcomes of IBMs are almost always emergent, according to our definition: all the IBMs we examine in this book have processes through which the interactions of individuals with each other and their environment affect system behavior. But most IBMs also have *intermediate* outcomes that are imposed by model rules: if mortality risks are assumed to be constant, then mortality rate outcomes are imposed; if the number of offspring parents produce is assumed unaffected by the parents' state, then fecundity relationships are imposed; if the distance that dispersing individuals move each day is drawn randomly from the same distribution, then dispersal speed results are imposed. How do we decide which results should or should not be imposed?

First, if the purpose of an IBM is to explain how a particular system behavior arises from individual traits, then of course it is essential that the system behavior emerges from adaptive mechanisms acting at the individual level. In fact, for this modeling objective we need to be especially careful: when we claim that an important outcome emerged from individual traits, critics will look for ways in which the outcome might have subtly been imposed.

A second reason to use emergence is to help make an IBM general and easily applied to a wide variety of sites and situations. In most IBMs there are one or several key individual traits from which the most important system behaviors emerge. Representing these traits in a mechanistic way can help ensure that individual decisions are realistic under a wide range of conditions. When we use traits that impose intermediate outcomes, we must be careful to make sure that the imposed outcomes are appropriate under all the conditions that could occur in the IBM.

For other intermediate system behaviors, it is generally best to limit emergence. A complete lack of emergence makes IBMs unrealistic and uninteresting, but if too many outcomes emerge from too many mechanistic processes, it can be very difficult to analyze and learn from an IBM. An IBM is most likely to be useful if it focuses on one or only a few of the most important emergent behaviors of the system it represents.

5.3 ADAPTIVE TRAITS AND BEHAVIOR

How do we design an IBM so that system behaviors of interest emerge from the adaptive traits of individuals? This is probably the most important question addressed by this book and by IBE, and it is addressed in different ways throughout the book.

5.3.1 What Adaptive Traits Are

Organisms typically have a variety of mechanisms for responding to changes in their environment and in themselves. Presumably, evolution has provided these mechanisms because they improve an individual's fitness—they produce behavior that, on average, increases the individual's success at reproducing and passing its genes on to later generations. We use the term *adaptive traits* for the decision-making rules used by the individuals in an IBM to select behaviors that improve their potential fitness (Zhivotovsky, Bergman, and Feldman 1996). An adaptive trait does not simply tell individuals what to do; it gives individuals a procedure for making situation-specific decisions. Adaptive traits produce adaptive behavior that varies with the situation or state an individual is in when it makes the decision.

The adaptive traits in an IBM are typically designed to model the real organisms' behaviors, which may be genetically programmed or learned. Of course, not all characteristics or behaviors of organisms are adaptive, and "adaptive" cannot be equated with "optimal." Some characteristics of organisms are simply constraints of evolutionary history and cannot be assumed to increase fitness; and organisms' limited ability to sense, predict, and calculate means that their decisions cannot be truly optimal. In designing adaptive traits, our objective should be to find traits that provide realistic adaptive abilities, not necessarily assuming individuals find optimal solutions to the problems they face.

Organisms exhibit a wide range of adaptive behaviors in response to different stimuli over different time scales. The following are *some of* the adaptive mechanisms that could be useful for explaining system behavior in IBMs.

- At very short time scales, animals choose among such behaviors as feeding, resting, and hiding. This choice may be driven by short-term motivations like hunger, satiation, and fear.
- Mobile animals adapt to changing conditions by moving. Movement may be in response to changes over time or space in food availability and mortality risks, or to seek habitat for specialized behaviors like mating.
- Both plants and animals can demonstrate plasticity in physiology and life history. Plants choose how much of their resources to invest in leaves, branches, roots, reproductive organs, or defense chemicals. Likewise, animals decide how much energy to store, commit to growth, expend for activity, or use for reproduction. An example life history adaptation is deciding when to reproduce; in many organisms delaying reproduction can result in higher fecundity but increases the risk of dying before reproducing. Changes in phenotype and life history state can be understood as determined by partly genetic constraints and partly by adaptive decisions considering the individual's state (size or energy reserves) and environmental conditions (Thorpe et al. 1998).
- Learning is an adaptive process that takes place over time scales from very short to an organism's entire life-span. Organisms may adjust their behavior very quickly in response to strong signals (e.g., the presence of a new predator), or may learn slowly from slow or weak signals (e.g., long-term or "noisy" trends in prey abundance).

In populations of real organisms, traits can of course vary among individuals: learned traits can vary due to the different experiences of individuals, and genetic traits can vary due to genetic polymorphism. Frequency-dependent fitness—when the potential fitness provided by a genetic trait depends on the trait's frequency and the frequency of other traits for the same behavior—is an issue discussed within the framework of evolutionarily stable strategies (Maynard Smith 1989). Questions about evolutionarily stable strategies could in principle be studied with IBMs (and probably will be in the future), but doing so requires simulating genetics and inheritance, which we do not address in this book. (However, section 6.9 discusses IBMs that are "calibrated" using artificial evolution and can have trait polymorphism.) The adaptive traits we consider represent an average over potentially polymorphic traits. However, we remind readers that a diversity of observed *behaviors* does not necessarily mean that there is a diversity of *traits*: even with the simplified traits we use in IBMs, seemingly small differences among individuals in state or environment can produce major differences in behavior that look like different strategies.

5.3.2   Advantages of Adaptive Traits to Model Behavior

Using adaptive traits in an IBM means that model individuals make decisions by attempting (directly or indirectly, as we will discuss) to improve their expected future reproductive success. Fitness-seeking adaptation as a basic design concept for IBMs has several advantages. First, and most important, the adaptation concept makes the theory of evolution a fundamental part of the IBM's conceptual basis. The assumption that individual traits act to improve fitness is one of the least controversial and most powerful concepts in biology.

A second advantage is that modeling becomes focused on real biological processes. If individual traits have evolved primarily to improve fitness, then building models of individual behavior around this assumption helps make the models realistic. The literature of the dynamic, state-variable approach to modeling behavior provides many good examples of this point. This approach (e.g., C. W. Clark and Mangel 2000; Houston and McNamara 1999) specifically assumes that decisions are made to maximize fitness, with fitness being defined as reproductive output at a future time (or something similar and closely related). C. W. Clark and Mangel (2000) illustrate a number of situations in which this fitness-based approach produces more general and realistic results than more abstract approaches to behavioral ecology.

A third advantage of using adaptive traits is that it facilitates using the often extensive autecology and natural history of organisms to design traits. Many kinds of observational and experimental information are often available, but it is often not clear how this information can be used to model behavior. This kind of information can be included in IBMs by putting it in the framework

82   CHAPTER FIVE

of adaptive traits. For example, often observations have been made of how individuals behave in response to a wide range of events (e.g., predator presence, changes in food availability, competition from other individuals, extreme weather). These observed responses at first seem unrelated and unuseful for modeling, but putting them in the framework of adaptive behavior—thinking about how each behavior affects the individual's probability of successful future reproduction—may help find a general model for these behaviors.

5.3.3 Directly and Indirectly Adaptive Traits

We define adaptive traits as being intended to improve fitness (i.e., to be fitness-seeking), but do organisms really make all their day-to-day decisions by trying to maximize probable future reproductive success? Almost certainly not: the decisions of organisms are often highly constrained by their morphology and senses, the range of their innate or learned behaviors, and their cognitive ability. Such constraints can be represented in the traits used in an IBM. In addition, the modeler may not have the information or the need to model explicitly how decisions affect fitness. Therefore, IBMs can have traits that represent fitness-seeking either directly or indirectly.

Direct fitness-seeking traits explicitly model the fitness consequences of each alternative behavior, using a specific fitness measure (discussed in section 5.4) along with a decision process to select one of the alternatives. Example traits that have been modeled as direct fitness-seeking processes are:

- Foraging behavior. The extensive optimal foraging literature is largely based on the assumption that behaviors such as the time spent foraging in various habitats optimize some measure of fitness.
- Habitat selection. Many models have assumed that animals choose which habitat to occupy, or when to leave their current habitat, by considering the consequences of their decision to fitness-related variables such as growth or survival.
- Changes in life history state. For example, Grand (1999) developed a model in which salmon decide when to move on to their next life history state in a way that maximizes their probability of surviving the transformation.

Many IBMs have used direct fitness-seeking without clearly stating it as such. Models in which individuals make decisions to improve growth, for example, implicitly assume that growth conveys fitness (section 5.4).

Indirect fitness-seeking traits are also common in IBMs. Typically, indirect fitness-seeking is used to model specific behaviors that are observed in the real organisms and assumed to contribute indirectly to fitness, but that would be very difficult to link to fitness directly. Often, indirect fitness-seeking involves following a set of simple rules instead of making a complex evaluation of alternatives. Some examples of indirect fitness-seeking are:

- The salmon migration models described in section 5.2. The individual salmon follow rules causing them to migrate toward their natal stream for spawning; the indirect fitness-seeking assumption is that the salmon's probability of reproducing successfully is highest if they spawn in their natal stream.
- "Boids" and related fish schooling models (sections 6.2.1–6.2.3). The simple movement rules used by individuals in these models are not directly related to fitness; however, the rules produce emergent flocking and schooling behaviors that are observed in birds and fish. Flocking and schooling are presumed to contribute to individual fitness, for example, by reducing individual predation risk.
- Traits that use probabilities or decision rules to reproduce observed decision-making behaviors. Examples include the lynx dispersal and beech forest IBMs described in sections 6.4.1 and 6.8.3. This approach assumes that reproducing decision-making patterns observed in real organisms provides fitness to model individuals.
- The case studies presented by Camazine et al. (2001). These studies found individual traits that explain many emergent phenomena observed in biological systems; and few if any of the traits involve direct fitness-seeking. Instead, these traits tend to be simple individual decision rules that give rise to complex system-level phenomena such as construction of elaborate nests by social insects. The structures that emerge from these simple traits appear to convey fitness to the individuals.

5.3.4   Design Guidance for Adaptive Traits and Behavior

Our primary recommendation concerning adaptive traits and behavior is that modelers think about and document the role of adaptation in their IBM's individual traits. In designing traits, it can be very useful to think about how individual behaviors affect the different elements of fitness (probability of survival, accumulation of energy, etc.; discussed in section 5.4) and how traits might be evolved or learned to increase fitness. Thinking about how each trait of model individuals affects potential fitness can help keep a model grounded in real biological processes.

The question of when to use adaptive traits in an IBM is straightforward: adaptive traits are used to produce emergence. When a modeler decides that a particular system behavior should be emergent instead of being imposed, then the problem becomes finding adaptive individual traits that give rise to the emergent behavior. However, whether a particular trait should be modeled as directly or indirectly fitness-seeking can be a more interesting question.

In general, direct fitness-seeking is useful for traits clearly and directly affecting key elements of fitness such as survival, growth, and production of offspring. Direct fitness-seeking has been useful for modeling complex decisions involving many inputs and many alternative outcomes. Direct fitness-seeking is especially useful as a conceptually clear and powerful way to model decisions that require trade-offs among fitness elements: choosing between behavior A that increases growth and gonad mass but reduces survival, and behavior B that reduces growth but increases survival. As we discuss in section 5.4, direct fitness-seeking using an appropriate measure of fitness provides a way to translate changes in fitness elements such as survival and growth into a common currency: potential fitness. This allows alternative behaviors to be compared by their effect on fitness potential. The cost of this power is, of course, increased complexity of the model formulation.

Indirect fitness-seeking is more likely to be appropriate for traits designed to reproduce observed behaviors that have indirect (but possibly important) effects on fitness. This approach is especially important for traits that appear to be highly "hardwired" in the real organisms and do not involve consideration or evaluation of complex decision alternatives.

The best way to represent adaptive traits will not always be clear. The following subsections of this chapter provide additional concepts for designing adaptive traits, and we provide more detailed guidance on modeling individual decisions in section 7.5. The IBE theory development and testing cycle in chapter 4 is ultimately the way to test, compare, and refine the traits used to model individual behavior.

5.4   FITNESS

In this section we explore one of the critical details of adaptive traits that use *direct* fitness-seeking: modeling how individuals estimate the consequences to their fitness that would result from each decision alternative. Unlike most other styles of ecological models, IBMs are well suited for modeling individual fitness directly; this section is designed to help modelers make the most of this advantage.

5.4.1   Fitness Concepts for Modeling Adaptive Traits

The word "fitness" has been used in many ways so it is important to define our terminology before proceeding. Fitness of an individual means its success in passing its genes on to future generations. Therefore, fitness is an outcome of the decisions an individual has made throughout its life and can only be evaluated after the individual has finished reproducing and the success of its offspring has been determined. The theory of natural selection allows us to assume that the important decision-making traits an individual has inherited genetically or learned are likely to be *fitness-seeking*: the individual has these traits because they generally increase the individual's eventual fitness. Therefore, many individual traits can be modeled by assuming they produce behaviors that increase the individual's *expected fitness*, the probable success the individual has in passing its genes on to future generations. Expected fitness is a current estimate, made by an individual, of its future fitness. A *fitness measure* is a specific, usually highly simplified and incomplete, model of expected fitness that is used in an adaptive trait. In other words, when an IBM assumes that individuals make decisions with the objective of increasing their expected future reproductive success, the fitness measure is the *internal model* of reproductive success that the individuals use to evaluate decision alternatives. A common example of modeling behavior using a fitness-seeking trait is assuming organisms make decisions to maximize their growth rate: growth rate is the fitness measure. This approach assumes that future fitness increases with growth, so growth rate is an adequate model of expected fitness.

It is important to understand that fitness is something that happens in the future: the fitness consequences of some decision an individual makes can occur long after the decision is made. Therefore, expected fitness is also a future concept and a fitness measure must be thought of as a prediction of the future consequences of a decision. (We therefore discuss prediction in section 5.5.)

In an IBM, an individual's expected fitness can depend on a number of events, such as survival until reproduction or growth to the size needed to reproduce. We refer to these events as *fitness elements*, which can be considered targets that must be achieved for fitness to be high. The extent to which these targets are achieved typically varies with several *driving processes* that are affected by the individual's decisions. Examples of fitness elements and processes affecting them are shown in figure 5.2. Driving processes are processes, often strongly affected by environmental conditions, that determine an individual's state with respect to the fitness elements. Energy intake is an important process driving many fitness elements, so it is no surprise that energy intake (or growth) has often been used by itself as a fitness measure.

5.4.2 The Completeness and Directness of Fitness Measures

A wide variety of fitness measures has been used in IBMs and related models. These fitness measures vary along two dimensions: *completeness* and *directness*. A fully complete, direct fitness measure would model how decision alternatives affect all of an individual's fitness elements, accurately predicting the individual's success at passing genes on to future generations under each alternative. Highly complete and direct fitness measures are biologically unrealistic, yet fitness measures that are too incomplete and indirect can provide model individuals with unrealistically poor adaptive ability.

The *completeness* of a fitness measure indicates how many fitness elements and driving processes are considered. A fitness measure could consider only

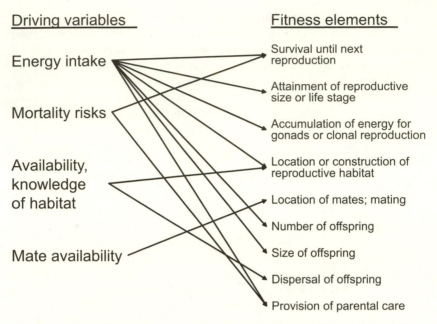

Figure 5.2 Examples of fitness elements and processes that drive them.

one element, for example, the probability of surviving over some future period, and consider only one process affecting survival, perhaps predation risk. On the other hand, the dynamic state variable modeling literature (Mangel and Clark 1988; C. W. Clark and Mangel 2000; Houston and McNamara 1999) includes such fitness measures as the expected number of female offspring, which considers expected survival to reproduction, accumulation of energy for reproduction, and number of offspring; with each of these fitness elements being driven by mortality risk and energy intake. The fitness measure proposed by Railsback et al. (1999) for modeling habitat selection considers two fitness elements: future survival and attainment of reproductive size. Attainment of reproductive size is driven by energy intake, and future survival is driven by energy intake (which determines survival of starvation) and other mortality risks such as predation.

The *directness* of a fitness measure refers to how explicitly the measure reflects the future fitness consequences of a decision alternative. A highly direct fitness measure is a prediction of how successful an individual will be at passing its genes on to future generations. In contrast, many IBMs (and related work in behavioral ecology) attempt to model direct fitness-seeking using very indirect fitness measures. The result is an IBM in which individuals make fitness-seeking decisions while lacking a good internal model of how their decisions really affect fitness. Probably the most common example of a very indirect fitness measure is assuming individuals make decisions to maximize their energy intake or growth rate. Even though energy intake is an important driver of many fitness elements (figure 5.2), future fitness is not a simple linear function of growth. For example:

- Survival depends in part on energy intake, which must be sufficient to avoid starvation. However, the importance of energy intake to survival can range from very high (if the individual is starving) to very low (if the individual has high energy reserves, if metabolic rates are low, or if risks other than starvation are high).
- Achieving the size or life stage necessary for reproduction also clearly depends on energy intake, but once the reproductive size or stage is achieved, growth may become much less important than other fitness elements.
- Obtaining energy for reproduction (e.g., for reproductive migration or gonad production) becomes important only as the reproductive phase is entered, not during nonreproductive life stages or seasons.

Ignoring these nonlinearities and simply assuming individuals make decisions to maximize energy intake therefore precludes a model from reproducing such behaviors as placing more emphasis on energy intake when starvation is imminent or size is small or metabolic rates are high, and placing less emphasis on growth when energy reserves are high, adult size is reached, metabolic rates are low, or mortality risks are high. These behaviors can be quite easily reproduced by using a more direct fitness measure—for example, by modeling how survival and achievement of reproductive status varies nonlinearly with growth (Railsback and Harvey 2002).

5.4.3 Examples of Fitness Measures

Two approaches have been developed for finding fitness measures that are direct and complete enough to produce realistic behavior, without being unrealistically complex. One approach used by some behavioral ecologists is to derive mathematically a simplification of the expected fitness consequences of decisions involving trade-offs between growth rate and survival probability. For example, Gilliam and Fraser (1987) derived the result that the expected fitness of juvenile fish is maximized if they select among alternative habitat patches in a way that minimizes the ratio of mortality risk over growth rate; Leonardsson (1991) derives a similar conclusion using a different set of assumptions. These derived fitness measures may at first appear very useful for modeling decision making in IBMs, but upon close inspection they can turn out to be quite problematic. The problem is the assumptions required to make these derivations: the conditions that must be assumed to make the derivation analytically tractable can be quite unrealistic and incompatible with an IBM. Deriving the "risk over growth" fitness measure required Gilliam and Fraser to assume, for example, that all habitat patches provide positive growth rates; and that the population is stable and has fixed, intrinsic rates of reproduction and mortality. Such assumptions not only are often violated in an IBM but are also incompatible with the fundamental assumption of IBE that population characteristics are emergent, not fixed.

The second approach is to develop fitness measures that specifically represent one or several fitness elements and how they depend on the driving processes, but are simplified enough to be computationally efficient and biologically believable. In section 7.5.3 we present an approach for developing fitness measures that are sufficiently complete and direct to reproduce a wide range of realistic behaviors, while still being realistically simple (see also the trout IBM example in section 6.4.2). This "state-based, predictive" approach borrows concepts from the dynamic state variable modeling literature (Houston and McNamara 1999; C. W. Clark and Mangel 2000) and assumes that organisms use simple predictions of future conditions to estimate roughly the fitness consequences of decision alternatives. The state-based, predictive approach can be used to develop fitness measures that directly estimate important, clearly defined fitness elements, such as the probability of surviving to some future time or the expected number of offspring, and consider how these elements depend on processes such as mortality risk and energy intake.

5.4.4   Design Guidance for Fitness Measures

The two primary considerations for designing fitness measures are how complete and direct the measures should be. A third consideration is whether different fitness measures should be used as the individual's state changes.

In designing fitness measures, we need to remember that they are *models*. Some fitness measures that have been quite successful in IBMs require elaborate assumptions and calculations: for example, they assume that individuals predict future conditions, estimate probabilities, and consider a variety of state and environmental variables in complex trade-offs. When we use such fitness measures, we of course do not think that the organisms actually make all these calculations; instead, we only assume that the organisms have genetic and learned behaviors that are modeled well by the fitness measures.

5.4.4.1   Completeness

Selecting which elements and driving processes to include in a fitness measure requires a trade-off. If important elements or processes are left out, then the decision-making trait will be too simplistic to reproduce some real behaviors;

but adding too many elements can make a fitness measure computationally difficult and possibly unrealistic. The following sequence of questions can help decide what fitness elements and driving processes should be included in a fitness measure.

First, what fitness elements are important to the IBM's purpose? Is the model intended to explain patterns in survival, growth, mating? If so, then it is likely important to include those elements in fitness measures. If an IBM is intended to explain patterns in fecundity, and a fitness measure is being designed for a decision that clearly affects fecundity (e.g., how much energy to allocate to growth vs. gonad production), then fecundity should be included as an element of fitness measures. If an IBM is designed to study how habitat fragmentation affects population persistence, then fitness measures may need to consider the effects of habitat connectivity on the ability of individuals to find mates and mating habitat. If, on the other hand, the IBM's purpose is to explain something less directly related to fitness (e.g., patterns of habitat selection or plant succession), then the modeler can ask: what fitness elements are so basic that they are likely essential for explaining any fitness-based decision? For most species and systems, survival to—and attaining size and energy for—future reproduction are probably the most fundamental elements of fitness and likely to be important in most decisions.

Second, what driving processes are directly affected by the decision that the fitness measure is used for? Does the decision alter the individual's energy intake, mortality risks, or ability to find mates? Third, considering the real system being modeled, what environmental and biological processes are the most important drivers of the selected fitness elements?

A final question is, What driving processes can the individuals be assumed to sense or "know" (section 5.7)? Remember that a fitness measure is not a model of how the environment affects individuals but represents the individual's *internal* model of how the environment affects its fitness. Therefore, variables or processes should not be included in a fitness measure if the individuals are not aware of them. For example, an organism may be assumed to make decisions by considering how mortality risks affect expected fitness, but which risks should be included in the fitness measure? It seems reasonable to assume that most organisms have some awareness of their current energy status, so they are likely to "know" their starvation risk. Likewise, many animals show instinctual knowledge of what conditions place them at high risk to predators and some ability to sense predator presence, and so could be assumed to "know" predation risk. On the other hand, it may not be realistic to assume the organisms "know" risks they have had little opportunity to adapt (genetically or via learning) to—perhaps harvest by humans or introduced predators. Population density is usually not useful for modeling individual fitness because individuals are unlikely to be aware of the density of their population.

5.4.4.2 Directness

The second major step in designing a fitness measure is explicitly simulating *how* driving processes affect the selected fitness elements. Once a modeler has decided which fitness elements and driving processes to include in a fitness measure, the problem becomes relating the two: how exactly does expected fitness depend on driving processes such as energy intake and mortality risk? This is the point at which we take advantage of the ability of IBMs to use more explicit and direct models of how expected fitness is affected by decisions. Representing, even simply, the nonlinear ways that processes such as energy intake and mortality risk affect future fitness can be key to the success of an IBM.

Among the resources available to help model the relations between fitness elements and driving processes are knowledge of the organism's natural history and physiology, energetics models, and probability theory. These relations are inherently nonlinear, often because physiology sets bounds such as a maximum rate of growth toward maturity and maximum fecundity. Expected fitness also often depends on the individual's current state—its current size, energy reserves, life history stage—as well as the driving processes. Finding the right level of detail for a fitness measure can be a challenge. The fitness measure needs to include essential relations and nonlinearities, but simplicity is also desirable. The dynamic state variable modeling literature (e.g., Mangel and Clark 1988; C. W. Clark and Mangel 2000; Houston and McNamara 1999) uses a similar approach and provides many helpful examples.

To emphasize biological realism and to guide model design, it is desirable (but not always completely feasible) for a fitness measure to have a specific meaning with clear biological significance. Examples include expected probability of survival over an upcoming n days, and expected number of offspring produced next reproductive season.

Another consideration important for many IBMs is designing the fitness measure so individuals can make good choices even when no alternatives offer high expected fitness. If an IBM sometimes simulates conditions when none of the decision alternatives offer high expected fitness (due perhaps to disturbance events or extreme competition), it can still be important for individuals to identify the least-bad alternative so they can attempt to survive until conditions improve. For example, a model might assume (as the dynamic state variable modeling literature often does) that expected fitness is zero when growth is negative. If this assumption is used, individuals will be unable to make good decisions during temporary situations where no alternatives offer positive growth, because all alternatives offer the same expected fitness: zero. If, instead, expected fitness is modeled in such a way that it decreases asymptotically toward zero as conditions get worse and worse (e.g., Railsback et al. 1999), individuals will always be able to select the least-bad alternative.

A final key consideration related to directness is selecting an appropriate way to model prediction: to evaluate expected fitness, an individual must predict how driving processes and fitness elements change in the future. This issue is discussed in section 5.5.

5.4.4.3   Changing Fitness Measures

As individuals progress through life history stages or make other changes in state, the relative importance of different fitness elements may change. Therefore, different fitness measures may be appropriate during different parts of an individual's life cycle. For early life stages, the most appropriate fitness measure may consider only survival and growth to the next life stage; when reproductive stages are attained, then growth may become unimportant, and a fitness measure considering reproductive output may be more appropriate. Thorpe et al. (1998) outlined a model in which salmon change fitness measures as they progress through life history states. Bull, Metcalfe, and Mangel (1996) developed a model, and supporting experimental data, in which overwintering fish use survival until the end of winter (during which they are largely dormant) as their fitness measure, but only until the spring feeding season begins. The design of an IBM's fitness measures needs to consider whether, and how, fitness measures should change with life history stage or other factors.

5.5   PREDICTION

The previous sections on modeling fitness and adaptation depend heavily on the concept that fitness is a phenomenon of the future, so prediction is necessary if decisions are to be made on the basis of how they affect fitness. Modeling prediction by individuals in an IBM is particularly fascinating. Prediction itself is a modeling task: organisms make predictions by applying internal models. In designing how to represent prediction in an IBM we are therefore trying to model the internal models that real organisms use. We thus have to take the individuals' perspective, which was one of the basic modeling heuristics in chapter 2.

5.5.1   The Importance of Modeling Prediction

The concept of prediction is rarely discussed in ecological modeling, yet many IBMs include some form of prediction and research on artificial complex adaptive systems has shown that the ability to anticipate the outcome of actions is critical for intelligent, lifelike behavior of individuals. For example, Holland (1998) showed that the skill of digital agents in playing games like checkers or chess is linked to their ability to anticipate the consequences of alternative game moves. Even rudimentary abilities to anticipate the future consequences of decisions confer a great fitness advantage, so we must assume that even the simplest organisms have some predictive ability (Levin 1999, p. 175). In fact, Zhivotovsky, Bergman, and Feldman (1996) described prediction of future *environmental* conditions as key to the adaptive ability of organisms, and it should be clear from the preceding section on fitness that prediction of future *internal* conditions is also critical. Consider trying to maximize our probability of survival considering only the immediate consequences of our behavior: we would simply maximize our instantaneous survival probability by locking the door and hiding under the bed. Intuitively, it is obvious to us that hiding is not sustainable behavior—sooner or later we must come out from under the bed and go to work to pay the grocery bill, or else we will eventually starve. Understanding the importance of feeding instead of only hiding requires consideration of future consequences of behavior, knowing that starvation will soon be our greatest threat if we continue only to hide. (Hunger is, in fact, a physiological mechanism resembling prediction: it reminds animals, long before serious declines in energy reserves occur, that future fitness requires eating.)

Holland (1995) discusses prediction via internal models that individuals use to anticipate outcomes of their actions. According to Holland, "tacit" internal models prescribe certain actions on the basis of simple implicit predictions. These implicit predictions are often so simple that they may not be recognized as a form of prediction. Holland provides the example of a bacterium that swims in a sugar gradient toward higher concentrations, under the implicit prediction that increasing concentrations lead to more food. More explicit are "overt" internal models in which the consequences of alternative decisions are evaluated using the information the individual knows about its habitat and itself.

Models of "tacit" prediction are common in IBMs, but rarely recognized as such by the modelers. For example, in a number of models (e.g., M. E. Clark and Rose 1997; Van Winkle et al. 1998), animals move to a different location if they have experienced a consistent decline in fitness potential. This approach to modeling habitat selection assumes that the animals implicitly predict that (1) if there is a consistent decline in fitness potential, then conditions are likely better at other locations (hence, fitness potential will be improved by moving), and (2) if there is no consistent decline in conditions, then conditions are not likely to be better at other locations (hence, there is no reason to move).

The approach to fitness-seeking traits we discuss in section 5.3 requires simulating how individuals predict future values of variables that represent the processes driving fitness elements. Growth, for example, affects many fitness elements so many fitness measures must include a prediction of future growth, which may in turn require predicting the ecological and physiological processes that control growth. All this prediction sounds daunting, but the little experience we have indicates that even simple models of overt prediction can produce useful, realistic behavior in IBMs.

5.5.2 Approaches for Modeling Prediction

Levin (1999, pp. 111, 195) discusses some of the issues in modeling prediction by organisms: abstracting important signals from noisy input, representing realistic limits on what information organisms have (section 5.7), and understanding the trade-offs between costs and benefits of obtaining more information. We have found little research on what models of prediction are realistic biologically or useful in modeling fitness-seeking adaptation. The dynamic state variable modeling literature (Houston and McNamara 1999; C. W. Clark and Mangel 2000) generally avoids this issue by assuming that conditions affecting fitness (habitat, competition, etc.) are static. There are, however, a few IBMs that addressed prediction fairly explicitly.

Huse and Giske (1998; section 6.9.1) developed a model of fish migration in which fish were allowed to base movement decisions in part on predicted seasonal conditions: fish were assumed to "know" the current day of the year and to have inherited traits encoding knowledge allowing them to predict environmental conditions from the date. In their model the prediction traits were artificially evolved instead of explicitly formulated, so we do not know the algorithm the model fish used to predict, for example, spatial distributions of temperature from the current date. Their study, however, is an important exposition of the importance of prediction to adaptive behavior in IBMs.

The trout habitat selection modeling study by Railsback and Harvey (2002; section 6.4.2) indicates that a very simplistic model of how individuals predict future habitat conditions can produce useful and realistic behavior. The model trout evaluate their fitness measure using the prediction that habitat and competitive conditions over a time horizon of several months are simply the same as those occurring at the present time. Not surprisingly, however, this approach produces much less realistic adaptive behavior when used with even longer time horizons (greater than one season) because the prediction method is not even roughly accurate over such long times.

Stephens et al. (2002a) developed a fitness measure used by marmots in a dispersal trait. The fitness measure estimates probable "lifetime fitness," essentially a prediction of the number of offspring (and surviving close relatives) under each decision alternative. This prediction is based on age- and sex-specific probabilities for a number of events such as the individual obtaining dominant status in a territory, mortality during several future periods, and number of offspring produced. Stephens et al. simply assumed that individuals know these probabilities, which were estimated by the modelers from field data.

Memory can be used as a basis for modeling prediction. An IBM could assume that an individual retains a memory of recent conditions and projects future conditions by extrapolating recent trends. Alternatively, predictions could use the assumption that short-term conditions will revert toward remembered longer-term average conditions. Although we know of no instances in which they have been applied and tested in an IBM, approaches of this nature have been proposed and seem promising. For example, Hirvonen et al. (1999) show how important memory can be in modeling prey choice decisions of foraging individuals. One obvious concern is using time scales for memory that are compatible with the prediction time scale: memory of weather over the past two days may be a useful predictor of tomorrow's weather but would likely be a very poor predictor of next month's weather.

Environmental cues appear to be a mechanism that some species use to predict changes and could readily be incorporated in IBMs. Changes in day length and temperature are likely used by many species to anticipate seasonal changes in weather and other conditions; the assumption by Huse and Giske (1998; discussed earlier) that fish know the date is therefore reasonable. Antonsson and Gudjonsson (2002) describe a more sophisticated example of prediction using environmental cues. They provide field evidence that juvenile salmon, before migrating from river to ocean, use river temperature variation as a cue to predict future thermal conditions in the ocean. Thus, it is not always absurd for IBMs to assume that individuals can predict events far away in both time and space.

5.5.3 Design Guidance for Prediction

As Holland (1995) points out, some kinds of prediction are actually common in models of individual behavior even though the modeler might not recognize the method as predictive. This is especially true of tacit prediction, traits in which the prediction underlying the algorithm is not explicit. Tacit prediction may be quite appropriate in an IBM and seems most likely useful as part of indirect fitness-seeking traits (section 5.3.2). However, our first guidance concerning prediction is that modelers should be aware of when they are using implicit prediction and carefully document the implied predictions. An IBM's developers and clients will be much better able to understand and judge the appropriateness of predictive traits once all the implicit assumptions are clearly stated.

Overt prediction is necessary in traits that use direct fitness-seeking to model adaptive decision making. In this approach, individuals base their decision on expected future consequences to their fitness. There is currently little experience in IBE to guide modeling of how individuals overtly predict the conditions they will be exposed to. The ability to foresee consequences of decisions can provide a tremendous fitness advantage, so we expect many adaptive traits of real organisms to be modeled well by assuming fairly sophisticated predictive abilities. We have not attempted to review the extensive literature on mental models and prediction in human cognitive psychology. Ecologists seeking alternative and innovative models of prediction might benefit from examining this literature.

The most appropriate approach to modeling prediction depends on what variable is being predicted and how uncertain that variable is. Seasonal changes in weather are quite consistent from year to year, so assuming a relatively accurate ability to predict seasonal weather may be reasonable for many organisms. However, for many fitness measures it is necessary to predict variables that are much less certain—for example, the degree of competition with other individuals. It seems less reasonable to assume individuals are able to predict such uncertain variables accurately; more approximate, "rule of thumb" approaches may be (but are not necessarily!) more appropriate. Similarly, different methods are likely to be appropriate for predictions of different time horizons: methods that may be reasonable in the short-term may be very inaccurate for long-term predictions.

Given the lack of research and experience in this field, our tendency would be to start with a very simple model of prediction, test it, and increase the sophistication of the prediction only as needed to reproduce the realistic behaviors of interest. This design process should start by examining the evidence concerning what the real organisms do, and follow the IBE theory development cycle of chapter 4.

5.6 INTERACTION

One of the characteristics setting IBMs apart from other population models is the ability to simulate interactions among individuals explicitly. The concept of interaction refers to how individuals in an IBM communicate with, or affect, other individuals. One way in which IBMs can resemble real ecological systems is by assuming that interactions are the mechanism determining what individuals "know" about each other and how information travels through a population.

Interaction is a key concept in CAS. In fact, one of the first concerns with classical modeling approaches addressed by the pioneers of CAS (especially in economics) was the assumption that all individuals have perfect knowledge of the variables driving their decisions. These pioneers realized that in real systems information is passed via interactions that are often local and uncertain (Waldrop 1992). Subsequently, representing interaction and its effect on system behaviors has been a major theme of CAS research (e.g., Nowak and Sigmund 1998; Axelrod, Riolo, and Cohen 2001; Cohen, Riolo, and Axelrod 2001; Gmytrasiewicz and Durfee 2001).

5.6.1 Approaches for Modeling Interaction

We describe here three general ways that interaction can be modeled. These are not necessarily alternative approaches with relative advantages and disadvantages. Instead, each is appropriate under some conditions.

5.6.1.1 Direct Interaction

Direct interactions involve an explicit encounter among individuals in which information is exchanged or the individuals otherwise affect each other. Communicating the location of food, contests for resources or dominance, and predation events are examples of direct interactions. Direct interaction may be either global (all individuals interact directly with all other individuals) or local (individuals directly interact only with neighboring individuals). Global direct interaction is unlikely to be useful or realistic except in very small populations. Direct interaction requires contact or signaling among individuals so that it may rarely be realistic to assume that individuals have the ability to communicate with large numbers of others over long distances simultaneously.

The ability to model interactions realistically as local instead of global is often considered a defining characteristic of IBMs, particularly in plant ecology (e.g., Huston, DeAngelis, and Post 1988). Local direct interaction can be useful in IBMs with temporal and spatial resolution similar to those at which individual interactions actually take place. A good example is the IBM presented by Camazine et al. (2001, chap. 20) to explain how dominance hierarchies in wasps emerge from direct contests among individuals; the dominance model discussed in section 6.2.4 is similar.

5.6.1.2 Mediated Interactions

Indirect interactions among individuals are often modeled as being mediated by some resource. Instead of interacting directly with each other, the model individuals indirectly affect others by producing or consuming a common resource. (If the interaction is competition, then direct and mediated interaction correspond to the classical concepts of interference and exploitation competition.) Interactions that are direct and local in the real world are often modeled usefully as mediated interactions. The real mechanism of competition for food, for example, may be numerous, short, direct contests among individual animals; but over longer time scales the average outcome of these contests may be modeled well as indirect competition for the food resource.

Several of the social insect systems examined by Camazine et al. (2001) include mediated local communication among individuals, with pheromones and building materials as the mediating resources. Army ants communicate by creating and following pheromone trails; termites create pheromone-scented soil pellets that stimulate other nearby termites to do likewise, leading to mound formation; and wall-building ants create structures by picking up stones and dropping them near stones dropped by other ants. Mediated interactions can be either local or global. The examples listed here are local interactions, but in some models (especially nonspatial ones) all the individuals may compete for a common resource pool (e.g., Uchmański 1999, 2000a, b; Grimm and Uchmański 2002).

5.6.1.3 Interaction Fields

A second alternative to modeling direct, local interaction is the assumption that each individual is affected by a local field of interaction created by other individuals. This approach is similar to the "independent-individual approximation" discussed by Auyang (1998) and exemplified by an individual investor's interaction with a stock market. Even though stock prices are actually determined by many interactions among many individual investors, the behavior of a single investor can be modeled by assuming the investor interacts with the market as a whole. The interaction field assumption can be a useful approach when (1) individuals really do base decisions on the cumulative or average influence of neighboring individuals, or (2) it provides a good approximation of multiple interactions that occur at shorter time scales.

Several IBMs have used interaction fields with notable success. The "field of neighborhood" approach to modeling competition for resources among plants (Berger and Hildenbrandt 2000; section 6.7.3) assumes each plant stem has a circular "field of neighborhood." The field's radius and strength of influence increase with plant size, and the growth and survival of each individual are affected by the total field of neighborhood exerted by all its neighbors. Huth and Wissel (1992; section 6.2.2) tested alternative assumptions about interactions in their fish schooling IBM. They found that assuming each individual fish interacts with (aligns its swimming with) the *average* of its neighbors produced more realistic results than did the assumption that each individual picks one neighbor to interact with.

Interaction fields must not be confused with the "mean field approximation" used in some analytical models. Mean field theory is borrowed from physics and assumes that each individual perceives the mean influence of all the other individuals. This approximation is used in analytical models to avoid the need to represent individual-level interactions (Bolker and Pacala 1997; Dieckmann, Law, and Metz 2000). With interaction fields, individuals are still modeled but their *local* interaction with other individuals is approximated by averaging the influence of neighbors instead of describing each interaction.

5.6.2   Design Guidance for Interaction

Interactions among individuals are a key part of many IBMs, so choosing how to model interaction is important. One of the most important potential advantages of IBMs is the ability to simulate the local, direct interactions that often control the movement of individuals, materials, and information through ecological systems. However, direct interactions can occur at scales that are inconvenient or inappropriate for a particular IBM. For example, direct competition among plants for soil nutrients can occur very slowly over large but dispersed volumes of soil where roots of competing plants neighbor each other. Consequently, integrative methods such as mediated interaction and interaction fields are often useful when it is not essential to represent direct interactions.

Answers to the following questions about the real system to be modeled should help make it clear which approach to modeling interaction is important:

- What are the actual mechanisms of interaction or communication? Are there mediating resources?
- Is it important to the IBM's objectives to simulate interactions explicitly? Are the system behaviors addressed by the IBM believed to emerge from *direct* interactions? If so, then it may be essential to represent interaction as direct. If not, then it may be better to use mediated interaction or interaction fields if these approaches are more efficient.
- Over what spatial and temporal scales do the real interactions occur? How do these scales compare to the resolution of the IBM? If the real interactions occur at much shorter times than the model's time step, or at much finer spatial resolution, then direct interaction will not be feasible.
- If mediated interactions or interaction fields are appropriate, what is the average effect of the real interactions over a model's time step and spatial resolution? What is a good way to represent, at the IBM's coarser resolutions, the real interactions?

5.7   SENSING

It is hard to imagine a very useful IBM that did not assume individuals have some information about their environment or the other individuals. Adaptation to changing conditions is impossible without some knowledge of those conditions, and sensing is how real organisms obtain information about their world. Any IBM that assumes individuals have some knowledge about their world includes at least implicit assumptions about how and what the individuals sense. We use a broad definition of the word "sense": few IBMs simulate the details of what can be detected via the physiological senses. More often, we consider the general ability of individuals to obtain information and what they "know" about their surroundings.

5.7.1   Modeling How Individuals Sense

The amount and accuracy of information individuals have are some of the important limitations on how well they can adapt and respond, and real organisms of course have limited abilities to sense and "know" their world. Consequently, designing an IBM requires consideration of how well the individuals can obtain the information they need for adaptive traits. Typical sensing-related questions that arise in designing IBMs include:

- What do individuals know about mortality risks and how they vary? For example, does an animal "know" what habitat has high predation risks or when predators are near?
- What do individuals know about resource availability and how it varies over space? Over what distances, and with what certainty?
- What do individuals know about themselves? Is it reasonable to assume an organism knows, for example, its energy reserves or disease status?

Designing how an IBM addresses the kind of sensing issues listed here requires asking three questions. First, *what kinds* of information does an individual have? What variables describing its environment or neighbors does an individual have values for? Second, *how much* information of each kind does the individual have? What range of distances, or how many neighbors, does it have information on? Finally, *how accurate* is the information? Does the individual have completely accurate values for the variables it knows? Or do the values contain random uncertainty? Or systematic bias?

One approach to representing sensing or information-gathering processes is to simulate, at least coarsely, the actual sensing mechanism. Often simulation of senses is simplified by using a Boolean "yes-no" approach: an individual either does or does not detect some signal. A number of the social insect models presented by Camazine et al. (2001) model whether insects sense pheromones from other individuals using a Boolean approach. The salmon migration model discussed in section 5.2 assumes salmon either do or do not detect the smell of their destination stream. For the salmon and insect models, the sensing approaches are supported by convincing studies of the real animals. A model of cowbirds (Harper, Westervelt, and Shapiro 2002) simulates how birds obtain information about habitat visually as they fly over it, also a believable representation of how real birds sense. Mechanistic simulation of sensing allows an IBM to represent each individual's knowledge as an outcome of the individual's experience.

Probably the most common way to model sensing is simply to assume that the individuals "know" certain information: assume that the individuals have

access to the value of selected variables describing their environment, themselves, or neighboring individuals. Knowledge of environment and neighbors
is usually limited to a specified spatial extent. For example, an individual in one
grid cell may be assumed to "know" food availability and competitor density
in its adjacent grid cells. This approach of assuming that individuals know specific, limited information is often reasonable because actual sensing processes
typically happen at very short time scales compared with a model time step,
at least for animals. A bear IBM may assume an individual bear spends each
one-day time step in one grid cell, whereas a real bear may spend part of that
day exploring and sensing conditions in nearby areas. It is unnecessary to model
all the bear's exploration within a day, so instead we simply assume that, over
a day's time, it has gathered sufficient knowledge about neighboring areas to
"know" conditions there.

Stochastic techniques (section 5.8) can be used to simulate uncertainty in
sensed information, and even how uncertainty varies with the state of the environment or individual. Stochastic techniques can be useful for simulating the
effects of incomplete or erroneous information on adaptive processes, effects
that real organisms must continually contend with.

5.7.2 Design Guidance for Sensing

How to model sensing and information gathering is likely to be an issue in
almost all IBMs. One of the key design questions is whether sensing needs to
be simulated explicitly and mechanistically, or whether it is more appropriate to
make simple assumptions about what individuals know. Mechanistic representation of sensing is necessary if the sensing process itself is expected to be an
important cause of the patterns and behaviors the IBM is intended to explain.
Such was the case in some of the studies of communication and interaction
among insects presented by Camazine et al. (2001). Mechanistic simulation of
sensing can also be useful if the sensing process can be modeled at the same
spatial and temporal resolution as the rest of the IBM, as in the cowbird model
of Harper, Westervelt, and Shapiro (2002).

If sensing is represented mechanistically, then the primary consideration for
designing the sensing trait will usually be the mechanisms actually used by the
organisms being modeled. The modeler can start by searching for information
on what individuals can detect, over what distances, with what accuracy; and
how these factors might vary with the individual's state or environment. The
study by Spencer (2002) of predator detection by turtles is the kind of literature
a modeler can at least hope for. This study showed that turtles could detect
the odor of native, but not exotic, predators—a key mechanism for explaining
effects of exotic predators on turtle populations. One of the most fascinating
new areas of plant ecology is discovering the mechanisms by which plants sense
(and actively adapt to) conditions such as light levels (Schmitt, McCormac,

and Smith 1995) and whether, and what, insects are attacking (Schultz and Appel 2004). IBMs are potentially of great value for understanding how these mechanisms affect population dynamics and viability.

Often, however, it is best to avoid detailed representation of sensing and instead use simple assumptions about what variables individuals know, over what spatial extent. These assumptions should be based on whatever information or understanding is available about what real individuals can sense or know within the IBM's spatial and temporal resolution.

Of utmost importance is making a good assumption about the distance over which individuals can sense during a time step: underestimating this distance can severely and unrealistically limit the ability of individuals to adapt. For example, when simulating movement of animals, modelers often automatically assume that individuals can only sense conditions in grid cells adjacent to their current location. However, if the real animals are actually capable of exploring a much greater distance during a time step, they can consider many more potential destinations than just adjacent grids. Stream fish, for example, are often modeled using grid cells of one to several square meters in size, but fish often explore tens or hundreds of meters per day, so a daily time step fish IBM can safely assume fish know conditions over many grid cells (Railsback et al. 1999). Specifying the distances over which individuals can sense can be one of the most important factors in an IBM's design.

5.8 STOCHASTICITY

By "stochasticity" we mean the use of random numbers and probabilities to represent processes in an IBM. ("Random" numbers are almost always pseudorandom numbers; see section 8.7.3.) Unlike many of the other concepts addressed in this chapter, stochasticity is widely used in other kinds of ecological model. The basic issue addressed here is what processes in an IBM should be modeled as stochastic.

One of the greatest misconceptions about IBMs is that they are inherently stochastic. The ecological literature often confuses *variability* with *stochasticity*, implicitly assuming that variability at the individual level is random. Because IBMs represent variability, some ecologists mistakenly believe all IBMs rely heavily on random processes. Camazine et al. (2001) classify IBMs as "Monte Carlo models"; Law, Murrell, and Dieckmann (2003) equate an IBM with "a stochastic process," while stating that birth and death events, and variation among individuals, are all random. In contrast with this belief, the state and behavior of an IBM's individuals can be variable in many ways even when few or no processes are modeled stochastically. In fact, one of the primary reasons for using IBMs is to understand how variability arises from deterministic processes (Huston, DeAngelis, and Post 1988). In this section we provide a framework for deciding what parts of an IBM should, and should not, be stochastic.

5.8.1 Stochasticity and Ignorance

Our colleague Glen Ropella once stated that "randomness is a way of injecting ignorance into a model." Representing a process as stochastic means that we either are indeed ignorant about the process or that we choose to pretend we are ignorant to avoid unnecessary detail (Turchin 2003). Consider, for example, how the BEFORE model represents the way forest canopy gaps are closed (section 6.8.3). If we knew the details of the spatial configuration of all the trees around a gap, their age, crown geometry, and other information, we could possibly predict with some certainty whether a neighboring canopy tree grows over the gap or whether a younger tree fills the gap by growing into it. Usually, however, we do not have this detailed information (we are ignorant); and even if we did have it, we may not need to represent all these details but only their typical outcomes (we pretend to be ignorant). We can instead model canopy closure stochastically, for example, by assuming that the probability of the gap being filled by a neighboring canopy tree is 0.7 and by a young tree is 0.3. Random numbers can then be used to decide the fate of each gap that opens during simulations: a random number is drawn from a uniform distribution between 0 and 1 and if this number is smaller than 0.7, the gap is filled by neighboring canopy, otherwise by a younger tree. Of course, for this stochastic approach to work well, we need some kind of empirical basis for the probabilities.

Modelers thus choose to represent a process as stochastic for one of two reasons. First is because too little is known about the process to model it mechanistically. Some kind of empirical model is needed in such cases, and if the process is highly variable a stochastic model can be appropriate. The second reason is that, even if the process is well understood, it is relatively unimportant and would require unnecessary effort to model mechanistically. When the effort or computational demands of modeling a process mechanistically are not justified by the process's importance, it can be appropriate instead to represent it as a stochastic process.

5.8.2 Uses of Stochasticity in IBMs

In this section we discuss three common uses of stochasticity in IBMs, and potential alternatives to stochasticity for these three applications. The choice between stochastic and alternative models can be viewed in the framework of empirical versus mechanistic approaches, with stochasticity being an empirical approach.

5.8.2.1   Representing Variability in Input and Driving Variables

Stochasticity can be used to represent variable model inputs, especially for environmental variables like weather. For example, M. E. Clark and Rose (1997) stochastically synthesized time series of flow and temperature input to drive an IBM of stream fish populations. A second common use of stochasticity to induce variability is in creating the initial population of individuals at the start of a simulation. The modeler can specify statistical distributions for the individuals' state variables, with the IBM then using these distributions to stochastically assign state variable values to the initial individuals. The initial weight of each individual, for example, can be drawn randomly from a log-normal distribution with mean and variance specified as input. Stochastic generation of variable inputs allows the modeler to generate replicate simulations (by using different pseudorandom number sequences) that can be used to examine the effect of input variability on model results.

One obvious alternative to using stochasticity to induce variability in input is to neglect the variability. Instead of synthesizing stochastic weather input, weather can be assumed constant or represented by monthly averages. A second alternative is to use observed data instead of a stochastic model to represent variability in inputs. Instead of synthesizing stochastic weather input, for example, a time series of weather observations can be used as input. Using observed data has the advantage of including natural patterns—trends, autocorrelation, periodic or rare events—that are not captured by simple stochastic models but possibly are important to simulation results. Different time series of observations can be used to generate replicate simulations. Of course, adequate time series of observations are not always available.

5.8.2.2   Reproducing Observed Behaviors

Many IBMs use stochastic processes to reproduce observed behaviors that have been described probabilistically. In the lynx model described in section 6.4.1, a lynx's decision of which neighboring patch to move into is a stochastic process with probabilities that depend on the habitat characteristics of potential destinations. These probabilities were selected to reproduce observed patterns of lynx movement. Markov processes and random walk models (Turchin 1998) are other examples of stochastic processes used to reproduce observed behaviors.

Stochastic methods designed to reproduce observed behaviors are an empirical approach to modeling individual traits. These methods often fall in the category of "indirect fitness-seeking" behaviors discussed in section 5.4: the modeler assumes that if the model's individuals use stochastic processes to reproduce observed patterns of behavior, the individuals' fitness will generally be increased. The alternative to stochastic models of behavior is, therefore, to attempt to model the mechanism underlying the observed behavior.

The advantages of the stochastic approach are those of empirical modeling. If the stochastic model of behavior is well-supported by observations, it is likely to be considered reliable within the range of conditions the observations were made in. Also, given sufficient observations, it can be easier to develop a stochastic model than a mechanistic approach. As with empirical approaches in general, however, stochastic models are subject to extrapolation uncertainty: stochastic parameters that reproduce one set of conditions may not be suitable for extrapolation to very different conditions. And, of course, the stochastic approach does not help explain the behavior it models.

It is important to understand that even if some process observed in nature can be described well using a random model, the process is not necessarily random. Modelers can be too quick to assume that *variable* processes must be *stochastic*. Highly deterministic processes can produce variable behavior that fits random models well, but the process is still deterministic, and modeling it as random will limit the IBM's ability to reproduce important dynamics. Tikhonov et al. (2001) present an example where a simple, mechanistic model of fish school movement, combined with dynamic habitat conditions, produced results that fit random and chaotic models well. Mechanistic traits may be viable alternatives to stochastic traits even for reproducing behavior that appears random.

5.8.2.3 Representing Complex Lower-level Processes

IBMs typically need to represent processes that have variable outcomes but are not important enough to model in detail. Modelers can first consider whether it would be best simply to ignore the variability; if not, then such processes can be modeled as stochastic. Modeling mortality of individuals is a very common example. In many models, the probability of mortality is a deterministic function of the individual's state or habitat, but whether the individual actually dies at any time step is determined stochastically. (A random number is drawn; if it is less than the mortality risk, then the individual "dies.") In such IBMs, the modeler has determined that the *probability of mortality* needs to be simulated explicitly as a mechanistic process, but that the actual mortality *event* would be too complicated to simulate explicitly so it is represented using a random number.

An alternative to representing such lower-level processes as stochastic is to model them mechanistically. Continuing the mortality example, the alternative to stochastic representation of mortality events would be to model explicitly the processes that kill each individual. In the case of predation, this would require simulating where predators are, how they hunt, and why an individual was attacked. In this example, it seems clear that representing mortality events as stochastic is a very useful and appropriate approximation.

5.8.3   Design Issues and Guidance for Stochasticity

In determining how stochasticity should be used, the first question to consider is whether the process being modeled really should be variable. Sometimes variability is considered necessary for realism, but too many kinds of variability may make an IBM difficult to understand. Too many stochastic inputs, for example, may make it difficult to analyze the more interesting variability that arises from adaptive individual traits. As we discuss in chapter 9, it is often useful to explore how an IBM's dynamics change as sources of variability are added or removed.

If some process in an IBM does need to produce variable results, the variability can be produced by deterministic or stochastic processes; often a deterministic process with stochastic components is useful. Stochastic processes can be useful for reproducing observed behavior, but the modeler must realize that such processes have the limitations of empirical approaches: susceptibility to extrapolation uncertainty and inability to explain (instead of only describe) behavior.

Stochastic processes can be used as part of an adaptive trait: a sequence of stochastic decisions can produce behavior that increases an individual's fitness if the probabilities are modeled appropriately. However, traits with stronger random components are likely to produce behavior less able to adapt rapidly to changing conditions. If an adaptive trait is stochastic, it is especially important to test whether the trait produces realistic behavior. (In section 7.4 we discuss the use of probabilistic rules in more detail.)

5.9   COLLECTIVES

Many organisms form aggregations that have strong effects on individual fitness and have behaviors and dynamics different than those of individuals. Familiar examples include schools of fish, flocks of birds, packs of coyotes or wolves, social groups of birds and marmots, and stands of trees—all of which are included in IBMs examined in chapter 6. Camazine et al. (2001, chap. 8) describe a particularly dramatic example, aggregation of amoebae into slime molds with fruiting bodies. Auyang (1998) used the term "collective" for such aggregations. According to Auyang, defining characteristics of a collective include that interactions among its individuals are strong, internal cohesion is strong while external interactions are weak, and a collective has characteristics and processes that can be understood independently of its individuals. Another defining characteristic of a collective in ecological systems is that a collective exists for longer or shorter times than do the individuals making up the collective. A collective can be treated as an additional level of organization between the individual and the population.

5.9.1   Representing Collectives

Collectives often must be considered in IBMs because the collective strongly affects the environment and behavior of individuals. For example, large collectives of animals such as schools or herds are assumed to reduce individual predation risk yet may also reduce the availability of food. Belonging to a collective may make food more available to predators that hunt cooperatively. Individuals belonging to a collective may behave very differently from lone individuals, so different traits may be needed to model individuals on this basis.

Three general approaches can be used to represent collectives; they differ in the extent to which the behavior of collectives emerges from traits of individuals.

5.9.1.1   *Collectives Emerging from Individual Traits*

The fish schooling models examined in sections 6.2.2 and 6.2.3 and models of slime mold and insect behavior presented by Camazine et al. (2001) are examples of modeling collectives as emerging from relatively simple traits of individuals. The IBM specifies traits for individuals, and these traits give rise to individual behavior that forms the collectives: model fish have rules telling them to stay near and align with their neighbors, and the clusters that emerge behave very much like fish schools. These models were developed specifically to explain how the collectives are formed, which of course requires representing the collective as emerging from the individuals. There are other benefits of this approach—*if* relatively simple individual traits that explain the collective behavior can be deduced: (1) emergence from individual traits is biologically realistic; (2) models potentially can be very general, reproducing wide ranges of behavior by the collectives in many conditions; and (3) the approach can be easy to implement.

There are at least two potential limitations to modeling collectives as emerging entirely from individual traits. First, of course, not all collectives are produced by individual traits as simple as those that explain fish schooling or slime mold formation. Clearly, collectives such as social groups of birds and mammals emerge from complex individual behaviors that are probably part genetic and part learned. It would be extremely difficult to model these behaviors—and completely unnecessary for models with objectives other than explaining formation of the collectives.

A second limitation is that it is often necessary to represent explicitly some characteristics of the collective, which requires modeling it as a specific entity instead of having it exist only as an emergent phenomenon. Consider mortality of schooling fish: "risk dilution"—the reduction in an individual's risk of being eaten that results from being in a large group—is presumed to be a reason why fish evolved the traits causing them to school. To model the effect of schooling on the mortality rate of a fish population, we must represent in our IBM how an individual's probability of being eaten varies with the size of the school. This requires modeling the school as a specific entity: that is the only way to know the school's size. Individuals cannot know the size of their school—they can only sense their nearest neighbors.

5.9.1.2 Collectives Imposed by Individual Traits

Collectives can also be modeled by giving individuals traits that force them to reproduce the collective behaviors observed in the system being modeled. Individuals are given traits telling them to belong to a collective and behave in a way that makes the collective persist and function. This approach is useful when we want the collectives to be formed by the individuals but are not interested in understanding the individual traits that lead to collectives and their behavior: the IBM is intended for problems other than explaining formation of collectives, or the individual traits leading to collectives are too complex to model. Such models can help us understand the consequences, but not the causes, of collective behavior.

This approach can overcome the first limitation of modeling collectives as emerging from individual traits but not the second limitation. Even when we force the individuals to form collectives, we still often need to represent explicitly the collective itself.

5.9.1.3 Explicit Representation of Collectives

In this approach, collectives are treated as explicit entities in the model, with state variables and traits of their own. The coyote (section 6.3.3) and lark (section 6.6.3) IBMs are particularly clear examples of this approach: these models include both individual animals and collectives—coyote packs, lark flocks— and both individuals and collectives have traits executed at each time step. In the species represented by these models, some things that are clearly important to individuals—information such as group social hierarchy, events such as reproduction—can only be understood at the collective level; but it would be very difficult to model the complex behavior of the collectives as emerging from individual traits. Instead, the modeler simply develops traits for the collectives.

Representing collectives explicitly does not mean that individuals are ignored. Instead, IBMs that include collectives can also represent how individual behaviors affect the collectives and how the state of an individual's collective affects individuals and their behavior. Even when the presence of collectives is hardwired, key states and behaviors of collectives can emerge from adaptive traits of individuals. For example, the IBMs we examine in section 6.3

were designed to study how individuals and collectives (social groups) interact to determine population dynamics. Individuals make decisions—especially,
when to disperse—that affect the formation and persistence of social groups,
while these individual decisions are based in part on the state of the social group.
The persistence and stability of the population in turn depends on formation
and persistence of its social groups. The population can only be understood
by modeling both individuals and collectives and the links between all three
levels.

5.9.2   Design Guidance for Collectives

Many modelers will face the question of whether and how collectives should
be represented in an IBM. Of course, if the purpose of an IBM is to understand
how collectives arise, then the IBM should not *impose* collective behavior but
instead represent the individual traits from which collectives emerge.

Many other IBMs are developed not to explain collectives but to study other
questions about species in which individuals form some kind of collective.
How do we decide whether, and how, the collectives need to be represented
in the IBM? First, does the presence and behavior of a collective have strong
effects on the individuals? Do the state and behavior of individuals depend
on whether they are in a collective and on the state of the collective? If so,
then the IBM probably needs to include collectives. Second, does the collective
have behavior that strongly affects individuals but cannot be predicted without
explicitly representing the collective? If the answer to this question is yes, then
it is likely necessary to represent collectives explicitly as a separate type of
model entity with its own state variables and traits.

When collectives are included explicitly as a type of entity in an IBM, they can
be represented by a combination of (1) collective-level traits and (2) characteristics that emerge from behavior of its individuals. When some key characteristics
of the collectives—for example, their size, when they go extinct or split into
new collectives—emerge from behavior of individuals, the IBM can still link
population-level phenomena to individuals.

How can we design traits of collectives? We can use the same general
approach as for individual traits: using literature and observations to propose
alternative traits, then testing the alternatives in an IBM to see which best
reproduces observed population-level patterns. However, there is an important theoretical difference between traits of individuals and traits of collectives:
while fitness-seeking is a powerful approach for designing individual traits, we
cannot use it as a basis for traits of collectives. Collectives convey fitness to
individuals, but we cannot assume that collectives themselves seek "fitness"—
that is, have traits acting to maximize the growth, persistence, or reproductive
rate of the collective itself. Without the theoretical basis of fitness-seeking, we
must instead depend on empirical information to design traits of collectives.

Consequently, collectives usually are given traits that simply impose behaviors observed in the real system.

5.10   SCHEDULING

In time-dependent classical models, time is treated either as *continuous*, with processes happening continuously at rates specified by differential equations; or as *discrete*, with processes happening in jumps over time steps as specified by difference equations. With IBMs, we are freed from the necessity of assuming events occur continuously (few if any IBMs use this assumption) or even discretely at regular time steps (although most IBMs do use this assumption). Instead, we have to think about and design the most useful way to represent time.

5.10.1   Scheduling: Designing a Model of Time

The real events that we represent in an IBM vary in the order in which they occur and how much time each takes. The timing of events in an IBM is most often modeled by assuming events are *concurrent*, all happening together, during each time step. A plant IBM might assume a daily time step and simulate a day's energy production, growth, and grazing damage as occurring concurrently each time step. All real changes within a time step are represented as discrete events in the model, discrete jumps in state. When we actually implement a model that assumes multiple events occur concurrently, we must decide on the order in which concurrent events are actually executed: the computer cannot execute multiple events concurrently, nor would we want it to. The execution order can strongly affect model results because the outcome of one event can affect simulation of the next event. The plant's daily energy production affects its daily growth calculation, and grazing damage could alter energy production; so the IBM would produce different results if the execution order of these events was changed.

Scheduling is the concept of modeling exactly how a model's events are represented in time. For most IBMs, scheduling is a matter of specifying the exact order in which events occur and how event execution is related to simulated time. In many IBMs all simulated events occur in a predetermined order once per time step, and each time step represents a specific length of time (an hour, a day, etc.). For some models, however, the timing of events is not predetermined and scheduling involves designing a process by which entities within the model decide what events occur when.

The concept of an *action* is useful for defining a model's scheduling; actions are the building blocks of a schedule. An action has three parts. The first part identifies a list of model entities and the second part identifies those entities' method (a specific trait or algorithm) that is executed by the action. When an action is executed, the model goes through the list of entities and executes the specified method for each entity. The third part of the action specifies the order in which the list of entities is processed. Some example actions are:

- Scheduling an IBM's weather simulator to execute a method that updates the current temperature. In this action, the list of model entities includes only one: the weather simulator. The method to be executed is the temperature update. Because this action acts on only one entity, no processing order needs to be specified.
- Scheduling all the habitat cells in an IBM to update their food production. The action's list of model entities is a list of all the habitat cells, and the method executed by the action is the cells' food production method. Assuming food production is independent in each cell, the order in which cell updates are processed is unimportant; the action can simply pass sequentially through the list of cells.
- Scheduling the animals in an IBM to move and feed. In this case, if animals move to find good feeding locations and they compete for food, exactly how movement and feeding are defined as actions becomes a more interesting and important issue. These events could be simulated using one action, in which each animal executes a method that includes first moving, then feeding. Or, two actions could be defined: first, each animal moves; then, in the second action, each animal feeds. These different action designs clearly could produce different results. And the third part of the action—processing order—now becomes important. An action processed from largest to smallest animal would represent a size-based dominance hierarchy, whereas an action processing animals in random order would assume no hierarchy (section 6.5.3 provides an example of how these two scheduling assumptions can affect an IBM).

After actions are specified, they must be *scheduled*: put at the desired spot in a queue of actions waiting to be executed (or, equivalently, in a loop that is cycled through; figure 5.3). From this perspective, designing an IBM's scheduling is seen as deciding what methods of what model entities are put together into actions, and deciding how the various actions are scheduled with respect to each other.

Scheduling can be organized as a hierarchy of actions: an action (e.g., any of the three preceding examples) can itself be treated as a model entity that appears in a higher-level action. An IBM could have three actions that execute the three basic traits of all its individuals: move, feed, and die ("die" simulating whether the individual lives or dies each time step). A higher-level action called "animal actions" could have a list of entities—the animals' move, feed, and die actions—and the order in which they are executed. Then the "animal actions"

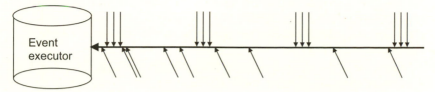

Figure 5.3 Timeline metaphor for scheduling. The large horizontal arrow represents time, with actions scheduled for later in time being further to the right. Actions scheduled at regular time steps occur at regularly spaced intervals, each interval being one time step. In this example, three actions are scheduled each time step, represented by the vertical arrows. Dynamic actions, represented by the slanted arrows, can be scheduled for any time—placed anywhere along the timeline. As simulated time progresses, events march left along the timeline and are executed when no other actions remain to their left.

could be included with habitat update actions and output-producing actions into the highest-level action. (This approach is exactly how several agent-based modeling platforms organize models and their software; see chapter 8.)

Actions need not be considered fixed once a model starts executing. It can be possible, and very useful, to add or remove actions from a schedule during execution.

5.10.2   Alternative Ways of Modeling Time and Concurrency

From the previous discussion, we see that the scheduling problem has three main parts. First we must decide in general how our IBM will represent time: with events happening concurrently at discrete time steps, with discrete events happening in continuous time, or both. Next, as we design the model and identify all the events (including those assumed to occur concurrently at each time step), we must decide how all the events that recur in the model should be aggregated into specific actions. Finally, we must decide the order in which concurrent actions are actually executed.

5.10.2.1   Discrete versus Continuous Time: Time Steps
          versus Dynamic Scheduling

The first major design decision for an IBM is selecting which (or both) of two kinds of scheduling to use. Most IBMs have used only discrete time with regular time steps. Discrete scheduling reduces a model's complexity by providing a simple, common way to model time, similar to the use of square grids to represent space. This approach keeps modelers from having to determine exactly when each simulated event occurs; instead, all events are assumed to happen once each time step. Time steps do not have to be of constant size; it is easy, for example, to use alternating time steps representing daytime and nighttime, with the number of hours represented by each step varying over the year.

When time steps are used for scheduling, time proceeds in chunks and the temporal relationship of events within a time step is ignored (see the discussion of selecting temporal and spatial scales in section 2.3). However, some eco- logical processes can be represented more naturally using *dynamic* scheduling, which assumes a continuous representation of time: each event has an exact time at which it is executed. Model time no longer proceeds in time steps but from event to event: the computer simply executes the next action in a queue that is ordered by time. Dynamic scheduling often works well when actions are created and scheduled by the model entities themselves, as simulation proceeds, instead of being predetermined and fixed.

Behavioral interactions seem like a natural application of dynamic schedul- ing. Consider a model of competition for space, containing many small habitat patches. Each patch is capable of supporting one or several individuals, depend- ing on the individuals' sizes. Dominance contests determine whether an intruder can stay in a patch and, if so, which individuals must leave. If an individual moves into a patch containing other individuals, then a dominance contest is dynamically scheduled to be executed immediately. If the contest results in individuals moving into other occupied patches, then additional contests are put on the dynamic schedule. Each initial movement could therefore dynami- cally trigger a whole chain of contests and moves, with contests from several chains being intermixed on the execution schedule.

5.10.2.2 Designing Actions

As an IBM is developed, the modeler identifies all the different things that the model's entities do. These things include not just the traits producing behavior of the individual organisms but also such "overhead" activities as updating habitat and producing output. Deciding how to aggregate all the traits and activities of a model's entities into schedule actions is a major part of the IBM's design, one that can have strong effects on results. Usually designing the first part of an action—the list of model entities it affects—is straightforward. Time- step actions usually act over all the entities of the same type: all habitat units, or all individuals. Dynamically scheduled actions may instead act on only one object: a single individual. On the other hand, designing the second part of an action—which of the entities' traits or activities are executed by the action—is often less straightforward.

Usually, the most interesting and troublesome issue in designing actions is whether several traits of individuals should be lumped into one action or executed as separate actions. If several traits are lumped in one action, then the first individual will execute each of these traits, then the next individual will execute each trait, and so on. On the other hand, if each of these traits is separated into its own action, then all individuals execute the first trait, then all individuals execute the second trait, and so on. It is easy to see that these two action designs could produce different model results.

In fact, *synchronous* versus *asynchronous updating*, a well-known design issue for IBMs and other bottom-up models (Ruxton 1996; Ruxton and Saravia 1998; Schönfisch and de Roos 1999), is a matter of how actions are designed. Under synchronous updating, the model's state is updated only once per time step, after all individuals have had their effects. This is achieved by letting all individuals sense the environment—including the state of other individuals—as it was at the beginning of the time step and by denoting the changes caused by the individuals in a temporary image of the enviroment. Only after all individuals have performed their traits, the state of the environment is—synchronously—updated by making the final state of the temporary image the new state of the environment. For example, when modeling feeding, one action tells all the individuals to execute their feeding trait, without updating the food availability. Then a second action causes the food availability to be updated, subtracting what all the individuals consumed. With asynchronous updating, the model's state is updated after each individual feeds: feeding and updating the food availability are combined in one action, so each individual determines how much food it consumes and then subtracts its consumption from what is available for the remaining individuals. There is a variety of ways to implement asynchronous updating, differing in the scheme used to select which model objects are updated when (e.g., Cornforth et al. 2002).

Asynchronous updating appears to assume that actions are undertaken one individual at a time (one individual feeds and reduces food availability; then the next one feeds; etc.). However, when we keep in mind that this is a way of modeling *concurrent* actions, it makes more sense to think of this approach as representing a priority hierarchy. The order in which individuals execute their actions reflects their priority for the resources being updated: individuals that go first have access to all resources, whereas individuals that go last get the leftovers (see the social spider model in section 6.5.3). Synchronous updating is more suited to representing situations in which the actions of one individual have little effect on the others, or in which no hierarchy among individuals is assumed. With synchronous updating, the order in which individuals execute their actions has little effect on what resources they have access to.

5.10.2.3 Scheduling Concurrent Actions

The third major scheduling issue is deciding how to schedule actions assumed to occur concurrently each time step, that is, to specify the order in which actions are executed. Most commonly, concurrency is modeled using *fixed scheduling*, under which actions occur in the same order each time step. The modeler specifies, as a very important part of the model design, the ordering of all the actions executed at each time step.

Randomized scheduling is an alternative that can be useful for avoiding artifacts of fixed scheduling. Randomization can take place at different levels within a schedule's hierarchy of actions:

- A low-level action (e.g., one that tells fundamental entities in the model such as habitat units and individuals what to do) can randomize the order in which these entities are processed.
- Multiple traits of an IBM's individuals can be included in one action, but the order in which these traits are executed can be randomized for each individual. The order of the traits changes from individual to individual within a time step.
- A higher-level action can include several low-level actions for one group of entities (e.g., the individuals), with the order in which these actions are executed randomized at each time step. The order in which the low-level actions are executed changes at each time step but is constant within a time step.

Typically, randomized scheduling is used for actions representing real events that really happen in an unpredictable order, or when the modeler wishes to avoid fixing the order of action execution. For example, actions that each represent one kind of mortality could be executed in random order because (1) there is no reason to believe any one kind of mortality would precede another within a time step, but (2) using a fixed order biases the frequency with which the different kinds of mortality occur. A number of studies have shown that how actions are randomized can affect results of simple agent-based models (e.g., Huberman and Glance 1993; Nowak, Bonhoeffer, and May 1994; Cornforth et al. 2002), although it is not clear that such artifacts are likely to be strong in IBMs rich in biological structure.

5.10.3   Design Guidance for Scheduling

Modelers must think about and document how they model time in an IBM, in the same way that we must think about how to model space. For most IBMs, the most important scheduling issue is how to group and order the execution of actions that are modeled as happening concurrently at each time step. There is extensive literature and experience concerning scheduling in the fields of discrete-event simulation, discrete mathematics, and communication systems. Readers are referred to sources such as Banks (2000) and Fishman (2001) for additional guidance. Software platforms designed for individual-based and discrete-event simulation (section 8.4) provide tools to make scheduling, and trying alternative schedule designs, easy. Following are specific issues that most IBM developers will need to consider explicitly.

5.10.3.1   Discrete versus Continuous Representation of Time

What actions should be discretized over time, if they occur concurrently once each time step? What other actions should be modeled over continuous time, using dynamic scheduling to model exactly when they are executed? Fixed time steps provide a common, simple way to represent processes that actually happen concurrently over long and variable time periods. Processes that actually occur nearly continuously (e.g., growth) are especially well represented by fixed time steps, as are events the exact timing of which is unimportant.

Dynamic scheduling can be used to model how specific individuals perform actions at specific times, usually within an overall time-step framework. Dynamically scheduled actions usually are only appropriate for modeling processes that (1) occur relatively quickly compared with the model's time step, and (2) produce results highly dependent on their order of execution. When dynamic scheduling is used, understanding *causality* in the model's results (section 8.3.5) becomes an especially important issue in designing the model's software and analyzing the model: data on what actions were actually executed when must be collected and analyzed to understand how the results arose.

5.10.3.2   Designing and Scheduling Actions

How should processes represented in the IBM be organized into actions? What things that the individuals and other model entities do should be lumped together so that they are executed together? And in what order should actions be executed, especially actions assumed to occur concurrently at each time step? There are no clear or general answers to these questions, only a few commonsense guidelines. The following sequence is often appropriate:

1. Actions updating the driving variables and environmental conditions that individuals must adapt to. (The events that individuals adapt *to* must be executed before the adaptive traits are.)
2. Actions executing the individuals' adaptive traits. (Individuals should be given the opportunity to adapt to new conditions before those conditions affect them.)
3. Actions through which individuals are affected by environment and each other, for example, feeding or energy intake, growth, mortality.
4. Actions by which individuals affect the environment, for example, by consuming resources such as food.
5. Observer actions that report the model's state to the user by updating graphical displays and writing file output.

However, the details within such a general organization can be very important. When it is not clear what scheduling is most appropriate, the best approach is to experiment with alternatives and see what effects they have.

5.11   OBSERVATION

Observation refers to collecting the information from an IBM needed to test and use the model. (Just to be clear, we are talking about observing what goes on *in the IBM*, not collecting field observations for comparison with model results.) From a software perspective (section 8.3.3), observation is a matter of producing the kinds of model output necessary to test the model and conduct the analyses the model was built for.

Observation is usually trivial for simple models because they only produce one type of output. For example, a Lottka-Volterra predator-prey population model produces only time series of population values, one for prey and one for predators. An IBM, however, produces many types of result: not only population values but also patterns of how individuals are distributed in space, plus the state (e.g., size, condition, location, behavior) of each individual. Our ability to test, analyze, and learn from an IBM depends on which results we observe, yet for many IBMs it is unnecessary or even impossible to output *all* the results. And specialized software is necessary to observe some IBM results.

Because IBMs produce many kinds of results, we must think about and design methods for observing simulations. This problem often resembles a field study design task more than a typical modeling task. We may need to think about ecological study design issues such as the resolution and frequency of observations, and we typically need observations at both the individual and population level. Ecologists are well aware that how we observe a system can strongly affect our understanding of it, and this is as true of IBMs as it is of natural systems.

There are at least three different perspectives for observing an IBM. Each gives a different view of simulation results that can be useful for different kinds of testing and analysis. The first perspective is the one typically used in computer simulation, the omniscient perspective. One of the primary benefits of using IBMs instead of studying natural systems is that we can observe whatever we want to in an IBM, without error or uncertainty and without altering the system. For example, we can output summaries of population status, spatial distributions of individuals and habitat, view individual behavior, at whatever temporal and spatial resolution we choose.

The second perspective is that of an individual in the simulation. We can attach "probes" to model individuals that report the world as the individual experiences it: what the individual senses about its habitat, its neighbors or competitors, and itself. This perspective helps us understand why the individuals behave as they do.

The third perspective is that of a "virtual ecologist," a simulated observer within the IBM that has the limitations of real observers in real ecosystems (Berger, Wagner, and Wolff 1999; Grimm, Wyszomirski, et al. 1999; Tyre, Possingham, and Lindenmayer 2001). A "virtual ecologist" models the process of collecting data from an IBM, producing observations that can be compared with observations collected by real ecologists in real ecosystems, or compared with omniscient observations from the same IBM. One application of this technique is to understand the importance and effects of bias and uncertainty in the data collection methods. Tyre et al. used a virtual ecologist that subsampled an IBM's habitat for the presence of individuals, as a real ecologist would sample some but not all habitats of a study system. A second application is to produce observations from an IBM that are comparable with observations of a real system that were collected using methods with a specific, known bias. Nott (1998) developed an IBM to study how dynamics of a songbird population depend on habitat fluctuations. Field data for testing the IBM consisted of counts of singing male birds during the breeding season, from helicopter surveys. These field data were known to represent the total population poorly because (1) the survey technique sampled only some of the habitat; (2) only breeding-age males were counted, not the entire population; and (3) the males do not sing if habitat conditions are not suitable for mating. Nott simulated when the breeding-age males would sing and a virtual ecologist that subsampled habitat in the same way the real helicopter surveys did. The IBM was then tested by comparing the virtual ecologist's observations with the real survey data.

Observation is covered only briefly here because chapter 9 considers how different kinds of observation are used in analyzing models, and chapter 8 discusses software tools to implement observer capabilities.

5.12   SUMMARY AND CONCLUSIONS

This chapter describes ten general concepts that help design, describe, and understand IBMs. These concepts were identified mainly from the literature on CAS, in which researchers search for ways to understand systems of interacting, adaptive individuals. Our hope is that these concepts will evolve into a general conceptual framework that serves some of the same purposes that differential calculus typically serves for classical models: providing a consistent way of thinking about models, a list of questions that guides the design of models, and an efficient way to describe and communicate models. These concepts are nowhere near as clear-cut and tidy as calculus, which is to be expected because IBMs are nowhere near as restricted in their ability to address complexity as differential equation-based models are.

Many of the issues discussed in this chapter are different ways of addressing the question of how mechanistic different parts of an IBM should be. In reality, everything that organisms are and do emerges from the interactions among their genes, their neurons, and their environment. It is never practical or necessary to simulate such low-level emergence, so many behaviors must be imposed at some level. The objective of this book is to develop a general approach for addressing how individual behaviors and system dynamics emerge from lower-level traits. Therefore, we cannot say exactly what dynamics should be emergent versus imposed, how detailed and mechanistic fitness measures should be, or what processes should be represented as stochastic. Instead, we hope to help establish a framework that helps modelers find appropriate answers to such questions for the systems and patterns they study.

We also hope that these concepts encourage modelers to think constantly about the real organisms and system they are modeling as they address model design issues. A great advantage of IBMs is that we can look to the real biology for answers to many modeling questions. Instead of mathematical tractability or data-fitting being primary modeling concerns, the ecologist's understanding of what really happens in nature can be (along with the problem the ecologist is trying to solve) a primary source of guidance in deciding what approaches to use in an IBM.

The conceptual framework presented in this chapter is summarized in the following checklist (section 5.13). We hope the checklist will have at least three major benefits. First, it can organize the process of designing IBMs and make the process more efficient. Thinking about how to address each of the concepts in the checklist should help modelers produce better IBM designs in less time. Each item in the checklist is an issue that will need to be considered in model design. The checklist should help the modeler identify and consider these important design decisions as early as possible, and explicitly.

A second benefit of the checklist is that its concepts can provide a common terminology and framework for communicating models. Documenting how each question in the checklist was answered can be an efficient way to describe the most important characteristics of an IBM. This benefit is very important because conventional ways of describing models (mainly by listing equations and parameter values) do not capture the essence of IBMs. Describing what outcomes of an IBM emerge from what underlying processes may, for example, be the single most important thing to understand about an IBM.

Finally, reviewers of IBMs, or proposals to develop IBMs, can use the checklist as an evaluation tool. Reviewers can evaluate modelers' understanding of IBMs by checking how thoroughly these concepts have been dealt with. Did the modeler recognize and explicitly address key concepts? Are the proposed approaches for dealing with key concepts appropriate? The checklist by itself cannot determine whether an IBM (or proposed IBM design) is good or bad, but it provides a framework for evaluating how appropriate an IBM's design is for the problems the IBM is intended to address.

5.13   CONCEPTUAL DESIGN CHECKLIST

Emergence

1. Which processes in the IBM are modeled as emerging from a mechanistic representation of adaptive traits of individuals? Do the system-level phenomena the IBM is designed to explain emerge from individual traits, or are they imposed by rules that force the model to produce a certain result?

Adaptation

2. What adaptive traits do the model individuals have to improve their potential fitness, in response to changes in themselves or their environment?

3. Which adaptive traits are modeled as direct fitness-seeking, with individuals making decisions explicitly to improve their expected success at passing genes on to future generations?

4. Which adaptive traits are modeled as indirect fitness-seeking, in which individuals make decisions to meet a specific objective that indirectly contributes to future success at passing genes on?

Fitness

5. For traits modeled as direct fitness-seeking, how complete is the fitness measure used to evaluate decision alternatives? The fitness measure is the individual's internal model of how its expected fitness depends on which alternative it chooses. Which elements of potential fitness—for example, survival to reproduction, attainment of reproductive size or life stage, gonad production—are represented in the fitness measure? Is the completeness of the fitness measure consistent with the IBM's objectives?

6. How direct is the fitness measure? What variables and mechanisms are used to represent how an individual's decision affects its future fitness? Is the choice of variables and mechanisms consistent with the IBM's objectives and the biology of the system being modeled? Does the fitness measure have a clear biological meaning? Does the fitness measure allow the individual to make appropriate decisions even when none of the alternatives are good?

7. How is the individual's current state considered in modeling fitness consequences of decisions?

8. Should the fitness measure change with life stage, season, or other conditions?

Prediction

9. In estimating future fitness consequences of their decisions, how do individuals predict the future conditions (internal as well as environmental) they

will experience? Do the simulated prediction methods produce realistic behavior while being biologically realistic? Are prediction methods appropriate for the time scales used to model fitness-seeking? Do the individual's predictions make use of memory? Of learning? Environmental cues?

10. What tacit predictions are included in the IBM? What assumptions are implicitly embedded in the tacit predictions?

Interaction

11. What kinds of interaction among individuals are assumed? Do individuals interact directly with other individuals? (With all others or only with neighbors?) Or are interactions mediated, for example, through competition for a shared resource? Or do individuals interact with a "field" of effects produced by neighbors?

12. What real interaction mechanisms, at what spatial and temporal scales, were the IBM's interaction design based on?

Sensing

13. What variables (describing both their environment and themselves) are individuals assumed to sense or "know" and consider in their adaptive decisions?

14. What sensing mechanisms are explicitly simulated? Does the IBM represent the actual sensing process?

15. If sensing is not simulated explicitly, what assumptions are made about how individuals "know" each sensed variable? With what certainty or accuracy are individuals assumed able to sense each variable? Over what distances?

Stochasticity

16. Are stochastic processes used to simulate variability in input or driving variables? Is stochasticity preferable to using observed values? Is it clearly desirable for these inputs or drivers to be variable?

17. What traits use stochastic processes to reproduce behavior observed in real organisms? Is this approach clearly recognized and used as an empirical model?

18. What variable low-level processes are represented empirically as stochastic processes? Is the variability important to include in the IBM?

Collectives

19. Are collectives represented in the IBM? Collectives are aggregations of individuals (flocks, social groups, stands of plants) included in an IBM because

the state and behavior of an individual depend strongly on (a) whether the individual is in a collective, and if so, (b) the state of the collective.

20. How are collectives represented? Do collectives occur only as phenomena emerging from individual behavior, or are individuals given traits that impose the formation of collectives? Or are collectives represented as explicit entities with their own state variables and traits?

Scheduling

21. How is time modeled in the IBM: using discrete time steps, continuous time, or both? If both are used, is dynamic scheduling used for events that happen quickly compared with the model's time step and are highly dependent on execution order?

22. What model processes or events are grouped into actions that are executed together? Do these actions produce synchronous or asynchronous updating of the model?

23. How are actions modeled as happening concurrently actually executed? What actions are on a fixed schedule, in what order? Are some actions executed in random order? What basis is provided for these scheduling decisions?

Observation

24. What kinds of model results must be observed to test the IBM and meet its objectives?

25. From what perspectives are observations of results taken: omniscient, model individual, or virtual ecologist?

Chapter Six

Examples

By operating the model the computer faithfully and faultlessly
demonstrates the implications of our assumptions and information. It
forces us to see the implications, true or false, wise or foolish, of the
assumptions we have made.

—Daniel B. Botkin, 1977

6.1 INTRODUCTION

In the preceding chapters we deal with concepts and strategies for developing
IBMs and conducting IBE, but now it is time to look at some IBMs in action.
So many IBMs have been developed in recent years that we were easily able to
assemble a mosaic of case studies that illustrate our ideas of how IBE is done and
the kinds of things to be learned by doing IBE. By presenting examples of IBMs
that have already been developed and used, this chapter also shows us what the
future should look like (Table 6.1). The example IBMs address a broad range of
ecological systems and questions but are not intended to provide a representative
overview of all existing IBMs. IBMs have been reviewed from a methodological
perspective by Huston, DeAngelis, and Post 1988; Hogeweg and Hesper 1990;
Breckling and Mathes 1991; DeAngelis and Gross 1992; Ford and Sorrensen
1992; DeAngelis, Rose, and Huston 1994; Judson 1994; Breckling and Reuter
1996; Grimm 1999; Grimm, Wyszomirski, et al. 1999; and Uchmański 2003.
Reviews of IBMs in specific disciplines include DeAngelis et al. 1990; Shugart,
Smith, and Post 1992; Van Winkle, Rose, and Chambers 1993; Dunning et al.
1995; Liu and Ashton 1995; Czárán 1998; Kreft, Booth, and Wimpenny 2000;
Wyszomirski, Wyszomirska, and Jarzyna 1999; Werner et al. 2001; and Huse,
Giske, and Salvanes 2002.

   The main focus of this chapter is on ecology: illustrating the kinds of questions typically addressed with IBMs and the kind of answers typically obtained.
However, we also examine many methodological issues. The examples illustrate the IBE framework of pattern-oriented modeling, theory development,
and modeling concepts that we laid out in chapters 2–5. None of the examples can, by itself, provide a comprehensive case study of our entire IBE
framework because this framework did not exist when the example models
were developed. Fortunately, however, most *elements* of our theoretical and

TABLE 6.1
Overview of IBMs Discussed in Chapter 6

Section	Ecological Question	Key References
Group and social behavior		
6.2.1	Flocklike behavior of "boids"	Reynolds 1987
6.2.2	Schooling behavior of real fish	Huth 1992; Huth and Wissel 1992
6.2.3	Directed movement of fish schools	Huse, Railsback, and Fernö 2002
6.2.4	Dominance behavior in primates	Hemelrijk 1999
Population dynamics of social animals		
6.3.1	Group size and spatial dynamics of a cooperatively breeding bird	Neuert et al. 1995
6.3.2	Group size and persistence of the group-living Alpine marmot	Dorndorf 1999; Grimm et al. 2003
6.3.3	Group size and population dynamics of group-living canids	Pitt, Box, and Knowlton 2003
Movement: dispersal and habitat selection		
6.4.1	Dispersal of lynx	Schadt, Knauer, et al. 2002
6.4.2	Habitat selection of trout	Railsback and Harvey 2002
6.4.3	Dispersal success in spatially explicit population models	Ruckelshaus, Hartway, and Kareiva 1997; Mooij and DeAngelis (1999)
Regulation of hypothetical populations		
6.5.1	Population regulation by unequal resource partitioning	Lomnicki 1978
6.5.2	Regulation and individual variability	Uchmański 1999, 2000a, b; Grimm and Uchmański 2002
6.5.3	Scramble and contest competition in a colony-building spider population	Ulbrich et al. 1996; Ulbrich and Henschel 1999
Comparison to classical models		
6.6.1	Analytical versus individual-based predator-prey models	Donalson and Nisbet 1999
6.6.2	Analytical versus individual-based logistic equation of population growth	Law, Murrell, and Dieckmann 2003
6.6.3	Separation of time scales in a model of nomadic larks	Fahse, Wissel, and Grimm 1998

(continued)

TABLE 6.1
(cont.)

Section	Ecological Question	Key References
Dynamics of plant populations and communities		
6.7.1	Fixed-radius plant IBMs	Overview
6.7.2	Zone-of-influence plant IBMs	Overview
6.7.3	Field-of-neighborhood plant IBMs	Berger and Hildenbrandt 2000, 2003; Bauer et al. 2002
6.7.4	Grid-based plant IBMs	Overview; Winkler and Stöcklin 2002
6.7.5	Individual-based forest models	Overview
Structure of communities and ecosystems		
6.8.1	Community model with trait-mediated effects	Schmitz 2000
6.8.2	Abundance distributions in a plant community	Pachepsky et al. 2001
6.8.3	Spatiotemporal dynamics in natural beech forests	Neuert 1999; Rademacher et al. 2004
Artificially evolved traits		
6.9.1	Horizontal movement in marine fish	Huse and Giske 1998
6.9.2	Vertical movement in marine fish	Strand, Huse, and Giske 2002
6.9.3	"Hedonic modeling" of vertical movement of marine fish	Giske et al. 2003

conceptual framework of IBE were applied more or less intuitively in the case studies we examine. We point out how each of the examples provides a good illustration of at least some part of the modeling cycle, the use of patterns to guide design and analysis of IBMs, or the development and application of IBE theory. Together, the case studies provide many good examples of all the topics addressed in previous chapters.

An especially important methodological objective is showing how the concepts developed in chapter 5 (e.g., *emergence, adaptation, fitness*; for a summary, see the Conceptual Design Checklist of section 5.13) can be used to describe and evaluate IBMs. These concepts are a more rigorous and coherent way to describe and think about the many essential characteristics of IBMs that cannot be captured just by listing equations. The concepts help us compare the essential characteristics of different IBMs so we can integrate what we learn from each into a higher understanding and theory of IBMs in general. And, although our objective in this chapter is not to critique the IBMs we examine, we show that the Conceptual Design Checklist clearly is a powerful way to identify conceptual weaknesses in IBMs. Especially, it helps us identify assumptions that were left unstated or unjustified and suggests alternatives to the traits—models of key individual behaviors—included in the IBMs. Therefore, at the end of some of the following sections we step through the checklist to describe and compare the sections' IBMs using the concepts and terminology of chapter 5. In later sections we simply use these concepts and terms throughout our discussion of example IBMs.

We do not describe the example IBMs in full detail, instead focusing on the problem each IBM addressed, the overall modeling strategy and model structure, and elements of the model relevant to the issues raised in the preceeding chapters. The following sections 6.2–6.9 each describe IBMs related to an ecological theme; within a theme some models are presented in greater detail while others are considered only briefly. The themes are in a sequence that loosely corresponds to increasing hierarchical levels of ecological organization.

6.2   GROUP AND SOCIAL BEHAVIOR

The models in this first theme do not fall into the typical realm of ecology, which studies systems at time scales greater than the life-span of individuals. Instead, these models address behavior (as do the IBMs presented by Camazine et al. 2001). We nevertheless include these models because they illustrate with particular clarity important issues of individual-based modeling, including imposed versus emergent behavior, the use of currencies and patterns for modeling, and the hypothesis-testing approach to developing theory. Examining how behavior can be modeled is important because the fundamental axiom of IBE is that population dynamics emerge from individual behavior (chapter 4). The models presented here are about flocking and schooling behaviors and the structure of societies in primates.

Fish schools are an example of collectives (section 5.9). Many animal species form such collectives or large groups (flocks, herds, swarms, schools, etc.), which move and behave as coherent entities (Krause and Ruxton 2002). Obviously, these collectives result from the behavior of the individuals, but how? One possibility is that collectives are organized by some master individual, or master plan. For example, fish schools typically move more or less linearly, so one could assume that all the fish follow one or several leaders, or that they all are directed by the same environmental cues, or that each fish knows about how the school should move and adjusts its behavior to cause the expected

movement of the school. These approaches assume that the school is either
organized by a leader, or follows an environmental template (Camazine et al.
2001), or that the behavior of the school is hardwired into the behavior of
the individuals. However, empirical evidence indicates that neither leaders nor
environmental templates account for the existence and behavior of the schools.
And the assumption that schooling behavior is hardwired into the individuals is
unrealistic because it requires unrealistic assumptions about what an individual
can sense: typically, an individual will have only local information about the
number, position, and velocity of the individuals in its immediate neighborhood, but it will not know the size of the entire school or how the school moves
as a whole.

Fish schools and similar collectives are thus prototypes of self-organized systems (Camazine et al. 2001, chap. 11). Their properties and behavior emerge
from the behavior of the individuals but are not directly coded into the individual's behavior. Each fish does not know or care about the properties of the
school but only about its own state and that of its immediate neighbors. Nevertheless, the school emerges from this individual behavior. On the other hand,
the relationship between individual and school behavior is not one-sided but
mutual. The behavior of the individuals causes emergence of the school, but the
school determines the structure of the neighborhood of the individual. The same
mutual relationship between individual behavior and system properties determines the structure and dynamics of ecological systems, although in ecology
this relationship usually is harder to identify and understand because ecological
systems are more complex and diverse. Modeling self-organized collectives is
thus a kind of test-bed for the methods for modeling ecological systems that we
propose in chapters 2 to 5.

6.2.1   Reynold's Boids

The "Boids" model of Reynolds (1987) is not only a very popular illustration of
emergence (numerous implementations of Boids are available on the Internet)
but also very useful: its algorithm has been used to generate animated herds,
flocks, and other effects in Hollywood movies. The Boids model is typical of
many agent-based models in that it does not attempt to represent any specific
system, but its purpose is instead simply to demonstrate that collectives resembling bird flocks or fish schools can emerge from simple behavioral rules of
individuals. The goal of Reynolds was to reproduce flocklike behavior with
as simple a model as possible. The model rules are based on biologically reasonable assumptions, which reflect both general properties of flocks (collisions
are very rare; most of the time, neighboring individuals move in more or less
the same direction, with more or less the same velocity) and the perspective
of the individual (see the modeling heuristic described in chapter 2: "Imagine
you are inside the system").

In Boids, model individuals ("boids") move in two-dimensional space by following three general rules (Reynolds 1987):

1. Avoid collisions with neighboring boids and environmental obstacles ("collision avoidance").
2. Attempt to match velocity with neighboring boids ("velocity matching").
3. Attempt to stay close to neighboring boids ("flock centering").

The two features making these rules biologically plausible are that they specify individual behavior, not the behavior of the flock; and that they refer to the local environment of an individual: only the nearby, easily perceived, neighbors of an individual have an influence on its behavior.

To implement the three general rules, further assumptions have to be made (see, e.g., Huse, Railsback, and Fernö 2002). An implementation of Boids must specify how an individual determines which other boids are "neighbors" to perceive and respond to; whether all neighbors have equal influence or whether influence is weighted by distance; how velocity matching and flock centering are implemented; how the three rules are prioritized and weighted (e.g., by parameters describing the relative "strength" of each rule in determining the boid's direction and speed); and what constraints are placed on the boid's velocity and acceleration.

The Boids model is a fascinating demonstration of emergence and self-organization, and such demonstrations are indispensible to inspire a new kind of thinking about complex adaptive systems (Waldrop 1992). However, from a scientific point of view, Boids is unsatisfactory for several reasons. Only one set of theories, or model rules, of individual behavior is used, instead of contrasting alternative theories to see which theories are best (chapter 4). No patterns of real flocks are specified as currencies by which one can assess the "quality" of the model output (chapter 3; see also section 9.4.1). The model provides only logical possibilities ("it may be that real flocks work like Boids") but not much evidence for whether Boids really does capture how real flocks work. And, finally, what are realistic assumptions and parameter values for the exact implementation of Boids's three general rules? For example, one could define a boid's neighborhood to be so large that the individuals know the centroid of the entire collective, which is unrealistic. Reynolds (1987) notes that with such an extreme neighborhood the behavior of the flock became unrealistic, but how is the range of perception that produces flocklike behavior in Boids related to the range of perception of real individuals?

All of these limitations arise not because Boids was poorly conceived; instead, they are a consequence of Boids being intended to demonstrate a concept, not to represent a specific natural system. Boids has been one of the most influential of all IBMs because it so clearly demonstrates how complex system behaviors can arise from simple, realistic traits of interacting individuals. But in IBE, demonstrations are only the first step from which we want to proceed and learn something about the real world— which is exactly what our next example, the fish school model of Huth and Wissel (1992, 1993, 1994; Huth 1992) does. Their fish school model is a pioneering work that anticipated many elements of IBM design, pattern-oriented modeling, and IBE theory development.

6.2.2 The Huth-Wissel Model of Fish Schools

The Huth-Wissel IBM of schooling behavior in fish is also described and discussed in detail by Camazine et al. (2001; chapter 11) as an example of self-organization in biological systems. Here, we focus on the research program, the modeling concepts and techniques underlying the model, and on the way the model was analyzed.

The modeling project started with an extensive survey of the empirical and theoretical literature on fish schools (Huth 1992). This survey had two main objectives: to identify alternative theories about the behaviors of individual fish that produce schooling, and to identify patterns which could be used as currencies for testing those theories upon implementation in an IBM. The theories tested by Huth and Wissel assume that individuals can sense only a limited neighborhood. As in a previous model (Aoki 1982), the range of perception of a fish is divided into three circular zones, characterized by their radiuses r_1, r_2, and r_3. If a fish senses another fish within its repulsion zone—the distance r to the neighbor fish is smaller than r_1—the fish turns by up to 90 degrees to avoid collision with the neighbor; fish turn to swim parallel to neighbors sensed in the parallel zone ($r_1 < r < r_2$); and fish swim toward neighbors sensed in the attraction zone ($r_2 < r < r_3$). These three rules correspond to the three general rules of Boids: collision avoidance, velocity matching, and flock centering. However, the behavioral rules in the Huth-Wissel model (unlike Boids) are stochastic: the turning angle and new velocity chosen by a fish at each time step is drawn randomly from a normal distribution, the mean of which is calculated from the three navigation rules. Because the degree of stochasticity in individual navigation behavior is unknown, the standard deviations of these two normal distributions are model parameters whose values must be calibrated by examining their influence on properties of the school.

Huth and Wissel identified several patterns observed in real fish schools to use as currencies in comparing alternative models of fish behavior. The main currencies are quantitative patterns: polarization (p) and nearest neighbor distance (*NND*; figure 6.1). Polarization is defined as the average angle of deviation between the swimming direction of each fish and the mean direction of the entire school (Huth and Wissel 1992); p is 0 degree if all fish swim in exactly the same direction, and p approaches 90 degrees as there is more and more variability in swimming direction. Values of p observed in real schools are in the range of 10–20 degrees. The value of *NND* reflects the compactness of the school, being the average distance between a fish and its nearest neighbor. In

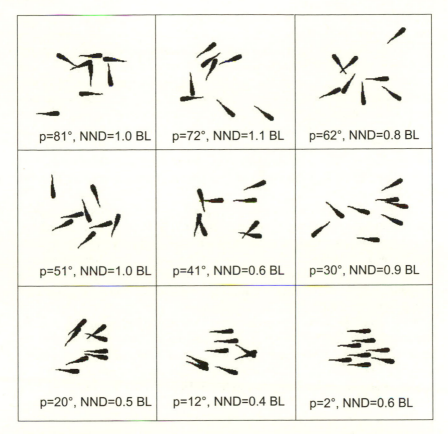

Figure 6.1 Visualization of the two "currencies," polarization (p) and nearest neighbor distance (*NND*), used in the Huth-Wissel fish school model; BL = body length. (After Huth 1992.)

real fish schools, *NND* is typically about one to two times the average length of a fish. In addition to these quantitative currencies, Huth and Wissel observed the model fish graphically (e.g., figure 6.1), allowing them to evaluate qualitatively whether the IBM produced behavior that looks like real fish schools.

In simulation experiments closely resembling the IBE theory development cycle (chapter 4), Huth and Wissel (1992) contrasted two theories of schooling behavior. One theory assumes that fish respond only to one neighbor (the closest neighbor, or the neighbor most directly in front; Aoki 1982); the second theory is that fish respond to the average orientation and swimming speed of all their neighbors. It would be very difficult to test these two theories convincingly using experiments on real fish; but Huth and Wissel, by implementing both theories in their IBM and comparing results to their quantitative (p and *NND*) and qualitative currencies, showed that the averaging theory was in all aspects more realistic than the one-neighbor theory (versions 10, 11 in figure 6.2).

In a second set of experiments, Huth and Wissel examined the sensitivity and robustness of their results, using methods similar to those discussed in chapter 9. First, robustness to uncertainty in parameter values was analyzed by varying all the IBM's parameters over wide ranges. Most parameters were found to have broad ranges over which their effect on school properties was negligible. Second, robustness to the model's structural uncertainty was examined by simulating more complicated alternatives to some of the IBM's simple, ad hoc assumptions. For example, the simplifying assumption that the response to neighbor fish changes abruptly as r exceeds r_1, r_2, and r_3 was replaced by smooth transitions between avoidance, parallel orientation, and attraction. This analysis found none of the more detailed theories to perform better than the simple ones (figure 6.2).

The important general lesson to be learned from these simulation experiments is that we should not be afraid to start an IBM with simplistic or ad hoc theories for adaptive traits of individuals—as long as we then use appropriate patterns and data as currencies for testing the theories. Huth and Wissel convincingly showed that their theory for how individual fish navigate by sensing neighbors is a simple yet robust model of fish schooling, and they rejected the "single neighbor" theory as a viable alternative.

This whole theory-testing procedure, however, could still be viewed as a kind of sophisticated fitting process: the theories are adjusted until the IBM's results fit the observations. The problem then is, as with any sort of fitting, that the fit could be good while the theories are still a completely wrong depiction of how real fish make navigation decisions in schools. How could Huth and Wissel make a strong case that their IBM has more to offer than a good fit to a few observed patterns? First is by reviewing the empirical literature on how schooling fish actually sense and interact with neighbors (also summarized by Camazine et al. 2001), which shows that there are physiological mechanisms corresponding to the IBM's assumptions. Real fish really do sense and interact with each other in different ways—avoiding, aligning, and attracting—at different distances. Second, the case can be strengthened considerably by looking for independent predictions of the IBM that provide evidence that the model is *structurally realistic* (chapter 3; section 9.9). If a model can reproduce a variety of patterns which were not used at all in model design and parameterization, then we can be much more confident that the model captures the essential structures of the real system. Huth (1992) made independent predictions of the following observed characteristics of real fish schools:

- Relationship between *NND* and size of the school. Larger schools tend to have smaller *NND*.
- Shape of the school. Schools often are ellipsoidal; herring schools, for example, have a ratio of 3 : 3.1 : 1 of length in swimming direction : width : height.

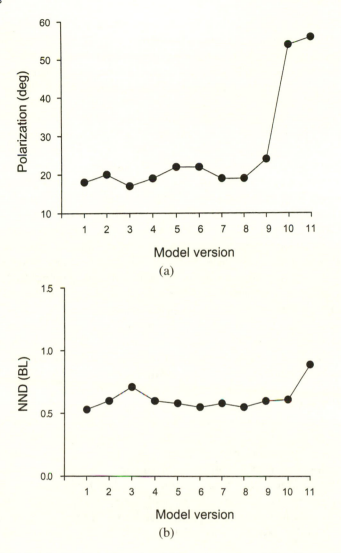

Figure 6.2 (a) Polarization (*p*, in degrees) and (b) average nearest neighbor distance (*NND*, in body lengths) of simulated fish schools for eleven versions of the Huth-Wissel model. $p = 0$ degree corresponds to a school with all fish oriented in the same direction. (After Huth 1992.)

- Position preference. Individual fish have no particular preference for a certain position in a school.
- Leading position time. Individual fish are at the front of a school for very short times.
- Distance to first, second, and third nearest neighbor. Herring, for example, have a relationship of these distances of $1 : 1.2 : 1.4$.

- Relationship between *NND* and average volume per fish.
- Relative position of neighbors, quantified as the angle between swimming directions of neighbors in both the horizontal and vertical dimensions.
- Existence of subgroups that show a strong correlation in behavior, whereas members of different subgroups show only a weak correlation.

For most of these independent predictions, the match between the IBM's results and observation was convincing (Huth 1992). For the other predictions there were either problems with interpreting the data or the observations were restricted to a certain experiment with a certain species so that it was not clear that the observed behavior was general. Together, the independent and the primary IBM predictions and the literature on real fish behavior make an overwhelming case that the Huth-Wissel model captures—despite its ad hoc nature and extreme simplicity—the essential individual behaviors giving rise to real fish schools. Note also that most of the independent predictions concern what we refer to as "weak patterns" (chapter 3). These patterns are not particularly striking, and each by itself could probably be reproduced by many model designs. The point of pattern-oriented modeling is that attempting to reproduce *all* these seemingly weak patterns *simultaneously* is in fact a powerful way to find the best model structures and theories. A further lesson is that the weak patterns found in different parts of a system are often interconnected: once we find an appropriate theory for key adaptive traits, many patterns emerge simultaneously because they are all a consequence of the same traits.

The Huth-Wissel model shows with particular clarity that the success of an IBM does not depend so much on *formulating* the model, as most beginners believe, but on *analyzing* it. Because this IBM (in contrast to Boids) addressed real systems, the modelers could pose alternative theories and then test them against observed patterns. Through their simulation analyses, Huth and Wissel were able to refine their theory for the individual navigation decisions that produce emergent schooling behaviors and make a strong case for its generality and usefulness. Next we look at a study that adapted the general theory of Boids and the Huth-Wissel schooling model to address a real ecological management problem.

6.2.3   The CluBoids Model of Huse and Colleagues

In chapter 4 we claimed that one advantage of the IBE theory development cycle is that once an adaptive trait has been rigorously tested in an IBM, it becomes part of the toolbox of IBE and can be used as a building block for other IBMs addressing different systems or questions. Here we describe an example of how the building blocks established by Boids and the Huth-Wissel model were used to tackle a related question about fish schools.

The model is about the Norwegian spring-spawning herring stock (Huse, Railsback, and Fernö 2002), which typically overwinters at the same location year after year. However, in the past fifty years there have been three abrupt changes in overwintering location. According to the "adopted-migrant hypothesis," juveniles migrating for the first time learn the location of spawning and overwintering areas by schooling with older individuals and return to these areas in subsequent years (McQuinn 1997). There is an important pattern in the three sudden changes in overwintering location: they occurred in years in which adult abundance was very low but recruitment of juveniles very high. Therefore, Huse et al. hypothesized that in these years the relative proportion of older individuals in the schools may have been so low that they were no longer able to "steer" the school to their learned overwintering areas. Instead, the school found new overwinter locations more or less at random and returned to them in subsequent years.

To test this hypothesis, an established fish school model was needed. Huse et al. decided to base their schooling simulation on Boids because of Boids's familiarity (in fact, Huse et al. developed their software by just modifying one of the many implementations of Boids) and simplicity. Huse et al. refer to their model as "CluBoids" (herring belong to the genus *Clupea*). To test the adopted-migrant hypothesis of why overwinter location changed when the relative abundance of experienced migrants was very low, Boids was modified so the modeler could make a small proportion of the fish (representing the older fish who learned their overwintering location) move in a directed way to a certain location while the other (juvenile) fish exhibit normal schooling behavior without any specific direction. Then the modeler could see how the tendency of the whole school to follow the "directed" fish depended on what proportion of the whole school was "directed." (The graphical interfaces used to control and observe behavior are shown in figure 8.1.)

Schools of 150, 300, and 450 individuals were analyzed. To standardize the initial conditions of the simulation experiments, the model was run until the school circled, a behavior also occasionally observed in real fish schools. While circling, the school has no net movement in any direction. Then, a selected number of fish was chosen randomly and made to stop following the Boids behavioral rules and instead move directly toward one specified position. The behavior of the remaining fish, which still followed their normal Boids-based navigation rules, was then observed. Only when the entire school followed the directed individuals was the behavior noted as a school response. In most cases the school responded either completely or not at all. The simulation experiment was repeated while varying the number of directed fish (figure 6.3).

The response of the CluBoids school to directed individuals showed a sharp threshold: when fewer than 3 percent of fish were directed, the school never followed, whereas when more than 7 percent of fish were directed the school always followed. This threshold was the same for all three school sizes

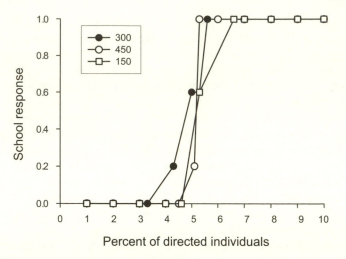

Figure 6.3 Response of CluBoids schools to directed movement of some individuals. Each plotted point represents five experiments in which the CluBoids aggregated into a school, and then a certain percent of the individuals (indicated by the x-axis) was directed to a specific location. The y-axis indicates the fraction of experiments in which the entire school followed the directed individuals. (After Huse, Railsback, and Fernö 2002.)

analysed. Further simulation experiments showed that the threshold at which the school followed the directed individuals varied with model parameters (especially parameters determining which fish are sensed as neighbors and how rapidly fish can move and accelerate), but the threshold-like response was robust to parameter changes.

What can be learned from CluBoids? Is it just a Boids-like demonstration model? Unlike Boids, CluBoids addresses a specific system and problem, so it does have the potential to provide ecological understanding. Huse, Railsback, and Fernö (2002) only took CluBoids for one quick loop around the modeling cycle, testing the IBM against only one pattern: the observed tendency for herring to change overwintering locations when the ratio of experienced to inexperienced migrants was very low. This first loop showed that the theory of herring school migration behavior embodied in the IBM is promising and suggests additional studies that could further validate the theory. At the individual level, the analysis could be strengthened by using a less-simplified, more-realistic model of schooling behavior, perhaps the Huth-Wissel model parameterized for herring. Environmental considerations could be brought into the IBM. First, studies could test the alternative hypothesis that environmental change (e.g., in currents) causes the herring to change overwintering locations (a question addressed inconclusively by Corten 1999). Second, environmental conditions could affect the schooling behavior; if herring migrate in darkness, for example, then schooling behaviors driven by vision will be affected. At the system level, further studies could focus on testing the theory that the herring schools usually are "steered" by experienced migrants: are there unique characteristics of "steered" schools that could be produced in IBMs and observed in real herring?

For IBE in general, the lesson from this example is that it is not necessary to reinvent the wheel: established theories of how system behaviors emerge from individual traits can be used as building blocks for new IBMs. The ability to use established theory was especially important for the Norwegian herring study because (like many ecological systems) it is extremely difficult to collect data on the individual fish and even on the entire schools. It would have been practically impossible to build an IBM of this system entirely "from scratch," yet Huse et al. were able to develop important understanding by applying established IBE theory. We expect to see many more building blocks similar to the fish schooling theory emerge as the theory development cycle of IBE is applied more routinely and consciously in the future.

6.2.4   Hemelrijk's DomWorld Model

A defining characteristic of fish schools is that they are "anonymous": individuals respond to the presence and activity of neighbors but not to their identity; in most cases the identity of neighbors is not even known. In contrast, living in social groups usually implies that group members know each other individually or at least know the social rank of other members. Therefore, modeling social behavior is more complex than modeling schooling behavior because the internal models used by the individuals for decision making are more complex. Here we describe the IBM of primate social groups of Hemelrijk (1999). The model addresses one important aspect of sociality, hierarchy—specifically, that individuals vary in "rank" and rank determines access to limited resources. The gradient of the hierarchy differs among species, or societies, so that Vehrencamp (1983) used the terms "despotic" and "egalitarian" to distinguish between societies with steep or flat hierarchies. Macaques are often used to study social hierarchy because some macaque species are despotic and some egalitarian. Thierry (1985, 1990) supposed that differences between despotic and egalitarian species are a consequence of the higher intensity of aggression and nepotism in despotic macaques. Hemelrijk (1999) studies an even more parsimonious version of Thierry's hypothesis: that differences between despotic and egalitarian societies might arise only from different intensities of aggression.

The rationale of Hemelrijk's modeling approach is identical to that used by Huth and Wissel (1992) for their fish school model: "patterns of interactions at a group level arise from local interactions between individuals and their environment. By interacting, individuals change each other and, therein, their social environment. In turn, the developing social structure feeds back to the individuals and shapes their interaction, etc." (Hemelrijk 1999, p. 361).

Hemelrijk thus assumes: (1) social structure is not imposed—coded directly in individual traits—but emerges from simpler traits for how individuals interact; and (2) several seemingly independent aspects of social structure, which previously were assumed to result from independent mechanisms, are all interconnected consequences of the same traits (just as many different characteristics of fish schools were explained by the same navigation traits in the Huth-Wissel model).

The design of Hemelrijk's model Dominance World ("DomWorld") was strongly influenced by the pioneering IBMs of Hogeweg and Hesper (1979, 1983). In Hemelrijk's model, the world is folded as a torus so it has no boundaries. The individuals, which are male or female, have a range of vision of 120 degrees and a maximum distance to perceive other individuals of 50 spatial units (parameter *MaxView*). The individuals move and interact with other individuals following a set of rules. Scheduling of the individuals' actions is asynchronous: each individual is "activated" (its movement and interaction actions are executed) and the world's state updated before the next individual is activated. To determine the order in which individuals are activated, the individuals each draw a "waiting time" from a uniform distribution. Individuals are activated in order of ascending waiting time; however, individuals are activated again sooner if a dominance contest (to be described) occurs nearby.

Each individual behaves according to the following rules. How an individual interacts with a neighboring individual depends on how close the neighbor is.

- If another individual is detected very close (within the "personal space"), the activated individual decides whether it will start a dominance contest, which it may win or lose (according to rules to be described). The individual that wins a contest moves toward its opponent by one distance unit and randomly turns by 45 degrees to the right or left to reduce the chance of repeated interactions between the same partners. The loser responds by fleeing.
- If another individual is detected within the near distance (but beyond the "personal space"), the activated individual just continues its movement.
- If another individual is detected at a far distance (within *MaxView*) the activated individual moves towards this neighbor.
- If no other individuals are within *MaxView*, the individuals turn by 90 degrees to the left or right to search for others.

The core of the model is the dominance interactions. Each individual has a variable (*Dom*) representing its dominance, the capacity to win a contest. When (according to the preceding rules) an activated individual must decide whether to start a dominance contest with an individual in its personal space, it executes the following additional rules.

- The activated individual first predicts the outcome of the contest by executing an internal simulation of the interaction: knowing the *Dom* values of itself and the potential opponent, it calculates its relative dominance as

$$\frac{Dom_i}{Dom_i + Dom_j},$$

with i and j denoting the activated individual and its opponent. This relative dominance is then interpreted as a probability of winning the contest. A random number is drawn and if it is smaller than the probability of winning, the activated individual predicts it would win and therefore starts a real contest.
- The outcome of the real contest is determined in the same way as in the individual's simulation. However, a new random number is drawn for the real contest—so an individual that decides to initiate a contest may lose it.
- Experiments on many animal species, including primates, have demonstrated that the effects of winning and losing are self-reinforcing (hence the stereotypes of "winners" and "losers"). Consequently, the *Dom* value of the winner is increased and that of the loser decreased:

$$Dom_{i,\,new} = Dom_i + \left(w_i - \frac{Dom_i}{Dom_i + Dom_j} \right) StepDom$$

$$Dom_{j,\,new} = Dom_j - \left(w_i - \frac{Dom_i}{Dom_i + Dom_j} \right) StepDom$$

where w_i equals 1.0 if the activated individual won and equals 0.0 if the opponent won.

The parameter *StepDom* represents the intensity of aggression: "In line with the larger rank differences in despotic rather than egalitarian societies, high values imply a large change in Dom value . . . and, thus, indicate that single interactions may strongly influence the future outcome of conflicts. Conversely, low StepDom values represent low impact" (Hemelrijk 1999, p. 363). In the simulation experiments, "fierce" and "mild" species are distinguished by their *StepDom* values: 0.8 and 1.0 for fierce females and males, respectively; 0.1 and 0.2 for the mild species. The higher values for the males reflect their physiologically superior fighting abilities. At the beginning of the simulations, all individuals of the same sex have the same *Dom* value.

Simulation experiments with *StepDom* values representing fierce species did indeed produce a broader distribution of *Dom* values than did experiments with *StepDom* representing mild species (figure 6.4). This result means that the intensity of aggression by itself can explain the emergence of egalitarian or despotic hierarchies in the IBM: when *StepDom* was high, the society diverged more strongly into winners and losers. A counterintuitive secondary result was

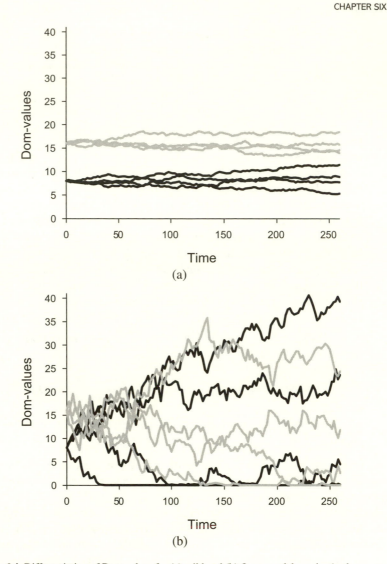

Figure 6.4 Differentiation of Dom values for (a) mild and (b) fierce model species (male = *gray*; female = *black*). (Modified after Hemelrijk 1999; data courtesy of C. Hemelrijk.)

that ranks overlap more between the sexes in fierce than in mild species, so that males are dominant over fewer females in fierce species (illustrated especially by the dominant female at the end of the simulation in figure 6.4b).

The most interesting feature of DomWorld is the emergent patterns that have a striking similarity to observations of despotic and egalitarian macaque species: simulations of fierce species produced larger differences in rank between contestants, less spatial cohesion, and therefore fewer dominance contests and more rank-correlated behavior. Because these patterns were not used for developing or parameterizing the model, they can be considered independent predictions indicating that the model captures key structures and processes of a real system. Other patterns produced by DomWorld are partially supported by observations.

DomWorld shows that not only ecology but also disciplines focusing on behavior—ethology, behavioral ecology, sociobiology—are likely to gain important new insights from IBE. The classical methods of behavioral ecology, which focus on single, isolated traits, may not be able to explain the emergence of group-level behavior as a consequence of one, or several interacting, individual traits. The DomWorld IBM and its analysis by Hemelrijk (1999) closely follows our IBE program to show how complex system-level phenomena— dominance hierarchies—can arise from relatively simple individual traits. Further articles about DomWorld and related models include Hemelrijk (2000a, b; 2002).

6.2.5 Summary and Lessons

What can we learn in general from these example IBMs addressing group and social behavior? First, these IBMs show that IBE can be a powerful way to develop theory for individual behaviors, especially behaviors that produce strong system-level patterns. Even simple IBMs can provide a richer environment for testing and falsifying alternative theories of behavior than we can often obtain in the laboratory or field. Of course, IBE is most powerful when it links patterns observed in the laboratory or field with individual-based simulation to develop and test theory for individual traits.

It is noteworthy that none of the four IBMs presented in this theme was designed for a specific species or ecological situation. Rather, they are more or less generic. Boids loosely represents flocking birds, but it has also been used (by ecologists and movie animators) for schooling fish, flocking bats, and stampeding wildebeest; CluBoids was motivated by observations of specific herring stocks, but contains no herring-specific details; the Huth-Wissel IBM is a generic model of fish school behavior and uses patterns from different species to validate the model; and DomWorld includes no species- or situation-specific elements, although it is designed to reproduce differences between egalitarian and despotic species of macaque.

Thus, these IBMs clearly show that the generality of an IBM is determined by the generality of the patterns it is designed to reproduce. Flocking and schooling behavior and social hierarchies maintained by dominance contests are common phenomena observed in many different species. Certainly, there are differences among species and situations in these phenomena, but the IBMs presented here focus on what is general, not on what is specific. On the other hand, these IBMs are not so general that they cannot be tested anymore: their "points of departure"

(Grimm 1994) are patterns that a model either does or does not reproduce, not general logical questions that do not clearly suggest testable hypotheses.

Another lesson from the models presented here is the power of the IBE theory development cycle: DomWorld is fascinating, but less convincing than the fish schooling model of Huth and Wissel because the theory development cycle was not applied to DomWorld (at least in Hemelrijk 1999). Huth and Wissel's tests of alternative theories and parameter values provided important insights. With DomWorld, the relative significance of the different aspects of the model remains unclear and we do not know how robust the study's conclusions are to changes in parameter values and model design. The way in which dominance interactions are modeled is fascinating but nevertheless completely ad hoc. Hemelrijk (1999) solved her research problem very well without needing to contrast alternative theories for the dominance interactions of social animals, but her analyses leave us itching to learn more about how the IBM could be refined and how well it could represent real macaque societies (see the subsequent section on fitness). Of course, DomWorld is the first model of its kind, and we expect that much will be learned from models inspired by it.

Another important issue illustrated by these case studies is communication of IBMs to the scientific community. An important measure of a model's credibility and success is whether it gives rise to "offspring": do other researchers reimplement and modify the model to perform their own experiments and address new questions? Boids has been amazingly successful in this way, as a quick search of the Internet will show. The Huth-Wissel model has likewise been successful, being reimplemented and modified by, for example, Reuter and Breckling (1994), Inada and Kawachi (2002), and Kunz and Hemelrijk (2003; for a review of schooling models, see Parrish, Viscido, and Grünbaum 2002). This reuse was possible because the model is completely and unambiguously described in Huth and Wissel (1992). DomWorld is, like most IBMs in ecology, more complex and cannot be fully described in a journal article. Chapter 10 discusses ways we can deal with this communication problem.

Finally, we apply the Conceptual Design Checklist of chapter 5 to the four models of this theme. This checklist is intended as a framework for describing key characteristics of IBMs, especially those characteristics that set IBMs apart from classical models and are not well captured in equations. We apply the checklist only to highlights of each IBM, not in a complete fashion.

Emergence.   All four models are good examples of what should be a universal characteristic of IBMs: that the system-level properties of interest (the patterns that the IBMs were built to explain) *emerge* from decision-making traits of the individual instead of being imposed by individual-level rules that force the properties to appear. What makes these four models all good examples of emergence is that each is sharply focused on one particular kind of system behavior and contains no individual behaviors except those from which the system behaviors of interest emerge.

Boids has become a classic illustration of emergence. The flocking behavior of a system of boids meets the criteria for emergence stated in section 5.2: flocking behavior is a system-level behavior, not just the sum of individual properties, and flocking behavior is of a different type (a spatial pattern) than individual properties (individuals cannot have spatial patterns, only locations). Most important, although we might anticipate some kind of flocking behavior from the three rules individual boids use to navigate, many characteristics of the flock's behavior cannot be predicted just by looking at the individual-level rules. In fact, the flock's emergent behavior depends on the number of boids and the characteristics of the space as well as on the traits of the boids. These same conclusions apply to the fish schooling model of Huth and Wissel, which differs from Boids mainly in the details of individual traits for navigation.

The CluBoids model allows the modeler to strongly impose a key behavior of some individuals: moving, when directed by the modeler, straight to a specified location. However, the system-level behavior of interest—whether the school does or does not follow the directed individuals—remains an emergent property.

Similarly, DomWorld was specifically designed so that the system-level behaviors of interest (social hierarchy patterns) emerge from individual decisions and interactions. DomWorld also provides an example rule that might be suspected of "imposing" important system outcomes: individuals execute their interaction behavior more often if other dominance contests occur in their neighborhood (see the section on *Scheduling*). It would be interesting to test how strongly this rule affects outcomes such as the emergence of cooperation documented by Hemelrijk (2000a, b).

Adaptive Traits. The three schooling and flocking models are excellent examples of adaptive traits modeled as *indirect fitness-seeking*. The "adaptive traits" of these IBMs, as defined in section 5.3, are the rules individuals use to decide which direction, and with what speed, to move. These rules constitute an adaptive trait in the sense that they are a model of an important decision-making behavior that contributes to individual fitness. The rules for movement are clearly not *direct* fitness-seeking, in which the individual makes a decision by considering its estimated probability of passing genes on to future generations. Instead, these adaptive traits are designed to explain and reproduce an observed behavior: forming flocks or schools. Flocking and schooling are widespread among many animals, so we assume this behavior has some fitness benefit to individuals (e.g., reducing predation risks or providing navigation to overwintering and spawning areas). By modeling how individuals form schools or flocks, we reproduce observed behavior that has indirect fitness benefits.

In the DomWorld model, the key adaptive behavior of individuals—the decision of whether to initiate a dominance interaction with another individual—is made explicitly to increase the individual's dominance. The model's author does not provide an explicit reason for assuming individuals make decisions with the objective of increasing their dominance, but most reasonable is an implicit assumption that more dominant individuals have higher expected fitness. Therefore, this adaptive trait can be considered *direct fitness-seeking*.

Fitness. The concept of fitness applies to DomWorld, which uses direct fitness-seeking to some degree. In the conceptual framework we develop in chapter 5, the fitness measure used by individuals in DomWorld is their prediction of whether they would win or lose a potential dominance contest. Individuals decide whether to start a dominance contest by considering only whether they expect to win or lose the contest.

This assumption is certainly reasonable for a minimal model to explain how social hierarchies arise, but thinking about the assumption within the fitness context suggests some alternative assumptions that could be interesting to consider. First, whether or not an individual expects to win a contest is not a direct measure of how the individual's *Dom* state variable changes as a result of the contest. The change in *Dom* depends also on the difference in *Dom* values between the two contestants—beating a more dominant opponent provides a greater increase in *Dom*. This dependence suggests an alternative fitness measure: the change in an individual's *Dom* value expected from a contest. Perhaps individuals should avoid contests unless they expect an increase in *Dom* above some threshold value. Further, what if an individual's expected fitness—its likelihood of passing genes on to future generations—is not a linear function of the individual's dominance? Perhaps only individuals with top *Dom* values get to reproduce—how would DomWorld's results be different if it assumed individuals made their decisions considering such nonlinearities? (Or, what if the relation between *Dom* and reproductive potential also emerges from individual interactions?) Such alternative fitness measures would be interesting to compare if DomWorld was developed further to explain more details of particular species' social structure.

Prediction. Boids and the fish school models do not include prediction: individuals make decisions only in reaction to their current environment. In contrast, DomWorld is one of the few IBMs that represent explicit prediction: individuals predict the outcome of a potential dominance contest when deciding whether to initiate the contest. Prediction is modeled by assuming individuals have an internal model of dominance contests that exactly matches the "real" contests—except that both the internal model and the real contest have a strong stochastic element (to be discussed) that causes predictions to sometimes be wrong.

Interaction. Interaction is an especially important concept for flocking and schooling models and for DomWorld. Direct interaction among individuals—dominance contests—is obviously a key characteristic of DomWorld. The individuals in Boids, CluBoids, and the Huth-Wissel fish school model all interact with a "field" of effect produced by their neighboring individuals. Boids was an important early illustration of how strong system-level patterns can emerge from local interactions among individuals. Huth and Wissel used their IBM to test alternative theories for interaction, showing that the "field" theory is better than the assumption that each individual interacts with only their nearest neighbor.

Sensing. The flocking and schooling IBMs assume that individuals adjust their speed and direction to match that of their neighbors and (in Boids) to avoid obstacles. Therefore, these models must assume that individuals can detect neighbors, identifying the other individuals within a specified sensing range. These models must also assume individuals "know" the speed and velocity of each neighbor and where obstacles are. The individuals in DomWorld are also assumed able to identify other individuals within a sensing range; and individuals can detect, without error, the *Dom* value of neighbors with which they could initiate a contest.

None of these models simulates sensing processes explicitly; individuals are simply assumed to "know" the characteristics of neighbors without error. However, the validity of fish schooling models has been greatly strengthened by other research (summarized by Camazine et al. 2001) that elucidates the physiological mechanisms real fish use to sense each other at various distances; this research shows that Huth and Wissel's assumptions about sensing are very reasonable. In fact, Huth and Wissel (1994) showed that the model could be made to reproduce schooling characteristics of different fish species simply by altering the parameters representing sensing ability.

Stochasticity. Boids and the fish schooling models provide an interesting contrast concerning stochasticity. The schooling model of Huth and Wissel includes a stochastic component in the decision fish make about which direction to swim, presumably to represent the imperfect ability of fish to determine and execute the "best" swimming behavior. In contrast, Boids and CluBoids have no stochastic component, yet still typically display considerable variability in velocity among individuals. How important it is to introduce stochastic "noise" into the navigation decision, for various model applications, is an unresolved question.

DomWorld uses a strong stochastic component in both the internal model of dominance contests that individuals use to predict contest outcomes, and in the "real" contests. Presumably, the stochasticity in the real contests is intended to represent important factors that (a) could cause a more-dominant individual to lose a contest, yet (b) are either unpredictable or unimportant to simulate explicitly. It is not clear what the stochasticity in the internal model used to predict contest outcomes is intended to represent. This stochasticity could represent uncertainty in the individual's ability to sense a potential opponent's *Dom* value (or in its knowledge of its own *Dom* value); however, the way in which random numbers are used in the prediction more closely resembles error in the individual's ability to predict contest outcomes when *Dom* values are known. This question is interesting because the high degree of stochasticity in the predicted and real contest outcomes undoubtedly has a strong effect on DomWorld's system behavior.

DomWorld also uses stochasticity to decide in which direction individuals turn when they change direction. Undoubtly, these decisions are stochastic to introduce a desirable level of variability in movement while avoiding the enormous additional complexity that would be needed to model how real animals make movement decisions.

All of these models illustrate perhaps the most common use of stochasticity in IBMs: to initialize some of the individuals' state variables (e.g., location, direction, speed). In DomWorld, the individuals' initial value of *Dom* was intentionally *not* randomized; this decision was made to ensure that the social hierarchy emerged only from interactions.

Collectives. Boids and the fish schooling IBM of Huth and Wissel are examples (along with others described by Camazine et al. 2001) of IBMs designed to explain how collectives emerge from adaptive traits of individuals. The collective, of course, is the flock or school: these are entities arising in the model that have strong internal interactions and cohesion, and characteristics (e.g., size, density, shape) that can be understood independently of its individuals.

CluBoids is a rare example of another representation of collectives: using emergent collectives in an IBM designed for a real ecological management problem. CluBoids was not designed to explain fish schooling but to understand how emergent schooling might affect herring migration. Although other IBMs in this chapter (see especially section 6.3) include collectives, none of the others except CluBoids uses emergent collectives to address some other problem.

Scheduling. The flocking and schooling models illustrate how scheduling details may or may not be important. These models simulate a continuous process—movement and adjustment of movement velocity—by using discrete time steps. The exact model results therefore depend on the exact order in which decisions and movement occur in each time step. CluBoids, for example, used one *action* executing over all individuals: each individual checks its neighbors, calculates its movement, and then moves; so the next individual's movement decision can depend on how its neighbor just moved (*asynchronous updating*; see section 5.10). Then, after the movement action has been executed for each individual, the graphical outputs are updated so the observer can see the new state of the model. To make these IBMs completely reproducible, their authors need to describe exactly their scheduling (which the authors did not all do).

DomWorld uses less typical scheduling (although this model's scheduling was also not thoroughly described by its author) because its dominance interactions are discrete events, not a continuous process. The model defines an action as one individual's full cycle of detecting neighbors, moving, deciding whether to initiate a dominance contest, and executing the contest. The order in which individuals execute this action is randomized. The schedule is also dynamic in one way: the next action for an individual is bumped up in the queue of actions awaiting execution if a dominance contest occurs in the individual's neighborhood (which was assumed to make it more likely that the individual will be encountering another individual soon).

Observation.   For Boids, CluBoids, and the Huth-Wissel model, the system characteristics of primary interest were spatial patterns of individual locations and movement. Therefore, it was absolutely essential to observe these patterns via graphical displays. The primary outputs of interest for DomWorld were distributions of *Dom* values among individuals, which are best observed from file output. DomWorld therefore could be (and apparently was) implemented without graphical output. However, spatial processes are critical to DomWorld and the model could be understood and tested more thoroughly if the locations of individuals were easily observable.

6.3   POPULATION DYNAMICS OF SOCIAL ANIMALS

In the preceeding theme we looked at IBMs addressing problems and time scales for which demographic processes—birth, death, immigration, and emigration—play no role. Now we turn to IBMs with full population dynamics: individuals not only behave but reproduce, die, and move among local populations, and the population's demographics change over time. These changes can include such striking phenomena as outbreaks, cycles, extinction, recolonization, and formation of spatial patterns. But why, to model these demographic phenomena, use IBMs instead of classical models, which represent population dynamics directly? The answer is that classical population models are of little use in either of two situations: (1) if we lack the long and variable time series of site-specific census data needed to fit the parameters of a classical model; or (2) if individual behavior is so complex and important that the assumptions of classical models—especially that demographic rates are constant or dependent only on population density—are obviously inappropriate.

Animals that live in groups with a social hierarchy are examples of the second (and often the first) of these two situations in which classical models are of little use. Group living often involves behaviors, such as territoriality and reproductive suppression by alpha animals, that strongly affect population dynamics. Obviously, individuals of such species are different from each other, interact locally, and base their behavior on complex decisions, so IBMs seem natural for studying their population dynamics. This theme examines three IBMs that represent populations made up of social groups having dynamics determined by individual behavior.

6.3.1   The Woodhoopoe Model of Neuert and Colleagues

This IBM addresses the effects of behavior on population dynamics of the green woodhoopoe (*Phoeniculus purpureus*), a territorial and group-living bird with reproductive supression and cooperative breeding (du Plessis 1992; Stacey and Koenig 1990). The modeling project addressed two problems. First, how do subdominant individuals decide whether they should undertake a scouting foray in hopes of finding a vacant territory where they can assume the alpha position and reproduce? On such forays, predation risk is much higher than in the home territory, but there is a chance that the subdominant individuals detect free alpha positions somewhere else. Second, how do spatial population dynamics depend on this individual scouting behavior? The need to link behavior with population dynamics makes IBMs an obvious approach; in fact we use this model in section 1.2 as an example of how IBMs can help us understand how population dynamics arise from individual traits.

In the IBM that Neuert et al. (1995) developed, individual birds are characterized by their sex, age, the territory in which they live, and their social rank within the territory. Space is divided into thirty territories arranged in a linear sequence (representing a long, narrow, riparian forest), and time divided into steps of one month. Within the territories, individuals die according to constant mortality probabilities, the alpha couple reproduces, and the surviving juveniles are assigned the lowest ranks (separate rankings and hierarchies apply to males and females). If a bird dies, all less-dominant birds in the territory move up the hierarchy by one position. Because only the most-dominant alpha couple reproduces, subdominants have a strong incentive to become alpha, which they can do in three ways. (1) Waiting until death of all the more-dominant individuals moves them up to the alpha position. This strategy is especially viable in small groups (some have only three or four birds). (2) Detecting and occupying a free alpha position in adjacent neighbor territories ("free" means that the alpha has died, and there is no subdominant of the same sex in the territory). (3) Undertaking scouting forays in search of free alpha positions in distant territories. The first two ways to become alpha involve merely waiting, but the third requires a decision to undertake a different behavior.

How to model this adaptive behavior—deciding whether to undertake scouting forays—clearly is key to the IBM's success. To devise a theory for this behavior, Neuert et al. used the modeling heuristic: *imagine that you are inside the system* (section 2.2). If you are a subdominant woodhoopoe, what information do you have and how do you use it to decide whether to go on scouting forays? Neuert et al. assumed that the woodhoopoes know their age and rank, and consequently use a simple decision trait: the older the bird and the lower its rank, the higher its prospensity to undertake scouting forays. This heuristic theory was contrasted with "null theories" where the prospensity to undertake forays was constant and independent of age and rank, or completely random. As shown in figure 1.1, only the heuristic theory was able to reproduce the group size distribution observed in the long-term field study of du Plessis (1992).

At this stage the modeling project stopped studying theories for the scouting behavior, simply because its resources were too limited. However, the IBM clearly presents opportunities for more analysis of this theory. For example, there is an opportunity to make and test independent predictions of the theory: the model was designed and parameterized for a riparian forest, with territories in a one-dimensional chain. Field observations were also available for a two-dimensional space (a savanna), where mean group size was significantly higher (du Plessis 1992). Could the simple heuristic theory for scouting behavior explain how this difference in group size emerges only from differences in spatial structure, or would additional factors such as predation pressure or food availability need to be added to the IBM? Similarly, more elaborate alternatives could have been posed and tested against the original heuristic theory, as we discuss under *Fitness* in section 6.3.4. Finally, the theory's assumptions about what individuals sense could have been tested: do birds really know the age and rank of themselves and others? How would the theory, and model dynamics, change if birds actually know less, or more? What if the birds could sense and respond to variation in predation risk when deciding whether to undertake forays?

All of these potential analyses could have been very interesting, and they illustrate how an IBM, once built, can be used as a laboratory for endless experimentation. However, they would have required considerable additional resources, especially field observations providing patterns capable of falsifying competing theories (section 3.3.3), and in some cases could make the IBM considerably more complex. With its original heuristic theory for scouting decisions validated at least by its ability to reproduce an observed group size distribution, the IBM was judged to be sufficiently valid to address the project's second ecological problem: understanding how spatial population dynamics depend on scouting behavior. This second problem does not require quantitative predictions but an understanding of the relationship between individual- and population-level behavior. Scouting forays are obviously important for the fitness of individual birds, but are they important for the population's distribution and persistence?

One of the most powerful and fascinating characteristics of IBMs is that these "what are the consequences of behavior X" questions are easy to answer. In our virtual population, we perform an experiment where behavior X is simply turned off—something we cannot do with real organisms (section 9.4.5). Neuert et al. examined the effects of scouting forays by simply turning off the ability of woodhoopoes to make forays, so birds can disperse only into adjacent territories. The consequence is that the territories lose their spatial coherence—that is, the population becomes fragmented by gaps of unoccupied territories (figure 6.5a). Further experiments showed that restoring even the smallest prospensity to undertake scouting forays restores the population's long-term spatial coherence (figure 6.5b). To understand this result, consider the fate of a single, isolated group: the mean lifetime of such a group is very limited—it goes extinct within

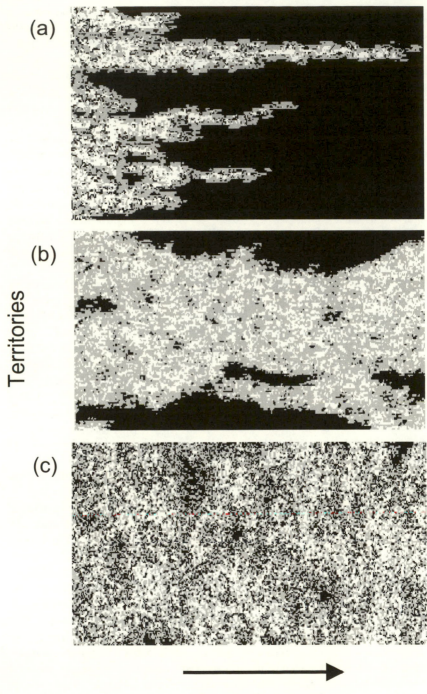

(a)

(b)

Territories

(c)

Time (years)

about ten years. Scouting birds recolonize territories where local extinction occurred, and only if at least about fifteen territories are linked via scouting birds is the entire population able to persist for longer times. The woodhoopoe population is thus a classic example of a metapopulation with local extinctions and recolonizations (Hanski 1999).

The simulation experiment shows that individual behavior is crucial for metapopulation persistence. Without long-distance forays, recolonization is too local and slow to keep up with the rate at which empty territories are produced by local extinctions. It is striking to see in figure 6.5 that this effect is more a qualitative than quantitative phenomenon: even the smallest number of "spatial correlators" (birds that link the fate of remote territories) is sufficient to produce an unfragmented, coherent population and long-term persistence.

There are several general lessons about IBE from the woodhoopoe example. First, it is a clear example of how we use IBMs to address the fundamental ecological problem of understanding how population dynamics emerge from traits of individuals (Levin 1999). We use IBMs to test theory for individual traits, then use the validated theory in simulation experiments to understand population dynamics. Second, a structurally realistic population model can be developed even if there is no time series of census data to fit the model to. Instead of being based on information encoded in a census time series, the IBM is based on the variety of information we have about the population's spatial structure (with group territories as the spatial units), the traits of individuals (with scouting decisions being the decisive behavior), and the subpopulation structure (the distribution of group sizes).

Finally, this example illustrates a key point about the modeling cycle (section 2.3): the critical importance of knowing when to stop elaborating and testing a model and start using it to solve problems. There were many ways to test and refine further the IBM's theory for how birds decide whether to scout for free territories, and it would have been easy for the project to get bogged down in finding the "best" or most "realistic" rules for individual behavior. But in this case, the modelers realized that solving their ultimate problem did not require more than the simple heuristic theory and that their resources were best spent, after validating the theory against the available pattern, by moving on to address their population-level problems.

Figure 6.5 Dependence of spatial coherence on dispersal behavior in the woodhoopoe model of Neuert et al. (1995). Each pixel column shows occupancy of 100 linearly arranged territories in one year: empty territories are black and lighter shading indicates higher occupancy. Panel (a) displays a simulation with no long-distance scouting forays; the population becomes fragmented and goes extinct over time. In panel (b) the individuals' prospensity for long-distance forays is only 2 percent of that in the full model; the population contracts in spatial extent but does not become extensively fragmented. Panel (c) displays the full model: the population retains its initial spatial characteristics. (Figure produced by H. Hildenbrandt.)

6.3.2   The Marmot Model of Dorndorf and Colleagues

The alpine marmot (*Marmota marmota*) is also a group-living, territorial mammal. The most striking behavior of this species is that it hibernates about half of the year. The long-term field study (fifteen years) underlying the marmot model (Arnold and Dittami 1997; Frey-Roos 1998) addressed questions of behavioral ecology, in particular the reasons for group living. But the data set from this study is so rich (687 marmots were marked individually and 98 were radio-tracked during dispersal) that Dorndorf (1999; Grimm et al. 2003) decided to develop an individual-based population model. (Stephens et al. [2002a, b] independently developed a similar IBM, based on the same data set but addressing different questions.)

The main question to be answered by the model was how group living affects the population's viability—that is, its ability to persist for long times (e.g., 100 years) with high probability (e.g., 99 percent). A comparison with other marmot species suggests that group living is an adaptation to long, cold winters: solitary-living marmots live in much more benign environments than the alpine marmot. Group living thus seems to somehow buffer the environment's harshness. In the model, environmental harshness is quantified as the length of a winter because long winters increase the risk that an individual's fat reserves are not sufficient to survive hibernation until spring. The IBM represents environmental variation by drawing each winter's length from a normal distribution with a specified mean and variance.

The marmot IBM is very similar to the woodhoopoe model, although the spatial structure and individual behavior are more complex. The marmots have state variables for sex, age, life stage (adult, yearling, juvenile), rank (alpha or subdominant), and territory in which they live. Territories within 500 meters of each other form a cluster because subdominant individuals can sense free alpha positions in territories up to 500 meters away. Subdominants can try to occupy such free alpha positions without having to undertake long-distance dispersal, which subjects them to high risk. The entire population consists of several clusters, which are linked by those individuals that do undertake long-distance dispersal.

Marmots often stay in their home territory for some years, but their reproduction is suppressed by the alpha individuals. Leaving the home territory corresponds to the scouting forays of the woodhoopoes, with the important difference that the dispersers do not return: either they find an alpha position somewhere else or they die the next winter because they cannot survive winters solitarily. Dispersing marmots can either detect and occupy a free alpha position, or they can try to evict an alpha individual—about 15 percent of changes in alpha position are due to the eviction, not death, of the alpha individual.

The most important difference between the marmot and woodhoopoe IBMs is in the trait individuals use to decide whether to leave the home territory.

The extensive marmot field data allowed the modelers to develop an empirical, stochastic model for this trait. From the data, age-specific probabilities of leaving could be estimated; for example, 21 percent of two-year-olds leave each year and 99 percent of five-year-olds leave, so two-year-olds were assumed to have a dispersal probability of 0.21, and so on. Moreover, the benefits of group living, an issue ignored in the woodhoopoe model, could be extracted from the field data. For example, the probability P of winter mortality for an alpha individual was determined, via logistic regression, to be:

$$P = \left[1 + \exp\left(6.82 - 0.286A - 0.028WS + 0.395SUBY\right)\right]^{-1} \qquad (6.1)$$

where A is the alpha's age, WS is winter length, and $SUBY$ the number of subdominants (including yearlings) hibernating in the group. (Similar equations were developed for yearlings and juveniles.) This equation states that winter mortality risk increases with the length of winter and with age, but decreases with the number of marmots in the group, a positive effect of group living on survival. The mechanism for this benefit of group living is energetic: the balance between heat production and loss in the hibernaculum improves with the number of individuals.

Once assembled, the marmot IBM was applied to the problem of how group living affects population persistence by comparing its predictions of mean time to extinction with theory. The theory of stochastic population dynamics predicts that if "demographic noise" (variation in abundance due to random variation in demographic processes—birth, death, immigration, and emigration) prevails, the mean time to extinction increases exponentially with the capacity of the habitat (Lande 1993; Wissel, Stephan, and Zaschke 1994; figure 6.6a). Therefore, extinction of the marmot population due to demographic noise is very unlikely beyond a threshold of, say, fifteen territories. If, however, environmental variation has strong effects on abundance, the increase in mean time to extinction with habitat capacity is much slower, in the extreme only logarithmic (figure 6.6a).

In the marmot IBM, group living buffers the population from environmental variation so effectively that population fluctuations are almost entirely determined by demographic noise. The relation between mean time to extinction and habitat capacity observed from the IBM shows that the population does not "perceive" the harshness of the winter and the variation in winter length (figure 6.6b). And, with further simulation experiments, it is easily shown that the positive effect of group living on survival is responsible for this buffering: if the effect of group size is removed from the equations for overwinter mortality risk, the extinction-habitat relation resembles the theoretical curve for strong environmental variation (figure 6.6b). The high persistence of alpine marmot populations thus appears to emerge from the winter group-living behavior of the individuals.

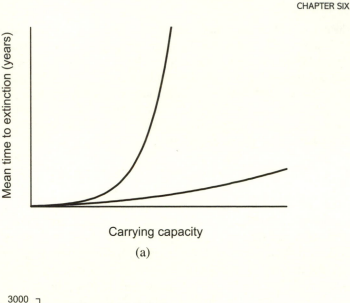

Carrying capacity

(a)

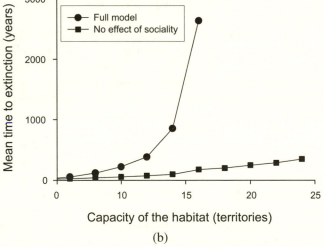

Capacity of the habitat (territories)

(b)

Figure 6.6 Relation between mean time to extinction (a measure of persistence) of small populations and habitat capacity. (a) Theory predicts an exponential increase in persistence if the population is affected by demographic noise but not environmental noise (*upper curve*); if environmental noise has strong effects, the increase in persistence with habitat capacity is much less (*lower curve*). (b) Results of the marmot model of Dorndorf (1999) and Grimm et al. (2003). In the full model, group living affects winter survival as described in equation 6.1; when this effect is deactivated persistence increased little with habitat capacity. (After Grimm et al. 2003.)

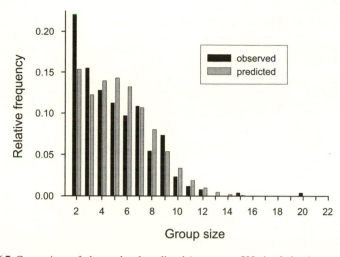

Group size

Figure 6.7 Comparison of observed and predicted (mean over 500 simulations) group size distributions in the marmot IBM. (After Grimm et al. 2003.)

As with the woodhoopoe model, the main validation criterion for the marmot model was reproducing an observed group size distribution (figure 6.7). Additional evidence of the model's credibility was provided by the observation that the simulated census time series was within the variation of the observed time series. Because of the buffering effect of group living, the correlation between winter length and population size was weak for both the observed and the simulated time series.

One lesson from the marmot example is that IBMs can be useful for conservation biology and population viability analysis (PVA; as defined by Soulé 1986). It has been argued that simple models are best for PVA because more detailed and complex models are too uncertain due to error propagation (Beissinger and Westphal 1998). However, simple models (e.g., matrix models with constant fecundities and mortalities) cannot capture the critical, but spatially and temporally variable, effects of individual behaviors like group living on population persistence. Because such effects are important for many species, IBMs are important tools for conservation biology (Burgman et al. 1993; Bart 1995; Matsinos et al. 2000).

A second lesson is that we cannot equate "emergence" with "instability." In this IBM marmot population dynamics emerge from individual behavior, and the benefit of group living on overwinter survival even provides a positive feedback to population growth: more marmots make overwinter survival higher, which makes more marmots. Emergence, and especially positive feedbacks, are often assumed always to make models unpredictable or even chaotic—and often they do. However, in this case the simulated population was more persistent with respect to environmental variability *with* the positive feedback than without it.

6.3.3 The Canid Model of Pitt and Colleagues

Now we look at a third group of animals that live in groups, are territorial, and have a social hierarchy with reproductive suppression: canids (wolves and coyotes). Vucetich and Creel (1999) and Pitt, Box, and Knowlton (2003) have concluded that analytical or matrix models with fixed demographic rates are not sufficient to support management (either preservation or suppression) of these species. Pitt, Box, and Knowlton (2003) developed an IBM for coyotes that is generic enough to apply to other canid species. We examine the model because it is a good illustration that IBMs, even for animals with complex social behavior, can be usefully realistic without being extremely complex. In fact, this IBM is quite simple. The model is intended to support management decisions, but the version described here addresses no particular question; the point is to show that the model captures essential characteristics of real coyote populations. This example also shows how a software platform specifically for IBMs helps describe models efficiently and unambiguously, as well as making them easier to implement.

Like the woodhoopoe and marmot IBMs, the coyote model characterizes individuals by sex, age, social status (alpha, beta, pup), and the group (pack) they belong to. The model considers packs but not territories, and is not spatially explicit. The number of packs is fixed at 100, but population abundance varies because the number of coyotes in each pack ("pack size") changes over time.

Also like the other IBMs in this theme, the coyote IBM's most important rules are for individuals leaving the group and mortality. Coyotes between one-half and two years old have a probability of leaving their pack (either by choice or involuntarily) that is proportional to the square of pack size, so pack size influences dispersal behavior. Coyotes two years of age or older do not leave their pack. Coyotes that leave their pack enter a pool of "transients." The mortality probability of adults was assumed to be a nonlinear function of age, and that of pups to be constant. Mortality of transients was assumed to increase with the total number of transients. The number of offspring produced is assumed to decrease with pack size. All of these rules and their parameters were based on extensive observations of coyote packs and individuals.

The Pitt et al. model was implemented using the Swarm platform for agent-based simulation models, which provides tools not only for implementing IBMs in software but also for organizing and describing IBMs in standard ways (section 8.4.3). The basic organizational structure of a Swarm model is a "swarm," a list of objects and a schedule of actions (section 5.10) the objects execute. Swarm's standard organization and terminology help describe the model efficiently and unambiguously, which is important for the communication of the model to the scientific community (chapter 10). Pitt et al. describe the schedule of actions in their model as follows. The model uses a monthly time step, and each action is executed once per time step.

Pack actions *(executed by all packs):*

- Check whether both an alpha male and an alpha female are present.
- If both alphas exist, and it is April, produce offspring: create pups, the number of which is stochastic but also depends on pack size, and add them to the pack.
- Check whether either alpha is replaced:
  - If it is December, and there is a contender (another adult of the same sex in the pack), both the male and female alpha coyotes are at risk of being replaced.
  - Replacement is a stochastic function with the probability of being replaced increasing with the alpha's age.
  - If replacement occurs, the alpha becomes a transient and the contender becomes the new alpha.
- Update the dispersal probability of each member according to its age and pack size.
- Force death of pups less than two months old if the pack has no adults.

Pack member actions *(executed by all individual coyotes*
that belong to a pack):

- If the age of two months is attained, leave the den.
- If the age of six months is attained, change from pup to beta adult.
- Update the age-dependent mortality probability and determine whether death occurs.
- If individual is a beta less than two years old, determine whether it leaves the pack, according to its dispersal probability.

Transient coyote actions *(executed by all individuals*
not belonging to packs):

- Update individual's mortality probability, depending on the total number of transients.
- Determine whether death occurs.

Pack alpha replacement actions *(executed by packs*
that lack an alpha individual):

- If there are beta individuals of the appropriate sex in the pack, promote the oldest beta to alpha.
- Otherwise, select a transient of the appropriate sex and promote it to alpha.
- If there are no available transients, select a beta from another pack.

TABLE 6.2

Comparison of Five Output Variables of the IBM of Pitt and Colleagues (2003) with Values from Published Studies of Coyote

		Literature Values	
Output Variables	*Model Results*	*Mean*	*Range*
Population size	525	500	420–560
Proportion of population transients	0.26	0.26	0.13–0.58
Offspring survival	0.41	0.41	0.32–0.73
Litter size	4.1	4.6	3.2–7.0
Proportion females breeding	0.43	0.44	0.33–0.7

This way of describing a model can be very clear, concise, and unambiguous. The woodhoopoe and marmot models use similar model rules ("If . . ."), but their rules and scheduling are harder to extract from the publications because the rules are described in an ad hoc way.

What we also see from this description of its schedule is that the coyote IBM is very simple conceptually: the packs and individuals do simple things and apply simple rules. The "complexity" is more in implementing and analyzing the IBM, not in its assumptions (true also for the woodhoopoe and marmot models). In some ways, IBMs can be more simple conceptually than analytical models: anyone can understand and discuss the coyote IBM's assumptions, whereas differential equation models include terms that can be unintuitive, difficult to relate to everyday observations, and even hard to understand without sufficient background in mathematics.

Pitt et al. looked at five output variables to validate their model: mean pack size, proportion of transients, average offspring survival rate, average litter size, and proportion of females breeding. These variables, which together provide a "fingerprint" of coyote populations, were compared with data from several field studies of different coyote populations, without any detailed calibration of the IBM's parameters. Considering the simplicity of the model, the match of observed and predicted values is surprisingly good (table 6.2). Further, a simple sensitivity analysis indicated that this validation success is relatively robust to parameter uncertainty. These analyses indicate that the model is valid enough for addressing many management problems, its intended purpose.

Among the conclusions drawn by Pitt, Box, and Knowlton (2003) from the IBM is that transient coyotes have an unexpectedly important buffering effect on population dynamics: when abundance increases, the density-dependent mortality of transients damps population growth; yet transients help packs maintain their breeding potential by replacing lost alpha individuals. Given this IBM's initial success, it is being followed up with a spatially explicit version that incorporates such additional factors as spatial variation in food supply (F. Knowlton, personal communication), expected to make the IBM applicable to even more management problems.

6.3.4 Summary and Lessons

Readers have no doubt noted that the woodhoopoe, marmot, and coyote IBMs are quite similar. All have similar basic model structures: individuals that belong to small social groups, dominance hierarchies within groups, and links among groups provided by dispersal of individuals. In each model the key adaptive trait of individuals is their rule for when to leave their group and disperse in hopes of reproducing. Even the list of individual state variables is similar among models: sex, age, dominance status, and group.

Why are these three models so similar, despite addressing different animal societies and ecological problems? Because they are also examples of the main points of chapters 2 and 3: that by focusing clearly on specific systems and study problems, and using specific patterns that capture the system's essence to guide model design, we can develop IBMs that include only the most essential structures and processes. For example, social hierarchies are essential to all three IBMs, but none of them contains the dominance contests that are the core of DomWorld, the hierarchy model we examined in section 6.2.4. The IBMs in this theme instead just assume the presence of hierarchies because the problems they address concern the *consequences* of hierarchies; DomWorld studied the *causes* of hierarchies.

One reason we chose the three IBMs in this theme is that they demonstrate both a strong and a—currently—weak point of IBMs and IBE. The strong point is that in all three cases we gained insights that would have been hard to be achieve without IBMs. Spatial coherence of metapopulations, the ability to buffer environmental harshness and variability, and internal structure of populations (e.g., group size, proportion of transients)—all these properties of populations emerge from the adaptive behavior of the individuals. In contrast to many classical models, these IBMs could be validated not by fitting them to census data but by comparing their various kinds of results with observed patterns of the population's structure. The models provided theoretical insights but were also realistic enough to tackle applied problems. In contrast to the belief that IBMs are always "complex," all three IBMs are conceptually simple: their rules for individual behavior are few, simple, and more intuitive than many classical models. The complexity arises from the interactions among individuals, which produce many kinds of output—for example, group size distributions, individual behaviors, spatial patterns—in addition to just abundance. But this complexity did not prevent successful testing and analysis of the IBMs.

The weak point illustrated in this theme is not a critique of any particular model but an indicator of how new and uncoordinated IBE still is. Although the three IBMs we examined (and many other IBMs of social species, e.g., Lankester et al. 1991; Verboom, Lankester, and Metz 1991; Letcher et al. 1998; Schiegg, Walters, and Priddy 2002) are similar in many ways, they were each designed from scratch, described using ad hoc and often incomplete terminology, and implemented in many different software platforms (section 8.4). The woohoopoe model was programmed in Pascal, the marmot model in C++, and the coyote model using Swarm. As a consequence, it is much harder for us, the "clients" of these IBMs, to understand them completely, compare them, or attempt to reproduce them. One of the most important things ecologists can do to help IBE mature rapidly is to adopt common modeling concepts and terminology (chapter 5) and to share software platforms that implement the common concepts and model designs easily and unambiguously (chapter 8).

These three IBMs help us see the potential benefits of the IBE theory development cycle (chapter 4). The IBMs all used theory for how an individual decides to leave a home group in search of the opportunity to attain alpha status elsewhere. Alternative theories were tested—at a very basic level—only in the woodhoopoe model; the other two models used empirical rules. The "problem" was the same in all three cases and, in fact, is a generic problem of group-living and territorial species: should I stay home, which has certain costs and benefits to my potential fitness, or should I leave, which has different costs and benefits? It seems quite likely that the concepts of fitness-seeking we discuss in section 5.4 could be used to pose general theories for this dispersal decision and that the theories could be tested using the IBMs and field observations presented in this section. We are unlikely to find one single theory that is applicable in all cases. But what we should be able to add to the toolbox of IBE theory (chapter 4) is a family of closely related approaches that are easily adapted to specific IBMs.

A final point illustrated by the models of this theme is how IBMs can indeed help integrate behavioral and population ecology. Clearly, behavioral traits were key to the emergent population dynamics. But the woodhoopoe IBM was used, along with observations of the population's group size structure, to test and compare alternative theories for this behavioral trait. Not only do IBMs let us explore the population consequences of behavior, they provide a rich environment for testing theories of behavior.

Now, we apply the Conceptual Design Checklist as a more formalized way to describe and compare highlights of the three IBMs of population dynamics of social animals.

Emergence.   In all three of the IBMs, characteristics of both the social groups (e.g., the number of individuals in each group, their extinction and reestablishment) and the total population (total abundance, persistence) emerge from individual traits for dispersal. However, these IBMs also illustrate how some system-level behaviors can be imposed by the rules for what individuals do, a common and often desirable characteristic of many IBMs. For example, in the marmot IBM the fraction of individuals dispersing, by age-class, is imposed by the stochastic trait for dispersal: the modelers gave two-year-olds a 21 percent probability of dispersing, so we can safely expect the IBM to predict that about 21 percent of age two marmots will disperse. Likewise, the mortality rates for some kinds of individuals (e.g., nondispersing woodhoopoes) are fixed by the individual traits instead of emerging.

Adaptive Traits and Behavior. There is an important but perhaps subtle difference between the woodhoopoe IBM and the other two in the key adaptive trait for deciding when to disperse. In the marmot and coyote IBMs, the dispersal behavior is modeled empirically, using stochastic rules with dispersal probabilities developed from observations of real animals. These rules are indirect adaptation: the traits simply cause the individuals to reproduce observed behaviors. This means the IBM should do a good job of representing dispersal behaviors as long as the populations being modeled are well represented by the empirical probabilities; but it also means that these IBMs cannot be used to study the dispersal behavior's causes, only its consequences.

The woodhoopoe dispersal trait is only partially stochastic: the probability of dispersing increases with the individual's age and its distance from the top of its group's hierarchy. This woodhoopoe trait makes sense only as a fitness-based decision (discussed further in the following paragraph): as individuals get older, or if they fail to approach the top of their hierarchy, their probability of reproducing before dying becomes lower and lower so dispersing becomes a better and better alternative. Consequently, we can think of the woodhoopoe's dispersal trait as direct fitness-seeking. Because the woodhoopoe model uses at least a partially mechanistic trait for dispersal, it can be (and indeed was) used to study how individuals make the dispersal decision.

Fitness. The concept of fitness applies to the only direct fitness-seeking trait of the three IBMs: the woodhoopoe's decision of how often to undertake scouting forays. The fitness basis of this trait was not explicitly stated by the model's authors, but we can easily infer it. This trait addresses two fitness elements. First is survival to reproduction, the apparent basis for the trait's assumption that birds become more likely to disperse as they get older. As the birds get older, their probability of surviving until they can reproduce decreases so they are more inclined to undertake the alternative strategy of dispersing. The second fitness element is attaining social status for reproduction: with reproductive suppression, the probability of reproducing within the natal group is lower for woodhoopoes of lower rank. This element apparently underlies the trait's assumption that birds with lower rank are more likely to disperse. The woodhoopoe dispersal trait is a very indirect fitness measure. Even though it apparently considers how survival and social status affect potential fitness, the trait includes no direct estimate of how age and rank affect an individual's expected probability of reproducing.

Thinking about the dispersal trait as fitness-based certainly suggests some alternative approaches that would more completely and directly estimate (perhaps using empirical information as well as theory) how an individual's expectation of achieving alpha status and reproducing depends on its current age, its social rank, perhaps other characteristics of its group and habitat, and how often it scouts for new territories (Stephens et al. 2002a in fact develop such a direct fitness-seeking dispersal trait for marmots). At its most complete (and complex), a fitness measure for this trait might consider kin selection: an individual can pass many of its own genes on by helping ensure the survival of related individuals. If individual woodhoopoes stay in their natal group, they may help raise the (probably closely related) offspring of the alpha pair; so kin selection could be an important component of the individual's potential fitness. Of course, testing these more complete and direct fitness measures requires observed patterns of individual and group behavior capable of distinguishing, via simulation experiments, which of the alternative measures are useful in what situations.

Prediction.   Explicit prediction is almost completely absent from these three IBMs, which is not surprising given their reliance on empirical traits. However, they use a subtle *tacit* prediction as the basis for dispersal: dispersal depends on individuals tacitly predicting that undertaking scouting forays or abandoning the natal social group increases their probability of reproducing.

Interaction.   These three models have the same type of interaction—social hierarchies—as a fundamental characteristic. The hierarchies are an indirect kind of interaction: alpha individuals interact with the other individuals by suppressing their ability to reproduce. Dominance is not simulated as direct interaction as it was in DomWorld. Instead, dominance in the population IBMs of social species is treated as an interaction field produced by the alpha individuals: the mere presence of an alpha prevents other individuals in the group from reproducing. What is the real mechanism behind this assumed interaction field? Social interactions at short time scales (of the type that DomWorld was designed to investigate) are widely observed and commonly believed to give rise to social hierarchies, even if the exact mechanisms determining and maintaining the hierarchy and reproductive suppression are not known.

Overwinter survival of the marmot individuals is also modeled as an interaction field among members of the hibernating group: the presence of marmots increases the survival probability the others. In this case the mechanism of interaction is clear: the rate at which each individual expends energy is lower when more marmots hibernate together.

The marmot IBM includes one direct interaction: a subdominant marmot can attempt to evict the alpha individual of another territory, a one-on-one contest. In real marmots this contest for the alpha position presumably involves aggressive interaction, but the IBM represents it simply as a stochastic event.

The coyote IBM has several interesting kinds of interaction fields. The assumption that young coyotes are much more likely to leave their natal packs when pack size is high can be interpreted as representing social interactions or competition that makes larger packs less desirable for (or more likely to eject) a young member. A second interaction field is the strong effect that adults have on survival of pups: if all the adults leave a pack or die, the pups are assumed to die. Presumably, the mechanism represented by this interaction is parental care: the pups, up to an age of two months, depend completely on adults for their survival. Third, the transient adults have a mortality probability that depends on the total number of transients—by joining the pool of transients, an individual changes the mortality risk of the other transients. The mechanisms behind this interaction stated by Pitt et al. are that transients (1) compete for resources and (2) often fight when they encounter each other, which is more likely when density is higher.

Sensing.   The three IBMs in this theme are quite similar in how they represent sensing—what information individuals have and how they obtain it. The woodhoopoes are assumed to "know" their own age and rank in their local group's hierarchy; whether an alpha position is free in a neighboring territory; and, when on a scouting foray, which territories have free alpha positions. In the marmot IBM, individuals are assumed to know their own age and status (alpha or not); whether there is a free territory within 500 meters; and, when dispersing, which territories they pass are free. Similarly, the coyote IBM assumes individuals know their own age and social status and their pack's size. In addition, coyote individuals have traits that depend on the month, so coyotes are assumed to sense the time of year. The IBMs are also similar in not representing sensing mechanisms explicitly; instead, individuals are simply assumed to "know" these variables.

Stochasticity.   One of the most obvious features of these three IBMs is their extensive use of stochastic processes in individual traits (and, in the coyote model, pack traits). Many key traits are modeled as partially stochastic: whether a particular behavior is executed is determined by drawing a random number, but the probability of executing the behavior depends on characteristics of the individual or its environment. Important examples include (1) the decision by woodhoopoes of whether to undertake scouting forays, which is stochastic with the probability depending on individual age and rank; (2) the number of pups produced by a coyote pack, a stochastic function of pack size; (3) whether marmots leave their group, a stochastic function of their age; and (4) mortality in all the IBMs being a stochastic event with probabilities depending on variables such as age and social status.

These stochastic processes clearly are used because they are a simple way to reproduce observed rates. But why were stochastic processes used instead of imposing the rates directly? For example, why assume that each two-year-old marmot has a 21 percent probability of dispersing, instead of simply selecting

21 percent of the marmots and making them disperse? Why not make the number of coyote pups produced each year a deterministic instead of stochastic function of pack size?

Although the authors did not always say why these traits were made stochastic, there are several likely reasons. First is simply so events like reproduction, dispersal, and mortality appear clearly as individual traits, not as being controlled by some model-dwelling demon with no counterpart in nature (e.g., so each marmot decides for itself whether to disperse, instead of some greater being selecting the marmots to disperse). Second, stochasticity is used to make traits variable without having to simulate the fine-scale (and, for these IBMs, unimportant) mechanisms causing the variability. In the marmot IBM, it was critical to include variability in demographic processes because one of the model's purposes was to examine relative effects of demographic and environmental variability on population persistence.

The marmot IBM illustrates another common use of stochasticity: representing environmental variability. Winter length, the driving environmental variable, is drawn randomly from a statistical distribution estimated from observed data. There are two reasons for doing so instead of simply using the historic record of observed winter lengths: the stochastic approach allows simulations to be much longer than the historic record and allows replicate simulations with different, but still realistic, sequences of winter lengths. Both of these capabilities are essential for the model's problem of explaining the population's high probability of persisting over long periods.

Collectives.   The social groups in these IBMs are clear examples of collectives, but a different kind of collective than those in the schooling and flocking IBMs examined in section 6.2. In this theme, collectives—woodhoopoe and marmot social groups, coyote packs—are explicitly represented as entities in the model. The collectives are "hardwired" and have their own state variables: location, number of members, and the like.

The coyote model is a particularly interesting example because several important model processes are traits of the collective, not of individuals. Beta individuals can replace their pack's alpha individual, but there is no individual trait for this process; instead, the pack has rules for when its alphas are replaced. Likewise, transient individuals can take over as a pack's alpha, but the pack selects a transient instead of the transients having traits for joining a pack. The pack is also assumed to have sensed information: to determine whether reproduction takes place, the pack is assumed to know the month and whether it has an alpha male and female. To determine whether and how one of its alphas is replaced, the pack must know whether it contains a contending beta and which beta is oldest; and, potentially, which transient individuals are potential alphas; and (if no internal betas or transients are available) which other packs have betas available to fill a vacant alpha position. And, if the pack senses that it no longer has any adults, its pups are assumed to die. Clearly, canid packs are just groups of individuals and have only the sensing abilities and behaviors of their individuals; but the model's pack traits are quite reasonable and appropriate. The IBM implicitly assumes that social interactions within and among packs, at time scales much shorter than its monthly step, produce behaviors that are best represented as characteristics of the pack itself. For the problems this IBM was intended to address, there is no need to represent explicitly the detailed interactions that give rise to these pack characteristics. While Pitt, Box, and Knowlton (2003) do not attempt to explain the real mechanisms and individual behaviors represented by the packs' traits, they make a strong case that these pack traits reproduce important behaviors observed in real coyotes.

Scheduling. All three of the IBMs modeled time using conventional discrete time steps—monthly for woodhoopoe and coyotes and yearly for marmots. However, only for the coyote IBM do we know from the published model descriptions how events were organized into specific actions and how these actions were scheduled. From the description in section 6.3.3 we can see the list of actions executed by each pack, then each pack member, each transient, and, finally, each pack lacking an alpha individual. Even so, this description is not a complete depiction of the model's scheduling; we do not know the order in which the packs and individuals are processed when their actions are executed.

Observation. The primary outputs of interest for all three IBMs are population dynamics of social groups and the entire population. These results are easily observed via output files reporting such variables as number of individuals in each group. However, spatial patterns of group size and persistence are intermediate results important for testing and understanding the dynamics of the woodhoopoe and marmot models. Normally, when spatial patterns are important it is essential to provide graphical observation tools, which was the case with the woodhoopoe and marmot model.

6.4 MOVEMENT: DISPERSAL AND HABITAT SELECTION

The key behavior of the social species described in the preceeding theme was dispersal, the individual's decision: shall I stay or shall I leave? Movement is a key adaptive behavior of almost all species. Mobile individuals can try to improve their fitness by finding new locations where their survival and reproductive success are better; but even most sessile species have at least one life stage that can move. Even if individuals have little control over their motion, their dispersal traits (transport mechanisms, timing) often include adaptive individual behavior as well as evolutionarily adapted genetic traits.

We distinguish two kinds of movement. *Dispersal* is usually considered a one-time movement to a new, permanent location; seeds disperse from parent plants, or animals disperse from their natal territory in search of their own territory. *Habitat selection* is a continual process of mobile individuals choosing what location to occupy while fitness benefits vary over space and time due to processes such as weather, food production and consumption, presence of competitors and predators, and the individual's size and life history state.

When traditional population ecology has considered movement of individuals, it has focused on the system-level outcome of movement, for example on the rates of dispersal or colonization among habitat patches in metapopulations. Dispersal has usually not been treated as a phenomenon emerging from the adaptive behavior of individuals, but instead models simply assume a dispersal rate, or empirical models of movement (such as the several varieties of "random walk") are fit to data (Turchin 1998). The main reasons for this "imposed" treatment of dispersal are the following perceived problems.

1. Multiple time scales. Movement occurs over days, hours, or even minutes, but classical population theory usually addresses demographic change over many generations.
2. Data requirements. Even if we want to include the details of individual movement, where would we get sufficient data to do so?
3. Complexity. If all the details of individual behavior are important, how can we ever understand system-level properties of ecological systems? Certainly we must simplify to learn anything.
4. Generality. Even if we could understand individual movement and its effects in some cases, everything we learn would be situation-specific with no chance of producing general insights.

These problems are, of course, exactly why we use IBMs. The examples in this theme illustrate that in IBE we can deal with these problems and build useful, and sometimes quite general, models of how habitat variability and individual behavior interact to explain movement and its population-level consequences.

6.4.1   The Lynx Dispersal Model of Schadt and Colleagues

This example illustrates how, despite sparse data, individual-based dispersal models can be developed, parameterized, and applied to real ecological management problems. The IBM addresses the Eurasian lynx (*Lynx lynx*). In the first half of the twentieth century, the lynx completely disappeared from all of Central Europe west of the Slovakian Carpathians due to persecution and the destruction and fragmentation of its habitat (Breitenmoser et al. 2000). However, both land use and public attitudes toward large carnivores changed in the second half of the twentieth century, allowing the slow recovery of lynx in several European countries. In Germany, natural immigration occurs in the Bavarian Forest, lynx are being reintroduced to the Harz Forest (Wölfl et al. 2001), and reintroduction to other locations has been vigorously debated (Schadt 2002; Schadt, Revilla, et al. 2002; Schadt, Knauer, et al.). The question is: can viable populations of lynx be reestablished in Germany? To answer this question we need to know where lynx habitat is, the connectivity among habitat patches, and the viability of both the local populations and the entire German metapopulation.

Schadt and co-workers addressed these problems in a series of papers. First, the suitability of the German landscape for lynx was assessed by developing a habitat model (Schadt, Knauer, et al. 2002). This model distinguishes between breeding, dispersal, matrix, and barrier habitat. In total, fifty-nine patches of breeding habitat—forested areas larger than 100 square kilometers, the average home range size of lynx—were identified; eleven of these patches were larger than 1,000 square kilometers and thus considered potential "sources" of dispersing individuals. Then a dispersal model was constructed (Kramer-Schadt et al. 2004; Revilla et al. in press) and virtual lynx were released from the source patches. They dispersed through the model landscape until they arrived at another breeding habitat patch. Finally, habitat and dispersal models were combined in a model of the population dynamics that predicted the viability and colonization success in the different patches, to support reintroduction decisions (Schadt 2002).

Here we only examine the dispersal model. Conceptually, the model is very simple: spatial patterns of lynx dispersal and the rate at which lynx successfully move among breeding patches emerge from the spatial distribution of habitat types and the individuals' adaptive traits for dispersal. Behaviors such as reproduction and even interaction among individuals were considered unnecessary. In fact, dispersal rules are the only individual trait in the IBM. Individuals decide (1) how far to move (s, the number of one square kilometer grid cells to step through) each one-day time step, and (2) which grid cell to move into next.

Individuals decide how far to move each day by drawing a value of s randomly from the probability distribution $P(s)$:

$$P(s) = \phi \left(1 - \frac{s-1}{s_{max}-1} \right)^x$$

with s_{max} the maximum number of steps, ϕ a normalization factor, and the exponent x determining the shape of the probability distribution: low values of x correspond to linear distributions (days with few vs. many movement steps occur with similar probability), whereas higher x makes days with few steps much more likely. This rule was made stochastic to reproduce the day-to-day variability in distance moved observed in real lynx.

For the second part of the dispersal trait, two theories for how lynx choose their next grid cell were compared. Both of these theories are partially stochastic but differ from the simplest "random walk" models by assuming that individuals have distinct preferences for specific habitat types. Habitat was depicted simply by categorizing each cell as one of three types: dispersal (habitat similar to that in which lynx have been observed), barrier (urban areas and water bodies lynx are assumed unable to cross), and matrix (other areas, mostly agricultural land). Both theories assume that lynx can always sense the habitat type of the nine grid cells surrounding and including their current cell.

Under the Habitat Dependent Walk (HDW) theory, lynx prefer dispersal habitat to matrix habitat. The individual's probability of moving into any adjacent cell is equal for all cells of dispersal habitat, lower for cells of matrix habitat, and zero for barrier habitat. The parameter P_{matrix} controls the relative probability of entering dispersal versus matrix habitat; it can be varied so the probability of entering matrix habitat ranges from zero to equal that of entering dispersal habitat.

In the alternative Correlated Habitat Dependent Walk (CHDW) theory, a correlation factor P_C for maintaining the same direction within a day is included. P_C is the probability that the next cell is in the same movement direction as the previous movement; alternatively, a random direction is chosen. (Staying in dispersal habitat, however, has stronger effect on choice of cell than does correlation in movement direction.) This theory's assumption that lynx tend to maintain a consistent direction through dispersal habitat can be interpreted as a tacit prediction: lynx predict that, when they are in desirable habitat, continuing in the same direction will likely keep them in desirable habitat.

The traits for dispersal also include an interesting component representing the lynx's spatial memory: if a model lynx moves for P_{maxmat} consecutive steps in matrix habitat, it is assumed to turn around and return to the cell where it left dispersal habitat. Without this rule, a lynx surrounded by matrix cells has little probability of finding dispersal habitat again. This rule can also be interpreted as tacit prediction: a lynx predicts, after being in undesirable matrix habitat for many steps, that it is more likely to find desirable dispersal habitat if it retraces its movement instead of continuing randomly. The parameter P_{maxmat} can be interpreted as describing the lynx's memory-related capability to return to dispersal habitat after an excursion into matrix habitat. A typical value assigned to P_{maxmat} was nine cells, but values up to forty were tested.

Why are the lynx IBM's traits for dispersal stochastic instead of assuming individuals automatically move into the "best" cell? As with previous IBMs we have examined, the reasons for using stochasticity are not all stated by the authors but we can make good guesses. One reason is that the lynx are very often choosing among equally good cells: if six of the nine available cells are dispersal habitat, there is no other basis for choosing among the six so the decision must be made randomly. A second potential reason for using stochasticity is to represent uncertainty in a lynx's ability to sense the type of the habitat lying in each direction: a lynx dispersing through unknown landscapes could wander into worse habitat because it does not know that better habitat is available in the opposite direction. Finally, stochasticity in the dispersal trait can reflect uncertainty in the categorization of habitat. A one square kilometer of habitat may be categorized (using remote sensing and geographic modeling; Schadt et al. 2002b) as matrix habitat while still having some characteristics making it desirable to lynx—perhaps a meadow that attracts roe deer or a water source.

The conceptual basis of the dispersal trait is another interesting question: does it represent *direct* fitness-seeking, with individuals moving at least in part to improve their fitness? Or is the trait *indirect* fitness-seeking, designed only to reproduce observed dispersal? While some components of the trait appear loosely based on fitness-seeking (individuals prefer the dispersal habitat type, which presumably provides lower risks and more food than matrix and barrier habitat; and individuals turn around and return to dispersal habitat if they get "stuck" in matrix habitat), the primary consideration in designing the trait appears to have been reproducing observed dispersal behavior. In fact, the categorization of habitat as "dispersal" and "matrix" habitat was based on observed behavior: Kramer-Schadt et al. (2004) defined dispersal habitat as the habitat types that real lynx prefer and matrix habitat as types that lynx tend to avoid.

The dispersal trait, including the two alternative theories for choosing which cell to enter next, clearly is very dependent on several key parameters. To compare the two theories and to calibrate parameters, the IBM was applied to habitat of the Swiss Jura Mountains, where telemetry data from six dispersing lynx were available (figure 6.8). Model lynx were released at exactly the same locations as the real lynx and followed for exactly the same number of days the real lynx were observed.

At first glance, parameterizing this landscape-scale model using telemetry data from only six individuals (a total of 303 location observations) seems hopeless. But Kramer-Schadt et al. (2004) followed a pattern-oriented approach, identifying multiple patterns in the data and using them to "filter" model results (section 9.8). None of the following four patterns is particularly "strong," that is, has high power to distinguish among alternative model structures or parameterizations; but used together, these seemingly weak patterns were able to reduce the uncertainty in model structure and parameters. The patterns, and the criteria for accepting model results as fulfilling the patterns, are:

- Habitat preference. At least 81 percent of movement steps must be in dispersal habitat.
- Average daily distance moved. Average daily distance the IBM individuals move must be within the range (mean $\pm$ SD: 41.7 ± 26.5 kilometers per day) of distances moved by real lynx.

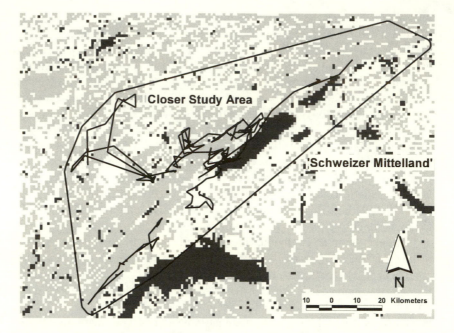

Figure 6.8 Dispersal paths of six subadult lynx in the Swiss Jura Mountains. The time lapse between location observations varies up to twenty days. Light gray grids are dispersal habitat, dark gray are barriers. (After Schadt 2002.)

- Avoided area. No individuals must cross the densely populated area between the Jura Mountains and the Alps, which was completely avoided by real lynx.
- Daily distance distribution. The statistical distribution of distances moved during a day must match the observed distribution, within a specified tolerance range.

All the parameters were varied over wide ranges, leading to 840 parameterizations of the HDW model and 8,400 parameterizations of the CHDW model. For each parameterization, the model was run 100 times and the average prediction compared with the patterns. For the HDW model, the percentage of parameterizations fulfilling the patterns 1–4 was: 76, 45, 58, and 26 percent; but only 11 percent of all parameterizations fulfilled all criteria simultaneously; the CHDW model was slightly more robust to parameter values, with 18 percent of parameterizations reproducing all four patterns.

Interestingly, this parameterization exercise reduced the "good" range only of P_{matrix}, the parameter to which results were most sensitive. Parameterizations fulfilling all four criteria occurred over the full range of the other parameters, indicating that interactions among parameters are important. Another interesting result was that in the "good" parameter sets, s_{max} was correlated with x, producing high probabilities of moving short distances, yet rare events of moving long distances. This could reflect a behavior of dispersing lynx observed by Breitenmoser et al. (2000): staying close to a prey carcass for up to a week and then moving far the next day.

Kramer-Schadt et al. (2004) also analyzed the IBM by comparing results to additional empirical data. The values of s_{max} in the good parameterizations (twenty-eight to sixty-two kilometers per day) reflect data known for the Iberian lynx (Revilla et al. in press) and Eurasian lynx in Poland (Jedrzejewski et al. 2002). Likewise, P_{maxmat} was restricted to 5 (SD ± 4) cells, corresponding to observations of the Iberian lynx (Revilla et al. in press).

While none of these validation analyses is particularly powerful, the combined evidence makes a strong case that the IBM captures the essence of observed lynx dispersal. The analysis also showed that the two alternative dispersal theories did not differ significantly in their power to reproduce the observed patterns. Because the CHDW theory did not substantially outperform the HDW theory, we can infer that habitat dependency is the more important factor for describing movement of the lynx; correlation in movement directions is compatible with the data, but not necessary to explain it. In the subsequent applications of the model, both versions were used (actually, the CHDW rule was used, but it is equivalent to HDW for parameterizations with P_C equal to zero). In applying the dispersal model to German landscapes, Kramer-Schadt et al. (2004) used 100 parameter sets randomly chosen from the parameter sets identified as "good" in the pattern-oriented filtering analysis.

Perhaps the most important lesson of the lynx model is that IBMs can help us address large-scale dispersal problems using two kinds of data that are often available: coarse, but spatially extensive, remotely sensed data on habitat; and high-resolution tracking data for relatively few individuals over relatively short time periods. The tracking data allow us to test traits for dispersal of individuals in the IBM, and then we can apply those traits to individuals moving through the huge expanses we can represent with satellite data. The habitat data may be coarse (perhaps in spatial resolution, but especially in the number and resolution of the habitat variables—the lynx IBM's habitat model was built from an existing database with seven values for one variable: land use), but using it as Schadt et al. did allow us to model problems of how dispersal is affected by spatial patterns in habitat.

6.4.2 Habitat Selection Theory in the Trout Model of Railsback and Colleagues

The trout IBM of Railsback and Harvey (2001, 2002) was designed to predict how populations (and communities of several trout species) respond to river management: if we change a river's temporal patterns of flow, temperature, or turbidity, or its channel shape or food production, how will populations respond? A wealth of literature and observation indicates that habitat selection is the primary way stream fish adapt to change. Stream habitat is highly heterogenous over short distances, and fish can rapidly find and move to better habitat when it is available. Therefore, the trout IBM was designed so that population dynamics emerge from the trait individuals use to select habitat in a diverse, dynamic world.

The primary model design problem was therefore to find a good trait for how trout select habitat from day to day as conditions change (Railsback et al. 1999). The literature (reviewed by Railsback and Harvey 2002) makes it clear that trout adapt their habitat selection in response to changes in both food intake and mortality risk; and that food intake and mortality risk are strongly affected by (1) habitat variables (depth, velocity, temperature, turbidity, predator types and density, and availability of cover for feeding and hiding—all of which can be affected by river management); (2) competitive conditions (larger fish exclude smaller fish from habitat); and (3) the individual's state (size, energy reserves, and life stage). Habitat selection in fish has often been modeled empirically, assuming that the habitat types most often used by fish offer the most benefits. However, among the many problems with the empirical approach (Garshelis 2000; Railsback, Stauffer, and Harvey 2003) is the virtual impossibility of developing empirical models that consider all these habitat, competitive, and individual variables. Therefore, Railsback et al. (1999) turned to theoretical approaches.

Habitat selection by trout and other salmonids has been studied extensively in behavioral ecology. Railsback et al. adopted the general approach of direct fitness-seeking: assume that individuals sense the fitness benefits of the habitat near their current location and then select the location offering highest benefits. The problem then was to find a useful fitness measure, the internal model that individuals use to evaluate the fitness benefits of habitat. Fitness measures used in previous studies were clearly inadequate for the IBM: assuming individuals simply maximize their growth rate ignores the importance of risk on habitat selection. One widely used measure that considers growth and risk—assuming individuals minimize the ratio of risk to growth—is based on assumptions that are clearly violated in the IBM, especially that all habitat offers positive growth (Gilliam and Fraser 1987). Instead, Railsback et al. (1999) developed the "state-based, predictive" fitness measure described in section 7.5.3. In summary, the IBM assumes individuals select the habitat that provides the highest predicted survival of both starvation and predation over a future time horizon, by using the following steps.

- Identify the potential destination alternatives: the habitat cells within a specified distance limit (200 times the fish's length) within which trout are assumed able to sense growth and risk conditions.

how?

Sense likelihood of occupancy or cost movement or permeability

- Sense growth and risk in the potential destination cells. Sensing is assumed to be error-free.
- For each potential destination, explicitly predict the probability of surviving risks other than starvation over a future time horizon (ninety days proved to be a good time horizon). This prediction is based on the simple and useful—though obviously wrong—assumption that current mortality risks would persist over the time horizon.
- For each potential destination, explicitly predict the probability of surviving starvation over the time horizon. This prediction is based on the assumption (also wrong yet useful) that the current growth rate would persist over the time horizon. First, the fish predicts its "condition"—its weight relative to the weight of a healthy fish of the same length—at the end of the time horizon; this prediction depends on growth rate (which may be negative), and the fish's current length and condition. Then the risk of starvation over the entire time horizon is estimated from the starting and predicted final condition.

or measure food availability

- If the fish is a juvenile, it also predicts how close to reproductive maturity it would come, over the future time horizon, in each potential destination. Reproductive maturity is defined as reaching the minimum length needed to spawn, so it is represented in the fitness measure as the fraction of spawning length that would be obtained at the end of the time horizon, assuming the current growth rate persists.
- For each destination cell, calculate the full fitness measure (referred to as "expected reproductive maturity," or EM): multiply the expected survival of nonstarvation risks by the expected survival of starvation and by the reproductive maturity term. Move to the cell with the highest EM.

Underlying this habitat selection theory are the IBM's submodels representing how growth, and survival of several specific kinds of mortality, depend on habitat, competition, and the individual's state.

The assumption that fish can accurately sense growth and risk conditions at distances up to 200 times their length is based on literature indicating that stream fish, at time scales much less than IBM's daily time step, thoroughly explore and become familiar with surrounding habitat.

Trout interact with each other via competition for food and cover. The IBM assumes trout have a size-based hierarchy within each habitat cell; this assumption is supported by field observations that such hierarchies (sometimes territory-based, but not always) exist and are maintained by aggressive interactions. Because these interactions occur at time scales much less than the IBM's time step, they are represented in the IBM as an interaction field: the food and cover available to an individual are reduced by the amounts consumed by larger individuals in the same cell.

The size-based hierarchy is maintained by the IBM's scheduling. Each daily time step, the habitat cells are updated with their current depth, velocity, temperature, and turbidity (other habitat variables are constant). Then, each fish executes its habitat selection trait. But habitat selection is executed in order from largest trout to smallest, and as each fish selects its cell the food and cover it uses are "depleted" and made unavailable to the following, smaller fish that use the cell (a kind of asynchronous updating). In subsequent schedule actions the fish execute their daily growth, then mortality, then reproduction.

To test and demonstrate the trout IBM's habitat selection theory, especially its potential to produce a wide variety of complex and realistic emergent behaviors, Railsback and Harvey (2002) conducted the analysis described in section 4.6 as an example of the IBE theory cycle. Subsequent analyses (Railsback et al. 2002) showed that different realistic population-level dynamics emerge from the IBM's habitat selection theory; and the IBM has since been applied to a number of river management and theoretical issues.

Perhaps the most important lesson from the trout IBM is that we can indeed find general theory for the most important adaptive traits of individuals, including habitat selection. Traits based on fitness-seeking adaptation are an important alternative to empirical models of behavior: once developed and tested, they can be applied to new sites and problems with confidence. This lesson is also illustrated by the work of Goss-Custard et al. (2001, 2002, 2003, 2004), Stillman et al. (2002, 2003), and West et al. (2002) on winter mortality of shorebirds. Their study has many similarities in history, approach, and results to the trout modeling project (section 4.6).

The trout IBM is a good illustration of some of the goals of IBE theory discussed in section 4.3, because theory for the habitat selection problem has also been addressed extensively using other ecological approaches. Implementing alternative habitat selection theories in the IBM (section 4.6) produced many testable predictions, quickly showing that some widely used theories (that animals select habitat to maximize growth or to minimize the ratio of risk to growth) are clearly inadequate in realistic settings. The IBM makes it easy to test the generality of theories, for example by simulating a variety of fish communities and habitats. And the trout IBM clearly is useful for applied ecology as well as the theoretical issue addressed here. In fact, this IBM is an example of how the need to solve an applied problem forces us to confront the weaknesses of conventional theory.

Finally, the trout IBM is another illustration of how a model's structure is determined by the patterns we use to define the essential characteristics of the system we are modeling. Chapter 3 emphasizes how we can use patterns to limit the complexity of a model; in this case, designing the IBM to explain observed patterns of habitat selection resulted in an IBM with more structure than preceding IBMs (i.e., M. E. Clark and Rose 1997; Van Winkle et al. 1998). For example, many IBMs of population dynamics treat mortality very simply, often (as in the IBMs examined in section 6.3) as a simple function of age and perhaps social status, with no consideration of what actually causes mortality. The trout IBM, however, needed to explain small-scale habitat selection that partly depends on specific mortality risks that vary sharply with characteristics of both habitat and individuals. Large trout are vulnerable to terrestrial predators, one reason they avoid quiet, shallow habitat where they are easily seen from the air; small trout are more vulnerable to predation by large trout and can avoid this risk by using shallow habitat. Reproducing this key pattern of how habitat selection changes with fish size therefore requires that the IBM represent not one but two types of predation risk and how these risks vary with depth and velocity.

6.4.3 Dispersal Success in Spatially Explicit Population Models

The term "spatially explicit population model" (SEPM) has become associated with a class of population models designed to "incorporate the habitat complexity of real-world landscapes" (Dunning et al. 1995). SEPMs are often applied to wildlife management, especially for animals that maintain home ranges in specific kinds of habitat. The management problems often concern the effects of habitat alteration on population viability, so the models must somehow represent habitat and its effects on the managed species. And SEPMs are most often, but not always, individual-based. The lynx dispersal model discussed in section 6.4.1 became part of the SEPM of Schadt (2002); other SEPMs have modeled owls (Franklin et al. 2000; McKelvey, Noon, and Lamberson 1993); woodpeckers (Letcher et al. 1998); sparrows (Liu, Dunning, and Pulliam 1995); bison and elk (Turner et al. 1994); and bear (Wiegand et al. 2003). Topping, Hansen, et al. (2003) developed a generic SEPM (ALMaSS), which was applied to voles (Topping, Ostergaard, et al. 2003), skylark, roe deer, carabid beetles, badgers, spiders, and other populations (see overview in Topping and Jepsen 2002).

SEPMs are not all identical, but most represent space as a grid of cells with an extent and resolution (cell size) relevant to the species, time scale, and problem addressed (section 7.3.1). Habitat complexity is often represented by categorizing each cell as belonging to one of a few discrete types, as in the lynx model. Most SEPMs assume habitat is static over time, but temporal variability can be included. Habitat data are often developed from remote sensing data analyzed via geographic information systems (GIS), but artificial landscapes have also been used to address theoretical issues (With 1997; Wiegand et al. 1999). Individual-based SEPMs have typically been relatively simple, depending mainly on stochastic traits to reproduce observed behaviors empirically. These SEPMs have made little use of direct fitness-seeking, and sometimes even lack basic characteristics of IBMs, such as resource dynamics, interaction among individuals, or variation in individual state variables such as size or energy reserves.

Dispersal is a key issue in many SEPMs, especially the problem of "dispersal success": how many individuals succeed in dispersing from their natal territory and establishing themselves in a new territory before dying? And dispersal success has been the topic of a well-known fracas about uncertainty in SEPMs versus simpler models. The widely cited study by Ruckelshaus, Hartway, and Kareiva (1997) analyzed a very simple, hypothetical SEPM that simulated what fraction of dispersers (moving randomly through a grid landscape) found a patch of suitable habitat before dying. They reported how dispersal success changed when error in the mortality parameter (probability of dying per movement step) was simulated, in model runs with mortality overestimated by 2, 8, 16, 24, and 32 percent. Dispersal success was reported to be extremely sensitive, dropping dramatically with even 2 percent change in mortality. This extreme sensitivity cast severe doubts on the credibility and usefulness of SEPMs because dispersal mortality is notoriously difficult to observe and quantify. Advocates of simpler, analytical models found their concerns about IBMs confirmed and even IBM users (including ourselves) accepted these results, as surprising (and, upon reflection, obviously wrong) as they were.

When Mooij and DeAngelis (1999) finally attempted to reproduce the model and results of Ruckelshaus, Hartway, and Kareiva (1997), they did not find high sensitivity of dispersal success to mortality probability. With help from the original authors (see Ruckelshaus, Hartway, and Kareiva 1999), Mooij and DeAngelis determined that Ruckelshaus, Hartway, and Kareiva (1997) had in fact simulated 2–32 percent errors in the per step *survival* probability (survival being one minus the mortality probability). This range in survival corresponds to errors in mortality probability of 665–10,635 percent, which are unrealistically large. (For example, reducing daily survival probability by 2 percent from 0.999 to 0.979 reduces the probability of surviving for a month from 97 to only 53 percent.) When Mooij and DeAngelis simulated 2–32 percent error in *mortality* probability, they found dispersal success not to be particularly sensitive and, in fact, found some evidence that SEPMs could be less sensitive to this parameter than simpler models.

Subsequently, Mooij and DeAngelis (2003) more directly examined the differences between simple SEPMs and even simpler, nonspatial models of dispersal success in vulnerability to parameter uncertainty. They determined uncertainty due to estimating parameters from a realistically sparse data set, for three models: (1) a model including neither time nor space, (2) a model including time but not space, and (3) a SEPM that used hypothetical spatial data. Each model has more parameters than the previous but uses more information from the data set to fit the parameter values. This analysis found that parameter uncertainty did not clearly increase with the complexity of the model, and in fact the SEPM had lowest parameter uncertainty. Mooij and DeAngelis concluded that the information provided by spatial data can more than make up for the uncertainty resulting from the need for additional parameters.

The interesting point of this story is how willing the ecological community was to accept the surprising and highly controversial results of Ruckelshaus, Hartway, and Kareiva (1997) until Mooij and DeAngelis finally tried to reproduce them. This kind of misunderstanding certainly hurts ecology as a whole and ecological managers as they struggle for ways to make difficult decisions; and it would be less likely if IBMs and their software were more transparent and communicated more thoroughly. In chapter 10, we address this problem of communicating IBMs so they can be easier to understand and reproduce.

It should also be noted that the basic message of Ruckelshaus et al.—that results of SEPMs may be very sensitive to dispersal mortality—is not necessarily wrong. Dispersal is a key process in many spatially distributed populations and metapopulations, and small changes in the time and risk involved in dispersal may indeed determine persistence or extinction of such populations. But this is an ecological problem that we can address with IBE, not an inherent limitation of "too complex" models. And we must remember why we are usually interested in dispersal success: because we are trying to understand how habitat loss and fragmentation affect population viability. Attempting to solve this problem without spatial information on habitat (especially now that such data are often easily obtained and analyzed) seems unlikely to be the most productive approach.

6.4.4   Summary and Lessons

Movement—both occasional long-distance dispersal and routine habitat selection—clearly is an important individual behavior that can strongly affect population dynamics (Lima and Zollner 1996; Turchin 1998), so it is a natural subject for IBE. Movement is especially important for one of the most common types of problem in ecology and ecological management: understanding how habitat change affects populations. The problem with movement is that it is even harder to observe and describe quantitatively than demography is: to quantify movement we must not just count individuals; we must identify each individual and track its location over time. Also, when we are studying movement it is much harder to pretend that individuals are unaffected by their habitat, and organism-habitat relations are often very complex.

So how can we model individual movement? Traditionally, SEPMs especially have relied on stochastic traits designed to reproduce empirical observations. This approach requires some spatial data on important habitat characteristics (often, habitat is simply categorizing as "suitable" vs. "unsuitable"); some simple, perhaps only intuitive assumptions about how individuals move among patches; and at least a few observed patterns for parameterization and validation. The IBMs that have been built this way are relatively simple and easy to parameterize (at least conceptually; the actual work may be tedious), and are certainly much more useful than nonspatial models for understanding relations between habitat and population dynamics.

However, SEPMs (and, consequently, IBMs in general) have been dogged by controversy arising from the deeply ingrained notion that ecological models are always "numbers driven" (Hengeveld and Walter 1999). Theoretical ecology has so focused on numbers—abundance, biomass, production, nutrient and energy fluxes, diversity indexes—that when we think about modeling we think about predicting the precise numerical value of some state variable. And when we think about building and parameterizing models, we think only about fitting parameters to data and naturally think of spatial IBMs as "data hungry" (Ruckelshaus, Hartway, and Kareiva 1997; Beissinger and Westphal 1998).

One way to avoid getting bogged down in numbers-driven controversy is, as we advocate throughout the book, to remain focused on predicting and understanding patterns, not just numbers. Once a model is able to reproduce a set of patterns, and perhaps even to be shown to predict successfully other independent patterns, we can be confident that we have done the best we can in modeling complex systems: capturing the essence—at least to some degree—of the system's properties and dynamics. At this point, but not sooner (section 9.3), we can fit parameter values to data and turn to quantitative predictions if needed to solve the problem we are modeling. And, even then, it is best to treat predictions not as absolute but as relative (Burgman and Possingham 2000; Grimm et al. 2004), for example, by ranking different management options (Turner et al. 1995). If this ranking is robust to changes in model structure and parameters (section 9.7), we can base management decisions on it. (Or the model and our analysis of it may show that the system is instead sensitive and unpredictable, also an important outcome of modeling.)

But the trout IBM illustrates a second way to avoid the limitations of numbers-driven modeling: basing our IBMs on well-tested theory and biological knowledge in addition to data. We can make IBMs that are "knowledge hungry" instead of "data hungry." For many species there is a wealth of information and even existing models of how habitat affects individuals; combining this information with simple assumptions about how individuals interact with each other and how they make decisions to increase their potential fitness shows great promise for producing general models of movement and other important behaviors. How can we make these models? The IBM design concepts from chapter 5 are a useful guide.

Emergence.   We need to think of movement and habitat selection as neither a fixed, inherent trait of a species nor (at the other extreme) merely random. Instead, we can think of movement as an emergent property of at least four processes (Railsback, Stauffer, and Harvey 2003): (1) the mechanisms by which habitat affects an individual's fitness, (2) the kinds of habitat that are available and how it is arranged spatially, (3) the ways that individuals interact and compete with each other, and (4) the population's abundance and structure. Information on any of these processes therefore should help explain movement.

Adaptation. Dispersal and habitat selection are, for very many species, extremely important to fitness. This has two implications. First, it is highly unlikely that natural selection would equip organisms with movement behaviors as simplistic as the models we often use. Second, direct fitness-seeking should be a productive way to represent movement traits.

Fitness. If we want to understand why individuals select or avoid different kinds of habitat, it is very helpful to look at how habitat affects the fitness elements shown in figure 5.2. For example, how does an individual's probability of surviving and achieving reproductive status depend on what habitat it uses? Even the simplest mechanistic models of how habitat affects processes like energy intake and mortality risk can be assembled into useful measures of expected fitness.

Prediction. What is reasonable to assume individuals can predict about temporal dynamics of habitat? In evaluating fitness benefits of alternative habitats, can individuals consider future seasonal changes, or changes due to resource depletion or increased competition? What can individuals predict about the consequences to themselves (mortality, growth, fecundity, etc.) of choosing one habitat or another?

Interaction. Competition for food and other resources, often mediated by territoriality or dominance hierarchies, is a kind of interaction especially likely to affect movement decisions.

Sensing. One of the most important assumptions involved in modeling movement is how much information individuals "sense" about the habitat alternatives available to them, and this assumption should depend on the IBM's spatial and temporal scales and on the individual's mobility. Especially important is representing how well individuals can explore habitat or sense gradients; many animals have remarkable abilities to sense food, predators, or habitat types at long distances.

Stochasticity. The degree to which movement decisions are stochastic should also depend on the IBM's spatial and temporal scales, and on the individuals' sensing abilities. The trout and lynx IBMs provide a very useful contrast. In the lynx IBM, individuals move rapidly (sometimes many steps per day) through coarse (one kilometer) grids, and the model lynx can only sense what is in the adjacent grids; so it seems reasonable to assume the lynx have little ability to explore and choose among the alternatives. Further, the grids are categorized into only three types, so lynx often must choose among alternatives that are equally beneficial. Therefore, it makes sense that the trait for lynx movement is highly stochastic. In the trout model, the individuals make choices over relatively short distances (which they could traverse within a few seconds or minutes) but long time scales (a day); and real trout are known to explore their surroundings continually. Therefore, assuming the trout move stochastically would greatly underestimate their real ability; the assumption that they sense and select the best cell is much more reasonable.

6.5 REGULATION OF HYPOTHETICAL POPULATIONS

So far in this chapter we have looked at models designed to understand what real organisms do, but there is much more we can do with IBMs. In this theme and the next we present IBMs developed more in the tradition of classical ecological theory instead of addressing specific systems. We start with IBMs that address the central issue of classical population models, population regulation: what keeps the size of a population from getting larger or smaller than it does? The questions addressed with these IBMs are: How significant is individual variability, and the modes of resource partitioning that give rise to individual variability, for regulation? Can individual variability be ignored, as in classical models, or is it a key element of regulation? And, most important, do IBMs including individual variability lead to the same results as classical models or do they raise serious doubts about the understanding of population dynamics and regulation delivered by classical models?

To make the theoretical IBMs of regulation comparable with classical models, many design elements of classical models were also used for these IBMs: the species modeled are generic and hypothetical; space is ignored so that interaction (competition for resources) is global, not local; trophic interactions with other species are ignored; and the environment—except resource production—is assumed to be constant.

6.5.1 The Łomnicki Model of Unequal Resource Partitioning

In the first study of the significance of individual variability for population regulation, Adam Łomnicki (1978; see also section 1.4) focused on unequal resource partitioning. In a population of N individuals, the food intake $y(x)$ of an individual of rank x, with $x = 1, 2, \ldots, N$, was assumed to be:

$$y(x) = a\left(1 - \frac{a}{V}\right)^x. \tag{6.2}$$

The parameter a is the maximum food intake of an individual and V the total amount of food available, assuming that $V > a$. The equation describes both unequal food partitioning among individuals of different rank and the increase in this inequality when food is scarce. The rate of food production per time step (one generation) is assumed to be constant.

This assumption that the inequality among individuals depends on the amount of food available is an important development of the classical concepts of scramble and contest competition (Begon, Harper, and Townsend 1990). Traditionally, the type of competition (scramble, or equal food partitioning, versus contest, with unequal partitioning) has been imposed as an inherent property of the system. In Łomnicki's model, as food level decreases, competition changes from scramble to higher and higher degrees of contest.

Although his depiction of food partitioning includes the concept of discrete individuals, Łomnicki did not model individuals. Instead, he simply calculated (using two coupled difference equations) the number of individuals, N, which receive enough food to reproduce and the amount of food consumed by the entire population.

The main result of the model is that unequal resource partitioning "stabilizes" population dynamics: both food level and population size inevitably reach equilibrium values. However, Łomnicki's main conclusion was that unequal resource partitioning provides an explanation of emigration that is compatible with the concept of fitness-seeking: individuals that do not receive enough resources in their home habitat still have a chance to reproduce if they decide to undertake the risk of emigration.

Łomnicki's model inspired a new way of thinking about regulation and its significance to other processes such as emigration. However, because he chose a classical difference equation framework to formulate his model, we cannot do what we like to do with IBMs: look "into" the population at lower levels or relate it to real populations. The model does not help us address such questions as: How does population structure change in the course of time? Is equation 6.2 a reasonable depiction of resource partitioning? What is the mechanism stabilizing the population? Certainly, monopolization of food during periods of low availability ensures that not all individuals starve simultaneously, but why does this mechanism stabilize population dynamics?

6.5.2 Uchmański's Models of Regulation and Individual Variability

Janusz Uchmański, another pioneer in the Polish school of individual-based modeling, was inspired by Łomnicki's work but, instead of simply assuming equation 6.2, Uchmański tried to find empirical evidence for the mode and causes of unequal resource partitioning. Is competition among individuals "symmetric," with the resources obtained by individuals proportional to their size; or "asymmetric," with larger individuals having disproportionate advantage over smaller individuals (Weiner 1990)? In his review of this question, Uchmański (1985) collected data on weight or size distributions as indirect indicators of the mode of competition, assuming that symmetric competition leads to more or less symmetric distributions, whereas asymmetric competition leads to distributions skewed so most individuals are very small and only a few are very large. The general pattern detected by Uchmański is that at low resource levels or high densities, weight and size distributions tend to be skewed, which indicates that under conditions of resource scarcity competition tends to be more asymmetric. This pattern may be less general than assumed by Uchmański (Latto 1992), but it certainly exists for many species and ecological situations.

Uchmański then developed an IBM to explore further how resource partitioning and the mode of competition affects population regulation (Uchmański 1985, 1999, 2000a, b). These models are a good example of how the patterns we want to explain affect the model's design (chapter 3). Uchmański's ideas concerned how resource partitioning affects weight distributions as well as population regulation so, unlike Łomnicki, he had to structure his IBM to produce weight distributions of its organisms. Individuals are treated as discrete entities characterized by weight and age. The model describes a population with nonoverlapping generations. The initial weight w_0 of each individual is drawn from a normal distribution, the mean and variance of which vary within a given range. Then, all individuals grow according to a simplified energy balance (bioenergetics) equation (e.g., Reiss 1989):

$$\frac{dw(t)}{dt} = a_1 w^{b_1}(t) - a_2 w^{b_2}(t). \tag{6.3}$$

The first term describes the assimilation (food intake) rate and the second term the respiration rate. It is assumed that under changing food levels V, the parameters b_1, a_2, and b_2 are constant, but the coefficient a_1, which controls assimilation, can vary with V and the initial weight w_0, as discussed later.

At the end of each generation, individuals reproduce only if they reach a threshold final weight. The number of offspring is proportional to the final weight. Food production is, as in Łomnicki's model, assumed to be constant. Instead of solving equation 6.3 numerically over time within each generation, Uchmański assumed that the resource level V is constant during each generation. Thus, he could directly calculate each individual's asymptotic final weight w_{end}:

$$w_{end} = \left(\frac{a_1}{a_2}\right)^{\frac{1}{b_2 - b_1}}.$$

The initial weight w_0 enters this equation through its effect on the assimilation parameter a_1. Uchmański (1999) explored the effects of unequal resource partitioning on population regulation by comparing different submodels for a_1:

$$a_1 = a_1 (w_0, V).$$

With a "null model" that assumes individuals are identical (i.e., w_0 and therefore food intake is the same for all individuals), the population increases in size exponentially and then goes extinct due to overexploitation of resources. Adding random mortality produced cycles of high and low abundance, but the cycles grow in amplitude so that extinction results after three or four cycles. Longer persistence was produced only by introducing individual variation in energy intake. Uchmański tested several submodels for a_1 (Uchmański 1999, 2000a, b), some producing quite long extinction times but all producing wide and growing abundance oscillations that eventually lead to extinction.

The results of Uchmański's IBM did not confirm Łomnicki's result that unequal resource partitioning led to population regulation toward an equilibrium. Uchmański's models did not produce equilibria but strong oscillations, with considerable risk of extinction during each cycle. And, unlike Łomnicki's findings, the relationship between population persistence and individual variability in initial weight was not monotonic. Instead, there was a range of intermediate variability that maximized persistence.

Uchmański's models are inspiring thought experiments that demonstrate the potential significance of individual variability to population regulation. However, their assumption of constant within-generation resource levels—so regulation only happens between generations—is a severe limitation. Resource depletion can occur in less than a generation, and within-generation regulation could produce different dynamics. Therefore, Grimm and Uchmański (2002) studied an IBM that is identical to the base version of Uchmański's model except that it simulates growth, resource depletion, and population regulation over subgeneration time steps. The following changes were made:

- An individual's assimilation rate each time step is limited by a maximum intake, and decreases as total food availability decreases. When food availability is high, all individuals have a near-maximum assimilation rate, but as food availability decreases, the assimilation rates of smaller individuals drop more quickly than those of larger individuals.
- Food production, individual growth, and food availability are calculated as in the original model, except at time steps shorter than one generation.
- Starvation is introduced as a mechanism for within-generation regulation. If an individual's weight drops to a specified percent of its previous maximum weight, it dies of starvation and is removed from the population before the next time step.

Within a generation, the model produces weight distributions that change as we expect with the abundance of individuals and availability of food. If initial food level is high and abundance low, all individuals grow rapidly and reproduce (figure 6.9a). If the initial food level is lower, there is a wider distribution of individual weights because smaller individuals get much less food and, therefore, grow more slowly than larger ones (figure 6.9b). The most interesting case is when both initial food level and abundance are high. Food levels drop quickly at the beginning of the generation, so many smaller individuals starve (figure 6.9c). This mortality reduces consumption so much that the remaining individuals—those with an initial advantage in weight—can continue growing.

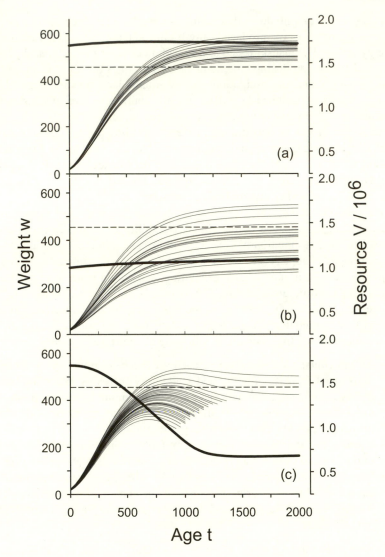

Figure 6.9 Individual growth (*fine lines*) and resource availability (*bold line*) within one generation. The dashed line denotes the threshold weight needed at the end of the generation ($t = 2000$)
to reproduce. See text for further explanations. (After Grimm and Uchmański 2002.)

When we look at this IBM's population dynamics over many generations,
we see a wider range of behavior than the earlier models of Uchmański. When
food production and vulnerability to starvation are both relatively low (individuals die only after losing 15 percent of their previous maximum weight),
the population shows diverging oscillations as in Uchmański's earlier models

(figure 6.10a). Starvation plays a role only during peaks in abundance. If food production is doubled (figure 6.10b), the oscillations are smaller in amplitude so the population can persist for longer times. If, in addition, starvation is more rapid (after weight loss of only 5 percent), the population is regulated between rather narrow limits and persists for long times (figure 6.10c). Finally, if the range of initial variability in weight is slightly reduced in such a way that slightly increases the total food consumption, starvation occurs in almost all generations and abundance is very steady (figure 6.10d). Grimm and Uchmański (2002) also found that if the time between generations is too short, starvation is unable to regulate the population well and cycles again become large enough to cause frequent extinction.

The IBM demonstrates that populations can be regulated by processes occurring at different time scales: in this case, reproduction rate at the generational time scale and starvation over shorter times. The relative importance of these two processes depends, in this model, on how vulnerable individuals are to starvation and how long generations are. And the two processes are linked by the current food level, which reflects both short-term consumption by the living individuals and consumption in previous generations.

The dynamics produced by the model of Grimm and Uchmański differ in one important aspect from classical models with nonoverlapping generations: if regulation is too strong, classical models can show chaotic fluctuations ("deterministic chaos"; May 1976). Grimm and Uchmański note that their model never produced dynamics resembling chaos. One of the preconditions of chaos in classical models turns out to be largely an artifact of model design: to produce chaotic dynamics, a model must not only have strong regulation but also a time delay between current population size and its effect on regulation. So chaos potentially can arise in models that use a full generation as the time step, but is unlikely to arise in models that consider within-generation regulation via mechanisms such as resource competition and starvation.

6.5.3 The Social Spider Model of Ulbrich and Colleagues

This IBM by Ulbrich et al. (1996) is inspired by the Uchmański models, but addresses a real species: a social spider occurring in Namibia. The species, *Stegodyphus dumicola*, occurs in colonies of several to hundreds of individuals (Seibt and Wickler 1988). Colony members share common nests and build large webs to trap both small insects that can be consumed by one individual and large insects that are shared. Colony members differ considerably in size even if they are of the same cohort and sex.

The IBM addresses essentially the same questions addressed by Uchmański: how does an individual-level mechanism—the mode of competition for food—affect individual variation in size and the colony's persistence?

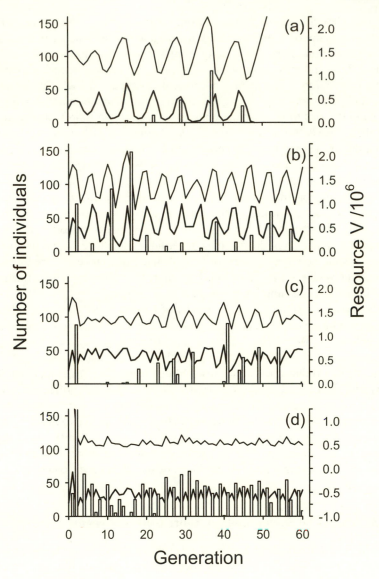

Figure 6.10 Typical time series of the number of individuals that survive until the end of the generation but do not necessarily reproduce (*lower, bold lines*), and dynamics of the resource, V, sampled at the beginning of each generation (*upper lines*). The vertical bars indicate the number of individuals that died of starvation. The scenarios are (a) the reference parameter set; (b) doubled resource production; (c) as (b) but individuals tolerate only 5 percent instead of 15 percent loss of their previous maximum weight before they starve; and (d) as (c) but with the range of individual variability in assimilation rate slightly reduced. (After Grimm and Uchmański 2002.)

Stegodyphus dumicola has an annual life cycle with nonoverlapping generations. Females only mature if they gain a threshold weight (120 milligrams), so
that individual variability in growth induces variability in the timing of reproduction and, in turn, variability in the weight of the following generation's
individuals—individuals born earlier are bigger than those born later. The IBM
uses a daily time step to capture these dynamics. Food availability each day
is assumed partially random but increases with the number of spiders because
each helps maintain the web.

The daily food resource is divided among the individuals either in contest or
scramble mode. In contest mode, the largest spider takes its full daily need, then
the second largest spider takes its need, and so on until no food is left. Scramble
mode is similar except that the order in which spiders feed is randomized each
day. The IBM then simulates, very simply, growth and reproduction from food
intake.

Experiments with the spider IBM (Ulbrich et al. 1996; Ulbrich and Henschel
1999) confirmed the basic results of Uchmański's models. Contest competition led to high individual variation in weight, whereas variation was small
with scramble competition (figure 6.11). When food was scarce, none of the
spiders reached maturity with scramble competition but contest competition
allowed some individuals—and the colony—to survive. Consequently, the
mean lifetime of colonies with contest competition is considerably higher.

6.5.4 Summary and Lessons

Readers have undoubtedly sensed that the models in this theme have a very different feel than the previous example IBMs. In fact, we have not even attempted
to fit these models into our IBE framework—the theory cycle, pattern-oriented
modeling, and design concepts. Why? The IBMs examined earlier in this chapter are what Grimm (1999) termed "pragmatic" models: IBMs motivated by
the need to solve specific problems of specific ecological systems. In contrast,
the IBMs in this theme address are "paradigmatic"—motivated by the desire to
explore and compare general paradigms of population ecology. What regulates
populations? How do modes of resource partitioning affect individual variation
and, consequently, population cycles and persistence? While some pragmatic
models clearly could be used to address such paradigmatic questions (e.g., the
marmot and trout models of sections 6.3.2 and 6.4.2), they rarely have been
(Grimm 1999). Therefore, the IBMs of this theme were built specifically to
address paradigmatic questions of the kind typically addressed with classical
models.

One problem with paradigmatic IBMs is that they inherit many of the
limitations of classical ecology along with its questions. Because the IBMs
of Łomnicki (1978), Uchmański (1985, 1999), and Grimm and Uchmański
(2002) address general ideas but not specific systems, the models cannot be

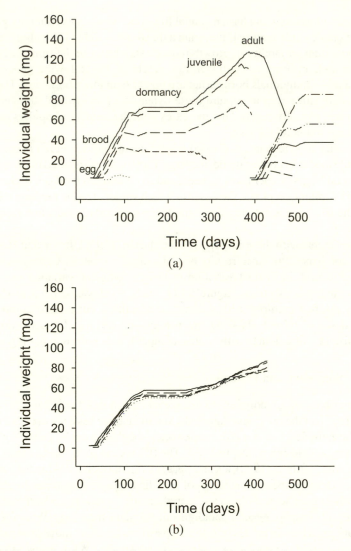

Figure 6.11 Growth trajectories for five randomly selected individuals (in each generation) for a low overall food level in the spider IBM of Ulbrich et al. (1996). (a) A colony with contest competition. (b) A colony with scramble competition; the colony was extinct after one generation because no females reached maturity. (After Ulbrich et al. 1996.)

tested and validated. The models ignore space and thus assume that competition is global, so it is not at all clear whether the regulation mechanisms occurring in these IBMs are significant in real systems where resource competition is local and spatially variable. Likewise, environmental effects are completely neglected. And the IBMs include no adaptive behavior: competition and resource consumption are tightly imposed by system-level rules instead of emerging from decisions made by individuals. Therefore, we can learn about the *consequences* of different kinds of resource partitioning but not about the *mechanisms* that might produce them. So these IBMs, like classical approaches to the same problems, may be general in the sense that they address problems of general interest and do not focus on any particular system, but the extent to which their results apply to real ecological systems remains an open question.

But these paradigmatic IBMs are still useful. They point us to questions and processes that may indeed be important in modeling specific systems. The clearest lesson from these IBMs of population regulation is that population dynamics and individual-level resource partitioning are tightly linked. We now know that addressing population regulation in an IBM requires careful representation of resource partitioning; and that when testing and comparing alternative traits for competition and resource partitioning in an IBM, it is important to look at their effects on population regulation and persistence. As the spider IBM of Ulbrich et al. (1996) shows, such generalities may indeed be important in specific populations.

6.6   COMPARISON WITH CLASSICAL MODELS

From the very start, the pioneers of individual-based modeling saw the primary benefit of IBMs as providing the ability to address questions that cannot be addressed by classical models (section 1.4). Yet the temptation to compare IBMs and classical models by applying both to the same problem is often strong (section 10.4.1). In this theme we look at several attempts to make this comparison to see what was learned by the exercise.

Why compare these two kinds of models that are so different? Several objectives are apparent from the literature. First is "validation": if some important result is produced by both individual-based and classical modeling, then perhaps we should have more confidence in the result. A second objective is testing classical models, because classical models are often very difficult to test against real systems. This difficulty is partly due to the data needed to parameterize and test population-level models, but also because classical models often make assumptions that are obviously violated in many real systems (e.g., global interactions; negligible environmental variability). Testing a classical model against output from a simple IBM allows the modeler to control the degree to which the classical model's assumptions are met. The third reason is simply to forget about real systems and understand the differences between the two types of model. What is it that we lose and gain when we switch from the individual-based to the classical descriptions of a system? Are the differences in model results fundamental, or does the individual-based description only add some details which could perhaps be captured by modifying classical models (chapter 11)?

The example studies we look at here address all three of these objectives but focus mainly on the third, comparing how the two types of model behave. The first two examples take the classical modeling perspective, using simple IBMs specifically designed for comparison with a particular classical model. The third example takes the opposite perspective, starting with a full-fledged IBM and looking at how it can be simplified toward a more classical model without changing its key characteristics.

At first glance, it may seem trivial to compare two kinds of population models, but most IBMs differ from classical models in so many ways that a truly direct comparison is impossible. Instead, comparison studies have usually created one or several models that add the elements of IBMs (space, discrete individuals, etc.) in a very simple way to a reference classical model. The models of Donalson and Nisbet (1999) and Law, Murrell, and Dieckmann (2003), which we examine here, follow this protocol. Our third example, the study by Fahse, Wissel, and Grimm (1998), is the only one we know of that starts with an IBM not developed for the purpose of model comparison and tries to understand its relationship to simpler mathematical models.

6.6.1 The Predator-Prey Models of Donalson and Nisbet

The goal of the models of Donalson and Nisbet (1999) is to explore the significance of two major limitations of the classical Lotka-Volterra (LV) predator–prey model: first, the LV model assumes that interactions have the same effects on all individuals, as if the individuals are well mixed so that interactions are global, not local. However, this assumption is clearly questionable in large systems because individuals interact only with their neighbors, a small part of the population, and because spatial variation in environment and density affect interactions (de Roos, McCauley, and Wilson 1991). Second, the LV model uses constant birth and death rate parameters, while in reality birth and death are discrete events and fecundity varies among individuals. For large numbers of individuals there may be little error due to using a rate-based description of demographics, but for small numbers variability among individuals (or over space and time) can produce demographics that deviate considerably from rate-based descriptions.

Donalson and Nisbet compared three models: the LV model, a stochastic birth-death (SBD) model, and a spatially explicit IBM. The baseline model for comparison is the density-independent LV model (e.g., Wissel 1989; Roughgarden 1998), which has the well-known property of neutral stability: both predator and prey abundance show cyclic fluctuations that have a mean determined by the equilibrium solution of the model's two equations and an amplitude determined by the initial abundances. The model's key parameters are the per capita rate constants for the three demographic processes: growth of the prey population, death of predators, and predation (death of prey, which also causes birth of new predators).

The SBD model was designed to represent the same processes as the LV model while adding temporal variability in demographic processes. This model represents individuals, but only in the sense that birth and death are discrete events so population size is an integer (see also Stephan and Wissel 1999). Stochasticity is used to represent temporal variability in the three demographic processes. Instead of modeling these processes as population-level rates (at the LV model's extreme of simplicity), and instead of modeling (at the opposite extreme) the detailed processes controlling individual reproduction and mortality, the SBD model represents birth, death, and predation as stochastic events. These events are modeled using exponential distributions parameterized to produce the same average rates of prey birth, predator death, and predation as the baseline LV model. The SBD model uses a continuous depiction of time (instead of discrete time steps) and dynamic scheduling: random draws from the three exponential distributions determine the time until the next event of prey birth, predator death, and predation (Renshaw 1991). Then, only the event with the shortest waiting time is scheduled for execution at the current time plus its waiting time. After each such event is executed, the population sizes are updated and the three exponential distributions reparameterized.

The third model is individual-based, but was designed specifically for comparison to the LV model instead of being intended to represent a real population. Nor does the IBM meet several criteria established in chapter 1 for being fully "individual-based": it considers neither life cycles nor resource dynamics, and there is no variability among individuals except in their location. The IBM differs from the SBD model only by adding spatial variability in demographic processes, which also requires representing the population as discrete and unique individuals. Individuals are represented in space by two vectors: position and velocity. Individuals move in a straight line until they hit a boundary of the square space; then they are given a new random direction.

Like the SBD model, the IBM is event-driven; however, because individuals have an identity now, the waiting time to the next event is calculated independently for each individual. The time to giving birth next (for prey individuals) or death (for predators) is stochastically determined for each individual via random draws from exponential distributions parameterized to reproduce the population-average rates of the LV model. Interactions are explicit predator-prey encounters. Predators are assumed to sense and capture prey that come within a specified distance; as a result the prey dies and the predator reproduces. Offspring of predator and prey are released randomly in the neighbourhood of their parents. Scheduling is again dynamic in continuous time: each individual determines the time at which its next events (birth, death, predation) will occur, and puts those events on a dynamic schedule to await execution. Understanding spatial effects is fundamental to the study's objectives, so the IBM's observation requirements included displaying predator and prey locations over time.

The three models are formulated in a consistent manner to make comparison as direct as possible: the rate parameters of the LV model are reinterpreted as probabilities in the SBD model, and the same probabilities are used in the IBM, which only adds individuality and the spatially explicit, local encounter behavior of individuals. This consistent design leads to similar dynamics in all three models: the well-known predator-prey cycles.

Donalson and Nisbet's study is an excellent example of using carefully designed simulation experiments to analyze models, as we discuss in chapter 9. The main "currency" they used for comparing the models was persistence (i.e., mean time to extinction), and they looked at how persistence varies with system size (area). In the LV model, persistence is infinite because extinction cannot occur. In the SBD model, persistence was found to increase linearly with system size. The reason for this is that, like the LV model it imitates, the relative magnitude of the cycle amplitude is constant and independent of system size. Therefore, the mean difference between the cycles' low point and zero increases as system size increases (a point also made by Stephan and Wissel 1999). The SBD model's stochasticity causes variability around the mean low point in the cycles, so the farther this mean low is above zero, the higher the mean time to extinction.

For small system size, the spatial IBM was found to behave similarly to the SBD model. Spatial effects are small in small systems, and individuals interact with all other individuals, meeting the assumptions of the nonspatial models. Surprisingly, at medium system sizes the IBM population was found to be *less* persistent than the SBD model but at large system sizes *more* persistent. Spatial effects thus destabilized the system at medium sizes but stabilized it at larger sizes.

To understand these spatial effects, Donalson and Nisbet used a variety of techniques: visualization, statistics, and simulation experiments that focus on the stabilizing effects of spatial patterns. Here, we only describe the visualisation and the simulation experiments. In figure 6.12 a series of spatial distributions of predator and prey is presented for three consecutive predator-prey cycles. In cycle 1 the spatial distribution seems only slightly nonrandom, and cycles 2 and 3 show the rapid emergence of spatial "waves" of high abundance. In cycle 3 there is almost complete spatial separation between predator and prey, as high densities of predators eliminate prey from large areas. Further experiments showed that the nonrandom spatial distributions such as those in cycles 2 and 3 were stabilizing: they caused subsequent cycles to have lower amplitudes.

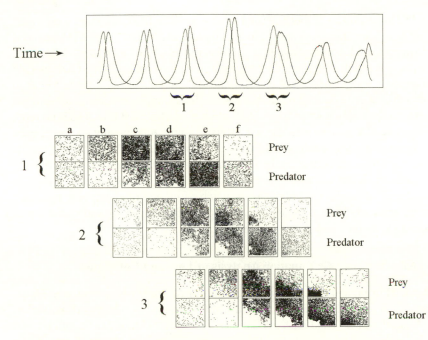

Figure 6.12 Time series and spatial patterns of prey and predator produced by the IBM of Donalson and Nisbet (1999). The spatial plots show the prey and predator spatial distribution at six events during each of the three cycles marked in the time series graph. These events are (a) total number of individuals is minimum, (b) predator population is minimum, (c) prey population is maximum, (d) total population is maximum, (e) predator population is maximum, and (f) prey population is minimum. (From Donalson and Nisbet 1999.)

The spatial effects in this model are subtle and the authors confessed that, although they were able to identify some stabilizing and destabilizing mechanisms, they were not able to explain fully why persistence was lower at medium system size than in the nonspatial model. But an even more fundamental concern about their spatial IBM is that it is based on numerous arbitrary assumptions such as homogeneous space, linear movement, and the absence of environmental effects. The authors noted that, indeed, small changes to these assumptions could cause significant changes in system dynamics. This observation casts doubts on the robustness of the model results and, in turn, the insights gained from the study. Further, even though complex dynamics emerge from the IBM, its individuals lack adaptive behavior: prey make no effort to avoid predators, and predators do not change their behavior as their need for food varies. The other IBMs examined in this chapter make us suspect that an IBM based more clearly on a real population, with adaptive traits of individuals, would be more robust than the Donalson and Nibet model.

If Donalson and Nisbet had used a more realistic IBM, however, they would not have been able to compare it so directly to the LV model. If an IBM designed to mimic, as closely as possible, a classical model based on ordinary differential equations (ODE) is inherently nonrobust, then how robust is the ODE model? Donalson and Nisbet (1999, p. 2506) make the point that: "By being forced to match explicitly defined implementations (for example, space and movement) against the implicit structure of ODE models, we find that the ODE is no more than one of the many possible selections from the total parameter space, even when using a set of interaction rules as simple as Lotka-Volterra. The results of this work cast doubts on the robustness of the ODE model results in many situations. If the individual-based spatial model is not robust with respect to a choice of model implementations (such as choice of movement patterns), then neither is [the] associated ODE model." The simplicity and generality of classical models may thus often be only apparent.

6.6.2   The Logistic Equation Analysis of Law and Colleagues

Law, Murrell, and Dieckmann (2003) published a study very similar in objectives and methods to that of Donalson and Nisbet, while addressing a different classical model. The logistic equation, one of the oldest elements of classical theory in ecology, models how a population grows from low density to densities at which competition slows growth and finally causes density to approach a fixed "carrying capacity." Like the Lotka-Volterra model, the logistic equation assumes that interactions—in this case, intraspecific competition—act at the population level, as if all individuals interact with all others. Law et al. examined the effects of this assumption by comparing logistic equation results to those of an IBM in which interactions are local and spatially variable.

The IBM used by Law et al. is similar to that of Donalson and Nisbet, as both were designed to mimic a classical model as closely as possible while adding temporal and spatial variability in demographic processes. Both use a continuous depiction of both space and time, and stochastic models of birth and death designed to reproduce the population-average rates of the logistic equation. However, the IBM of Law et al. is loosely based on plants instead of mobile animals. Dispersal occurs only at birth: new individuals are given a random location in the neighborhood of their parent, after which there is no movement. Interaction occurs only as a local density effect on mortality: an individual's mortality risk increases with the density and nearness of neighboring individuals.

In their analysis of the IBM, Law et al. used population trajectory over time as their currency for comparison: How closely does the trajectory of population density in the IBM follow the familiar rise and leveling of the logistic curve? And does density reach the same "carrying capacity" as the logistic curve? Law et al. repeated this comparison while varying the parameters controlling the average distances over which newborn individuals disperse and over which competition affects mortality; these are the only parts of the IBM not specified by the rate parameters of the logistic equation.

The analysis found that the dispersal and competition parameters have strong effects. At one extreme, when both dispersal and competition occurred over short distances, the population actually declined to zero instead of following a logistic trajectory: individuals formed small clusters where mortality was very high. At the other extreme, when dispersal occurred over long distances and competition over short distances, the population reached an equilibrium well above the carrying capacity predicted by the logistic equation. As expected, when dispersal and competition both occurred over long distances (approximating global interaction), the IBM matched the logistic equation very closely. These results support the same conclusion that Donalson and Nisbet reached: that the classical model itself is nonrobust in the sense that processes ignored in it have strong effects on results—changing, in the logistic equation case, not only the shape but the direction of the population trajectory.

6.6.3   Separation of Time Scales in the Model of
Fahse and Colleagues

The IBM of Fahse, Wissel, and Grimm (1998) simulates nomadic larks in the Nama-Karoo, a semiarid grassland biome of South Africa. The model was constructed for two reasons: to test which searching and flocking strategy is optimal for these birds, and to design reserves for the larks (Dean 1995; Fahse, Wissel, and Grimm 1998). The larks usually move in small flocks to find patches of grassland habitat that are suitable for reproduction (Dean 1995). These small patches (on average about 3.5×3.5 kilometers) are rare and ephemeral, appearing only after rainfall, which is typically patchy. The grassland patches provide food and shelter for nestlings. However, successful breeding must occur within two weeks after a patch appears; older patches dry out before the nestlings mature. The environment of the larks is thus an ever-changing mosaic of suitable and nonsuitable breeding areas (figure 6.13). If a flock finds a suitable area, it will start a breeding session. The flock sizes vary over space and time due to mortality, reproduction, and behavior: flocks split or combine according to decisions made by their member birds. Larger flocks are assumed more likely to find breedings areas but competition for food is believed to cause per capita breeding success to decrease with flock size.

The IBM represents both space and time discretely, and uses a one-day time step. The space has a grid of cells, fifty by fifty, each the size of an average grass patch. Which cells turn into usable grass patches, and when, are determined stochastically. New grass patches are either found by a flock within two weeks or revert to unusable status because they are no longer suitable for successful breeding (figure 6.13).

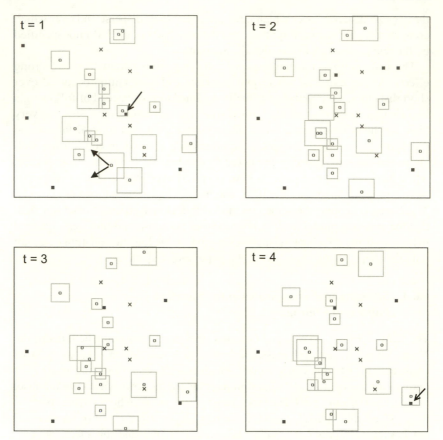

Figure 6.13 Spatiotemporal dynamics of randomly created grass patches suitable for breeding (*filled squares*), flocks of larks searching for these patches (*circles*), and locations where birds are breeding (*crosses*). The frames around the circles indicate the flock's range of vision. At the bottom of the $t = 1$ display, arrows denote the splitting of a flock into two smaller ones that appear at $t = 2$. At $t = 1$ and $t = 4$, arrows also show a grass patch being found by a flock. Time proceeds in steps of one day; the entire area is 50×50 cells, and each cell represents 3.5×3.5 kilometers. (From Fahse et al. 1998.)

Individual birds are distinguished by age, life stage, and the flock to which they belong; and mortality occurs at the individual level. However, the larks live in flocks their whole lives and flock-mates interact and cooperate intensively. Therefore, the IBM explicitly represents the flocks as collectives, and adaptive behavior is modeled at this collective level. Flocks have traits allowing them to adapt their location and breeding behavior in response to the availability of grassland patches that can support reproduction. A flock is assumed to sense the presence of a breeding patch within a radius that increases with flock size. The reason for assuming bigger flocks are better able to detect patches was unstated, but it could be that each individual explores neighboring habitat somewhat independently, or bigger flocks may be more likely to have individuals that are better explorers, or perhaps bigger flocks *must* spend more time exploring to find adequate food. When a patch is detected, the flock moves to it and breeds. Flocks also have traits for splitting into two smaller flocks and for combining with other flocks they encounter.

The technical problem with this model was that it combined a daily time step with a large space and population, so it ran too slowly (on a 1997 personal computer) to perform even the most basic analyses. In general, options for dealing with this problem include simplifying the model, shrinking the space and population, or using software engineering techniques (section 8.7.4). But Fahse et al. turned to a more fundamental approach. They wondered if it would not be possible to extract a per capita growth rate:

$$r(N) = \frac{f(N)}{N}$$

where N is the population size and f is the population growth rate, that is,

$$f = \frac{dN(t)}{dt}$$

of the population from the IBM.

In the IBM, of course, r is not determined directly by N but instead by mortality, the actual number and size of flocks, flock behavior, the structure of the landscape, and so on. Therefore, finding a function f at first seems very difficult, but on the other hand if one process in the IBM dominates demographics a simple function might be found by focusing on that process. This possibility seemed likely for the lark IBM because Fahse et al. found that the number and mean size of breedings flocks reached a relatively steady equilibrium value much faster than population size reached its equilibrium (figure 6.14). In fact, the processes determining the size distribution of breeding flocks dominates the rate of population change because in each time step the number and size of breeding flocks determines the number of new progeny; mortality explains little variation in abundance because survival rates are assumed to be constant. Thus, once we know the breeding flock size distribution and how it depends on N, we can extract the population growth rate f.

However, there is still a problem finding the population growth rate: f is a function of breeding flock size distribution, but breeding flock size distribution is undoubtedly a function of N. To overcome this circularity, Fahse et al. applied a technique well known in physics but only occasionally applied in ecology (but see Ludwig, Jones, and Holling 1978): the separation of time scales. This means treating processes that change over fast versus slow time scales separately: while

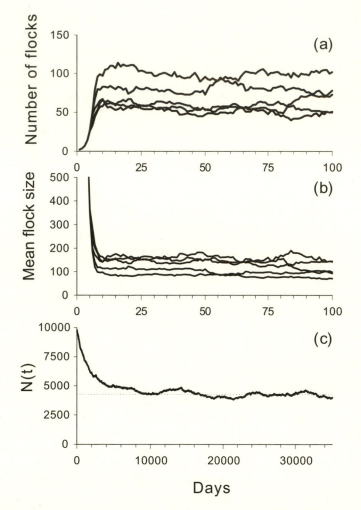

Figure 6.14 Separation of the behavioral and the population dynamic time scales in the IBM of Fahse, Wissel, and Grimm (1998). The daily number of flocks (a) and daily mean flock size (b) adjust very quickly from initial conditions, typically reaching relatively constant values within 10 days. (The different curves correspond to different values of the parameter *Nsplit*, which describes the readiness of individual birds to split a flock.) However, (c) a long-term trend in the total number of individuals $N(t)$ remains (here, for over a decade). (After Fahse, Wissel, and Grimm 1998.)

describing the fast processes, the "slow" variables are assumed constant. In the lark IBM, N can be treated as a slowly changing variable, whereas the variables controlling the flock size distribution are fast, because they are based on behavior occurring on daily time steps. Consequently, to extract the function f from the simulation model, the slow variable, N, was kept constant by deactivating all demographic processes: the IBM was run with birds executing all their behaviors except dying or producing offspring. After this version of the IBM was run long enough to reach equilibrium, the size distribution of breeding flocks was observed from it. The number of nestlings that this distribution would produce could easily be calculated from the IBM's assumptions concerning fecundity, and this number divided by N equals r. Equilibrium population sizes predicted by this "behavioral" version of the IBM matched results of the full IBM very well.

Repeated simulations using a wide range of initial values of N allowed Fahse et al. to find how per capita growth rate r varies with N. This function turned out to be a negative linear relation, which means that population dynamics in the lark IBM can be simplified to exactly the classical logistic equation that Law, Murrell, and Dieckmann (2003) analyzed (sections 6.6.2 and 11.4).

This example clearly shows that under certain conditions the population dynamics of an IBM can be described (but not explained) by a classical model, even when the IBM was not intentionally designed for direct comparison to the classical model. However, the discovery that the IBM could be approximated by a logistic equation does not mean that the lark study could have been addressed using only a logistic equation model. The problem is parameterization: without the IBM, the only way to parameterize the logistic equation is fitting it to a time series of sufficiently variable observations, which was not available for lark in the Nama-Karoo. The logistic model also does not lend itself as the IBM does to the lark study's second problem, examining effects of preserves that alter habitat conditions over space.

The general protocol of Fahse et al. for extracting a function for population growth rate from a detailed IBM can be used in other IBMs; when successful, it can facilitate the analysis (using methods discussed in chapter 9) of IBMs which are too big to run many, many, times. This protocol can be used whenever there is a clear "interface"—separation of time scales—between behavior and demographic variables. If so, birth and death processes can be turned off to determine a function for how r varies with N. The results of the behavioral population IBM can then easily be tested by comparing them with those from the full IBM.

6.6.4 Summary and Lessons

Comparing classical models and IBMs seems like a natural project: knowing all the limitations of both, we want to learn how they are similar or different, and which might be "best" in various situations. And the project initially seems fairly clear and easy: apply both types of model to the same system and compare their results. However, the conclusions we can draw from the comparisons in this theme are not nearly as satisfying and clear-cut as expected.

One conclusion is that classical models simply cannot be translated verbatim to the individual level. We cannot translate population-level rates (population increase, predation, competition) into individual events (birth, death) without introducing behavior to the model, and the examples show that behavior affects results as much as the rate parameters copied from the classical model.

Does an IBM produce the same results as the classical model it was carefully designed to mimic? The examples show that the answer to this question is sometimes—depending on parameter values and what kind of behavior is used—but not always, even when the IBM contains as little behavior as possible.

But this ambiguous result is not the most dissatisfying aspect of these comparisons. Translating a classical model into an IBM is the tail wagging the dog: the resulting IBM is too simplistic to be interesting. Donalson and Nisbet and Law et al. found their IBMs to be nonrobust, and much of this nonrobustness is no doubt because the IBMs lack many characteristics of real IBMs: individual variability, life cycles, resource dynamics, and (especially) adaptive behavior. Real IBMs, as we propose them in this book, start with an ecological question, with patterns in real systems, and with theories about the adaptive behavior of individuals. All these characteristics are excluded when we design the IBM only to mimic a classical model.

The third example examined in this theme goes the other way: Fahse et al. started with a real IBM of a specific population and then found a simple, classical model that reproduced one of the IBM's key outputs, equilibrium population size. Several worthwhile objectives can be addressed by this approach. First, we can see whether classical models can reproduce results of a useful IBM; this is still a comparison of model versus model, but at least the comparison is grounded in the characteristics of a real system that are captured in the IBM. Comparing classical models with "real" IBMs of real systems seems likely to tell us much more about the value and limitations of classical models than does comparing them with highly simplified IBMs. In general, simplifying full IBMs is an important method for analyzing IBMs (chapter 9). We can remove elements from an IBM and see how its capabilities change as the simplicity of an analytical model is approached (section 11.5.2).

A second objective of trying to find a classical model that "fits" an IBM was the motivation of Fahse et al.: the need for a simpler model more amenable to extensive analysis than the IBM. Other approaches for generating simpler versions of an IBM that have been explored (but rarely published, as far as we know) include using IBM output to parameterize matrix models and statistical models (including time series and spatial models). These approaches all have limitations. The IBM is still needed to validate and parameterize the simpler models, and the simpler models must be reparameterized each time new assumptions or parameter values are tried in the IBM (which happens often during analysis). Certainly, many of the IBMs we examine in this chapter are too complex for these approaches to be practical or even feasible. Often we must just rely on software techniques to speed up an IBM.

6.7 DYNAMICS OF PLANT POPULATIONS
AND COMMUNITIES

Instead of focusing on specific examples in this theme, we provide an overview of general approaches used in many plant IBMs. We discuss general approaches because plant IBMs developed along quite a different pathway than animal IBMs did, partly because of basic differences between animal and plant ecology. In animal ecology, population ecology is a dominant discipline and was the first discipline in ecology to become "theoretical" by using mathematical models. Classical models operate at the population level in animal ecology, but even the notion of a population being the natural unit for modeling is much less prevalent in plant ecology. Classical modeling's fundamental assumption that change in population size is a simple function of current population size seems much less plausible for plants (Crawley 1990). The key process in plant population dynamics often appears to be recruitment, which often appears most strongly influenced by factors other than current population size: disturbance, weather, soil conditions, local competition for unoccupied sites, and the like. The effect of such factors on recruitment seems less amenable to modeling, so many plant IBMs focus on within-generation processes: growth, local competition, and density-dependent mortality.

Another difference between animal and plant ecology is that "behavior" is more readily associated with animals: their movement, feeding, interactation, and other behaviors are easier for us to observe and understand. Because we are like animals, it is natural to us to develop and apply theories about adaptive, decision-making traits of animals. Plants also have adaptive behaviors (section 6.7.6), but these behaviors are of different types, and often occur at different time scales, than ours. Without an intuitive understanding of how plants make decisions, we tend to make adaptive behavior less an explicit issue in plant ecology than in animal ecology. Instead, because plants cannot adapt to local conditions by moving, ecologists have focused on local competitive interaction as the key concept in individual-based plant ecology. Because of this focus on interaction, this theme mainly concerns how competition among neighboring plants is modeled (a topic reviewed by Czárán 1998; see also Kenkel 1990; Czárán and Bartha 1992).

In addition to having population dynamics that emerge primarily from competitive interactions among individuals, many of the modeling approaches in this theme share a major simplifying assumption: the use of space to represent interaction. Plants interact with neighboring plants in generally negative ways: one plant can block sunlight and rainfall before it reaches others; roots of one plant use up nutrients and moisture that otherwise would be available to others; and some plants use allelopathic chemicals to suppress neighbors.

(There are also positive interactions among plants, e.g., the risk of a tree being killed by wind is reduced by neighboring trees.) Modeling these interaction mechanisms explicitly can involve considerable complexity and uncertainty. Some models do represent some interactions explicitly: Pacala, Canham, and Silander (1993) simulate competition for light; the BEFORE model examined in section 6.8.3 simulates effects of neighbors on wind mortality. But many successful plant IBMs assume all interaction is negative, ignore mechanisms, and lump interactions together as "competition," and further assume that the degree of competition is a simple function of how space is divided, or shared, among individuals.

Three general classes of plant IBM are reviewed in the following subsections. First are *distance models*, a class distinguished by Czárán (1998); such models represent the interaction between individuals as a function of the distance between them. We examine three kinds of distance model that use different functions for how interaction varies with distance: fixed-radius neighborhood, zone of influence, and field of neighborhood. Next are grid-based models, which use a discrete representation of space. Finally, we consider IBMs widely used in forest ecology and management: gap and growth-yield models. We provide specific examples of some of these classes.

6.7.1   Fixed-radius Neighborhood Models

In fixed-radius neighborhood (FRN) models each plant is the center of a circle of a fixed radius. Other individuals within a plant's circle are, by definition, neighbors that interact with the plant. The influence of neighbors on a plant may simply depend on their presence or, in more sophisticated models, on species, age, or other state variables. The functional relationships representing this influence may be assumed ad hoc or—as in many detailed forest models (e.g., Pretzsch, Biber, and Dursky 2002)—determined empirically by using regression methods.

The plant population models of Pacala and Silander (1985) and Pacala (1986, 1987) use the FRN approach. These were pioneering plant IBMs, but their design was constrained by the authors' objective of comparing them directly with analytical, classical models. These IBMs assume that individuals, as soon as they exist, have a fixed radius within which they interact with neighbors; very little about change in individual state is represented. This assumption parallels classical models in ignoring the life cycle of individuals (Uchmański and Grimm 1996): instead of representing how individuals grow and develop, only the presence of plants and their neighbors is represented. As with the models examined in section 6.6, the objective of comparing these models with classical models resulted in their being so simplified that they retain few of the characteristics or advantages of IBMs (Czárán 1998).

6.7.2   Zone of Influence Models

Like FRN models, zone of influence (ZOI) models assume a circular zone around each plant, but the meaning of this zone is more explicit: the zone specifically reflects the area over which a plant obtains resources such as light, nutrients, and water (Ford and Diggle 1981; Wyszomirski 1983, 1986; Weiner 1982; Czárán 1984; Hara 1988; Wyszomirski, Wyszomirska, and Jarzyna 1999; Weiner et al. 2001). If the ZOIs of two plants overlap, they therefore interact via competition for resources within the overlapping area. The effect of competition is reduced growth. Resources are usually not modeled explicitly in ZOI models, but it is assumed that the area of the ZOI represents the plant's resource intake and therefore its performance. A plant with no neighbors has maximum performance (e.g., a size-dependent growth rate). The more a plant's ZOI overlaps the ZOIs of neighboring plants, the lower its performance.

In contrast to FRN models, the radius of the ZOI is not fixed but depends on the size of the plant, usually quantified by biomass or stem diameter. A plant thus may start out with no neighbors when its ZOI is small, but as its ZOI grows and overlaps with others, local competition increases. Growth slows down, as a consequence, and in some IBMs plants under severe competition die.

Although ZOI models are conceptually simple, their implementation is not straightforward because calculating the area of overlap becomes cumbersome as soon as more than two zones overlap. Wyszomirski (1983) used an elegant algorithm to cope with this problem: sampling forty-four regularly distributed points within the ZOI of each individual to see if they overlap ZOIs of other individuals. Czárán and Bartha (1989) simplified the ZOI concept by considering the distance between plant stems instead of overlap areas.

Most ZOI-based IBMs were designed to study growth in even-aged monocultures (single-species cohorts planted all at the same time; see the review of Wyszomirski, Wyszomirska, and Jarzyna 1999, and the overview in Weiner et al. 2001). Typically, the objective was to study how various factors influence the distribution of individual plant sizes as a cohort grows. For example, empirical studies (reviewed in Uchmański 1985) indicate that higher densities lead to more positively skewed weight distributions: many small but few large individuals. Other studies addressed the effect on size distribution of the spatial arrangement (regular, random, or aggregated) of the plants.

Still other studies attempted to relate size distribution to the mode of competition, so such relations could be used to infer whether competition among plants is "symmetric," with the negative effects of competition proportional to the size difference of the competitors, or "asymmetric," with effects more than proportional to size difference (Weiner 1990). The general approach—another instance of the pattern-oriented approach to modeling described in chapter 3— was to see whether IBMs assuming symmetric or asymmetric competition best reproduce observed distributions of plant sizes (section 6.5.2 describes other studies of this problem). However, one result from these studies was learning that the mode of competition cannot necessarily be inferred from the skewness of a population's size distribution. If density is too high, all plants may be so reduced in growth that large size differences simply cannot emerge, so asymmetric competition may not have detectable effects (Wyszomirski, Wyszomirska, and Jarzyna 1999; Uchmański 2003; Bauer et al. 2004). A limitation of most of these studies is ignoring mortality, which at higher density (and, therefore, higher competition) certainly occurs in nature. Mortality reduces local density and therefore affects size distributions. Another limitation is that these studies focused on one narrow problem in an unnatural context: size distributions of monocultures. Consequently, they have so far made relatively little contribution to our general understanding of plant population and community dynamics.

6.7.3 The Field-of-Neighborhood Approach of Berger and Hildenbrandt

A recent extension of the ZOI approach is the field-of-neighborhood (FON) approach of Berger and Hildenbrandt (2000). Originally it was developed to model long-term dynamics of mangrove forests, which means that effects of competition on mortality and recruitment had to be considered as well as growth. The size of a plant is represented by the radius of its stem. As in ZOI models, individuals have a circular zone of influence around their stem. The radius R of this ZOI is assumed related to the radius r of the stem:

$$R = ar^b. \tag{6.4}$$

The parameters a and b can be determined empirically or simply assumed (Berger and Hildenbrandt 2000; Grimm and Berger 2003). An individual's growth rate is not limited by competition if the ZOI does not overlap with any other plant. However, growth may be assumed also to depend on the individual's state (e.g., size or age) or on environmental variables other than competition. In the mangrove model of Berger and Hildenbrandt, growth was also affected by groundwater salinity and nutrient availability.

While the ZOI only defines the geometry of interactions—whether there is an effect of neighboring individuals at any point—the FON approach also represents how the strength of interaction varies over space. On a plant's ZOI a scalar "field of neighborhood" is defined; the FON quantifies, at each point, the strength of the plant's effect on potential neighbor plants (figure 6.15). This influence is assumed to be strongest ($= 1.0$) at the stem and decreases exponentially toward the border of the ZOI. The total FON at a point is the sum of fields from all plants having the point within their ZOI. This FON concept provides, among other advantages, a simple way to model the effect of competition on seedling establishment: it can be assumed that if the total FON

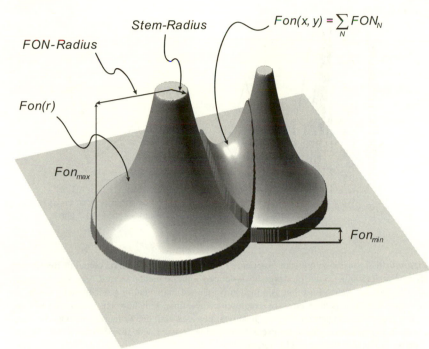

Figure 6.15 Visualization of the field of neighborhood (FON) of two plants with overlapping fields. Stem radius and FON radius correspond to r and R, respectively, in equation 6.4. Between the two plants, a part of the total field $F(x, y)$ is drawn. FON_{max} and FON_{min} are the maximum ($= 1.0$) and minimum values the field of neighborhood can assume. (Figure courtesy of H. Hildenbrandt.)

at a seed's location, $F(x, y)$, is above some threshold (which may be zero), the seed cannot develop.

The FON is also used to quantify the competition among neighboring plants. The strength of competitive interaction affecting a plant k due to n neighbors is calculated by first integrating $F(x, y)$ over the areas where the neighbors' ZOI overlaps the ZOI of plant k, then dividing by the area A of plant k's ZOI:

$$F_A = \frac{1}{A} \int_A \sum_{n \neq k} F_n(x, y) \, da.$$

Because the interactions with different neighbors are assumed to be additive, the summation and integration in the preceding equation can be reversed, which simplifies the calculation of F_A. Finally, the plant's current growth rate is determined by multiplying the growth rate in the absence of competition by a correction factor C assumed to decrease linearly with F_A:

$$C = 1 - 2F_A.$$

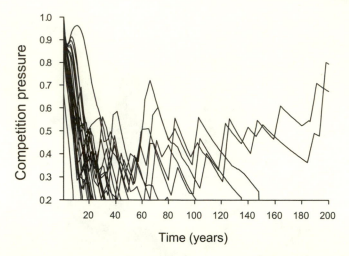

Figure 6.16 Competition pressure experienced by individuals in the KiWi model of Berger and Hildenbrandt (2000). Competition pressure is defined as the correction factor C (see text) averaged over the past five years. Each line is a randomly selected individual in a self-thinning cohort of mangrove trees. Competition pressure is represented over time, with the x-axis indicating years. As a plant grows, the competition pressure it suffers increases until one or more neighbors die; then competition pressure drops and the cycle of increasing pressure with growth starts again. (Figure courtesy of H. Hildenbrandt and U. Berger.)

Without neighbors, F_A is zero and therefore $C = 1.0$. If F_A is larger than 0.5, C is assumed to be zero and therefore growth is completely suppressed. A completely suppressed plant can start growing again if one or more neighbors die so that F_A is decreased.

Mortality is also assumed to depend on local competition: if a plant's growth is completely suppressed for more than a certain time (e.g., five years), the plant dies. As a result, plants with too many neighbors that are too large successively die or have their growth suppressed. Consequently, growth rates in dense stands typically are highly variable over time: a plant's growth jumps up when a neighbor dies but then decreases as its growth, and that of neighbors, again builds up the value of F_A (figure 6.16).

The FON approach fulfills the criteria for modeling local competition formulated by Stoll and Weiner (2000): each plant has an explicit spatial location, a basal area where no other individual can exist, and a zone of influence where it influences, and is influenced by, neighboring trees. The number, size, and location of neighboring plants all affect the level of competition affecting a plant. This last criterion is not fulfilled by the ZOI approach, in which location does not matter for smaller plants having their ZOI completely overlapped by the ZOI of a larger plant.

The FON approach is similar to the so-called ecological field approach (Wu et al. 1985; Walker et al. 1989), in which a field—not necessarily a circular

ZOI—describes the influence of a plant on a specific resource. However, the mechanistic details of how plants influence resources in their neighborhood are often unknown, and including them can make an IBM very computationally intensive. The FON approach provides a level of mechanistic resolution between the complexity of ecological field methods and the simplicity of ZOI models.

In conjunction with developing the FON approach, Hildenbrandt (2003) addressed a basic problem of implementing distance models. Interaction is represented as a variable function of distance between plants in continuous space, so identifying the neighbors that each plant interacts with each time step is a difficult implementation problem. The conceptually easy approach—checking each pair of plants in the IBM to see if they are neighbors—is computationally burdensome for large populations because the number of such pairs increases with the square of the number of individuals, N. Distance IBMs must therefore use sophisticated algorithms and data structures to keep track of neighbors. Hildenbrandt (2003) implemented a FON IBM using a data structure called Hilbert R-Trees (Guttman 1984; Sellis, Roussopoils, and Faloutsosj 1987; Beckmann et al. 1990); this technique resulted in computation time increasing only linearly with N.

To illustrate the potential of the FON approach, we now look at its application in two IBMs.

6.7.3.1 Cyclic Population Dynamics of Plants

Inherently complex population dynamics (cycles, chaotic fluctuations, etc.) are a key issue of theoretical animal population ecology (e.g., Turchin 2003), but plant populations are typically assumed to be inherently "stable" unless affected by disturbances, environmental fluctuations, or pathogens (Crawley 1990; Krebs 1996). However, a few empirical studies indicate that stability is not necessarily inherent (Symonides, Silvertown, and Andreasen 1986; Thrall et al. 1989; Silvertown 1991; Tilman and Wedin 1991; Crone and Taylor 1996). Therefore Bauer et al. (2002) studied nonequilibrium plant population dynamics using a full-life-cycle IBM and the FON approach. Reproduction was modeled by assuming that plants above a minimum size produce seeds; the number of seeds increases with plant size up to a maximum number. Seeds were distributed randomly around the parent plant at distances drawn from a negative exponential distribution. Growth was assumed to be a sigmoidal function of size, the same assumption used in the so-called gap models of forests (section 6.7.5).

The IBM describes perennial plants, using a one-year time step. Initially it was assumed that seedlings cannot emerge anywhere the local FON from neighboring plants is above zero. This assumption could represent extreme shade-intolerance of the seedlings or allelopathy. The IBM produced fluctuations of population abundance between 550 and 900 individuals in its

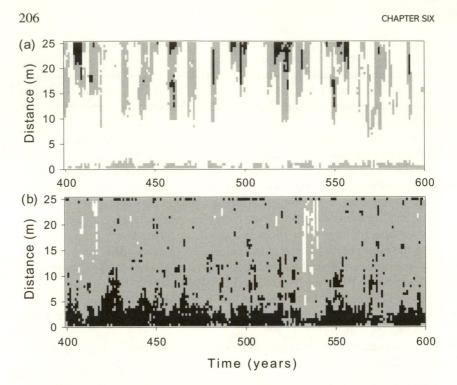

Figure 6.17 Patterns of spatial distribution in the perennial plant IBM of Bauer et al. (2002). Ripley's L-function was used to classify distributions as regular (*black*), random (*gray*), or clumped (*white*). For each year (x-axis), the classification was applied at increasing spatial scales (y-axis). Different distribution patterns occur for small individuals (a) and large individuals (b). (After Bauer et al. 2002.)

fifty-square-meter area. Autocorrelation analysis found significant, long-period cycles in abundance. The population's age structure also changed over the cycle: at peak abundance the population was dominated much more by young individuals than at minimum abundance. Analysis of the individuals' spatial distribution revealed that larger individuals were regularly distributed at smaller scales—distances of up to four to five meters (figure 6.17). At larger scales, the distribution of large individuals was random. In contrast, smaller individuals tended to occur in clumps. Robustness analysis (section 9.7) found the IBM's cycles to be robust to parameter values and most model assumptions. However, the abundance cycles disappeared when seedlings were allowed to establish where the total FON was less than 0.5 instead of only where FON is 0.

Bauer et al. (2002) concluded that the cycles are generated by both *self-thinning* and *monopolization of space* by older and larger individuals. Existing plants monopolize space by preventing seedling establishment wherever FON is greater than zero. Seedlings can only appear in gaps created by the death of older plants and therefore have a clumped distribution. After a clump of seedlings is established and starts to grow, competition causes most of these individuals to die; this self-thinning (discussed further in the following example) results in the gap again being dominated by one or a few individuals. The importance of self-thinning is indicated by the regular distribution of larger individuals: self-thinning is known to generate regular distributions (Leps and Kindlmann 1987).

This understanding of the cycles and their cause leaves several interesting questions unexplored. For example, would cycles also occur in a larger system? Would the cycles persist if seeds were dispersed over greater distances? So far, only one field study confirming the relation of cycles to space monopolization and self-thinning has been found: Tilman and Wedin (1991) describe cycles in a population of perennials in which leaf litter prevents establishment of young plants. The same phenomenon of monopolization of space leading to cycles— though for immobile animals (corals)—has been discussed theoretically by Iwasa and Roughgarden (1986) and Roughgarden and Iwasa (1986).

6.7.3.2 Self-thinning in Monocultures

Self-thinning can be quantified by the relation between the density d and average biomass $\bar{w}$ of individuals as a population ages. For monocultures of higher plants, self-thinning relationships show a striking pattern: in diagrams of $\log(\bar{w})$ versus $\log(d)$ plotted over time as the cohort ages, there is a long linear section with a slope of about $-3/2$. This pattern appears to be independent of the species and initial density of the cohort (Yoda et al. 1963; Harper 1977; Westoby 1984; Silvertown 1992). A linear relationship in a log-log diagram means that the relationship between density and average biomass follows a power law:

$$\bar{w} = Cd^{-3/2}.$$

Power laws indicate a scale-invariant or "self-similar" process: no matter what density we start at along the linear section of the self-thinning trajectory, a decrease in density of, for example, 10 percent coincides with the same increase in average biomass of 17 percent. Thus, along the linear section of the trajectory local competition, mortality, and growth affect each other the same way and independently of the average size or density of the plants.

Early theoretical explanations of the "$-3/2$ self-thinning rule" were based on geometric considerations: the area occupied by a plant scales with r^2, with r being the radius of the circular projection of the plant, whereas the individual's biomass scales with r^3 (Yoda et al. 1963). Later, this and all other explanations of the self-thinning rule were hotly debated and the existence of the pattern itself was questioned (Lonsdale 1990). Currently, allometric theories favor a power law exponent of $-4/3$ instead of $-3/2$; however, the allometric theory does not try to explain the cohort's trajectory but instead predicts the upper limit of biomass in mixed stands of given densities (Enquist, Brown, and West 1998).

The shift from $-3/2$ to $-4/3$ is an interesting instance of how theory filters our perception of data (Fagerström 1987): the empirical evidence supporting the theory of Enquist et al. is impressive, but the empirical evidence supporting the $-3/2$ power law is no less impressive and was presented in virtually every ecology textbook for decades.

Considering the obsession of plant ecologists with self-thinning, it is surprising that so few attempts have been made to understand it mechanistically—to simply simulate the process of self-thinning. A mechanistic analysis could help identify the conditions under which the logarithmic biomass-density trajectory has a linear section and determine how the slope of this linear section depends on biological processes (Li, Wu, and Zou 2000). Previous attempts to model self-thinning with IBMs include Firbank and Watkinson (1985); Adler (1996); Li, Wu, and Zou (2000); Weiner et al. (2001); and Stoll et al. (2002). The FON approach can also be used because it includes a simple mortality rule. Berger and Hildenbrandt (2003) used their mangrove IBM to simulate biomass-density trajectories for two different initial numbers of plants (500 and 1,000) and different values of the parameter b (equation 6.4; b ranging between 0.4 and 1.0). Smaller values of b make the radius R of the ZOI grow faster with the radius r of the stem so that for the same stem sizes and spatial configuration of plants competition strength is higher. Berger and Hildenbrandt found two things: first, the slope of the linear segment of the trajectory depends on b: the stronger the overall competition strength (i.e., the smaller b), the larger the slope; for weak competition strength (b larger than, say, 0.8), a linear segment hardly emerges. Second, Berger and Hildenbrandt did what is possible only with IBMs: they "looked into" the population during self-thinning. They observed the skewness of the size distribution of the plants and found that—independent of initial density and b—the linear segment starts when the skewness of the size distribution is maximum and ends when skewness becomes zero. The linear segment thus occurs over a period when the size distribution changes from positively skewed to symmetric. Both of these findings of Berger and Hildenbrandt clearly indicate that understanding local competition among neighboring trees, not just allometry, is key to understanding the extent and slope of the linear biomass-density trajectory.

Another use of IBMs for understanding self-thinning mechanistically is explaining a phenomenon that is simply assumed, not explained, by the aggregated theories of self-thinning. This assumption is that during self-thinning space is more or less evenly divided among the existing individuals so that the average area A occupied by a plant is inversely related to the number of plants, N, or: $A \propto N^{-1}$. This assumption was made explictly by Yoda et al. (1963) and more implicitly by Enquist, Brown, and West (1998) and Enquist and Niklas (2001). Westoby (1984) suggests that the relationship $A \propto N^{-1}$ occupies a central place in our understanding of ecosystems, and Zeide (1987, 2001) claims that it constitutes the "core" of ecology because it links two main branches, production ecology and population ecology.

What mechanisms might enforce $A \propto N^{-1}$ as a cohort of plants proceeds along the linear segment of the biomass-density trajectory? In IBMs based on the FON or similar approaches, $A \propto N^{-1}$ is not assumed a priori; instead, we can see whether and how it emerges from local competition, growth, and mortality (figure 6.18; Berger and Hildenbrandt 2000, fig. 3). It turns out that in the FON-based mangrove IBM of Berger and Hildenbrandt (2000), the emergence of $A \propto N^{-1}$ is extremely robust to changes in model parameters and structure (H. Hildenbrandt, personal communication). This density-area relationship thus appears to be valid and indeed decisive for explaining the robust self-thinning patterns in plant cohorts.

6.7.4 Grid-based Plant IBMs

Now we leave behind the distance models and look at some other ways IBMs have represented plant interaction. In grid-based models, space is discretized into uniform cells (usually square, but sometimes triangular or hexagonal). Spatial effects within the cells are ignored—for example, interaction among individuals in the same cell does not depend on their location within that cell—but spatial interactions are modeled as occurring among cells: the state of a cell is influenced by the state of its neighbor cells. This basic design of grid-based models (section 7.3; Czárán 1998; Wissel 2000) is widely used in ecology and other disciplines, not just for IBMs.

In grid-based plant IBMs, cell size is often chosen to approximate the average or maximum size of an adult plant. Within a cell, one or more individuals may exist, but their location within the cell is not considered explicitly. Moreover, interactions among individuals within a cell are not explicitly represented, and only their average result considered. For example, in grid-based models of single-species Pacific forests, Jeltsch and Wissel (1994) assumed that each cell is occupied by one adult individual. In reality, if this individual dies the opened space will be occupied by a cohort of seedlings, which then grow and self-thin until again the cell is occupied by only one adult. While the IBMs based on the FON approach (section 6.7.3) represent recruitment and self-thinning explicitly, Jeltsch and Wissel ignored these processes and simply assumed that after a tree dies it is replaced sometime later by another tree of the same size.

The grid-based approach has several advantages: the state of small spatial cells is easily described, and changes in the state of a cell's plant(s) in response to neighboring cells are easily described using "if-then"-rules (section 7.4). Grid models have an important computational advantage: while identifying a plant's neighbors is a difficult problem for distance models (section 6.7.3), neighbors in a grid IBM are easily identified by looking one cell (or several cells) in each direction on the grid.

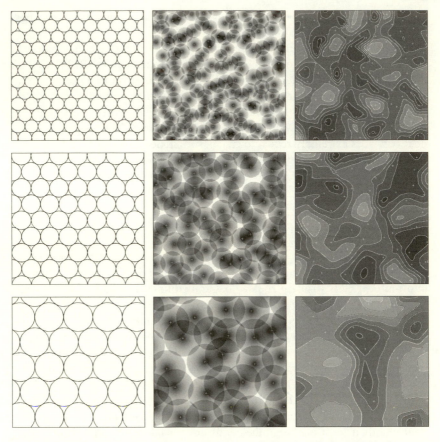

Figure 6.18 Visualization of the areas influenced or occupied by plants during self-thinning, comparing classical assumptions with the FON-based model KiWi. Time and self-thinning proceed from top to bottom. The left column illustrates the classical assumption that the average area of a plant is proportional to the inverse of the number of plants. The center column displays the location and field of neighborhood of trees as simulated by KiWi. The right column is a contour plot of competition pressure from the KiWi simulation. The value of competition pressure is assigned to the stem location of each tree and interpolated among trees; darker shading indicates stronger pressure. (Figure courtesy of H. Hildenbrandt.)

Many grid-based plant IBMs describe individuals in a very coarse way, often considering only one or two state variables: location (grid cell) and perhaps size. The plant's life cycle is often ignored because the assumption that one cell holds only one individual makes no sense for small seedlings. Even these coarse descriptions of individuals can be sufficient for modeling some problems because many plants are close to full-sized for most of their life-span. Other IBMs have treated grid cells as units similar to collectives: processes such as reproduction and recruitment of new plants are modeled as internal functions of the grid cell, not as behaviors of individual plants; yet full-sized plants are treated as individuals. Useful examples of grid-based plant IBMs, which in fact are only partly individual-based, include the beech forest model BEFORE (sections 1.2 and 6.8.3); a model of a shrub community in a semi-arid region (Wiegand, Milton, and Wissel 1995); a model explaining large-scale, stand-level diebacks in single-species forests of the Pacific region (Jeltsch 1992; Jeltsch and Wissel 1994); a savanna model addressing the coexistence of trees and grass (Jeltsch et al. 1996; Jeltsch, Milton, et al. 1997a); an IBM of competition between two tree species along an environmental gradient (Groeneveld et al. 2002); a theoretical model of the coexistence among annual plant populations (Silvertown et al. 1992); and further examples described by Czárán (1998).

6.7.4.1 The Winkler and Stöcklin Model

A grid-based representation of space can be combined with a truly individual-based approach to modeling plants. In such models, a grid cell no longer represents the size of an adult individual. For example, Winkler and Stöcklin (2002) developed an IBM of the perennial herb, *Hieracium pilosella*. The IBM addressed the question of how sexual and vegetative reproduction, competition with a grass species, and disturbance—mainly by cattle trampling—influence the spatial distribution of *H. pilosella* along an environmental gradient that favors the grass species on one side and *H. pilosella* on the other side (figure 6.19). The model area was 200×50 square centimeters. One grid cell of 1 square centimeter represents the initial size of an individual *H. pilosella* rosette, but larger individuals are represented by clusters of grid cells. Every year, a rosette's diameter increases by a certain amount. Larger individuals (diameter of 3–5 centimeters) will, with a probability depending on their size, reproduce sexually and also vegetatively by producing stolons with a juvenile rosette at the apex. The competing grass was modeled using a similar description of tussock growth, but was assumed to reproduce only vegetatively.

Winkler and Stöcklin analyzed the IBM by comparing its predicted spatial patterns of *H. pilosella* to corresponding patterns observed in real populations. This analysis indicated that two additional processes were needed in the model: long-distance seed dispersal and facilitation of seedling establishment in the vicinity of grass tussocks (figure 6.20). Interestingly, one of the few adaptive traits that we found in the plant IBM literature turned out to be decisive for the maintenance of *H. pilosella* populations: phenotypic plasticity of stolon length. The IBM assumes that when a stolon reaches a grid cell that is already occupied it can, with a certain probability, search for unoccupied sites in up to four additional cells. Without this adaptive behavior of the stolons, the population was not able to persist in the presence of disturbances. Winkler and Stöcklin also concluded that a mixture of vegetative and sexual reproduction is necessary

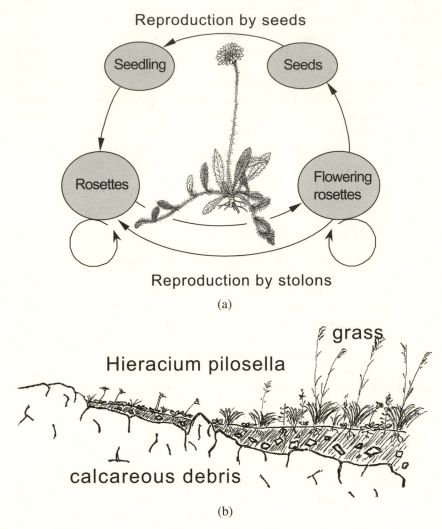

Reproduction by seeds

Seedling

Seeds

Rosettes

Flowering rosettes

Reproduction by stolons

(a)

grass

Hieracium pilosella

calcareous debris

(b)

Figure 6.19 (a) Life cycle of *Hieracium pilosella* with four stages. (b) Schematic profile of the distribution of *H. pilosella* and competing grass in a gradient of soil quality in calcareous grassland. (Modified after Winkler and Stöcklin 2002.)

for this species to maintain populations in the presence of high interspecific competition and a shortage of open space.

6.7.5 Individual-based Forest Models

Forest IBMs are certainly the most important and successful class of plant IBMs, and of IBMs in general. There are hundreds of forest models and many

(a)

(b)

Figure 6.20 Vegetation pattern results from the grid-based IBM of Winkler and Stöcklin (2002). Fictitious *Hieracium* species with only (a) sexual or (b) vegetative reproduction are simulated. Soil fertility (increasing from left to right) affects *Hieracium* seedling establishment and grass regeneration. *Hieracium* rosettes are shown in dark gray with black spots indicating their origin; grass patches are light gray. (From Winkler and Stöcklin 2002.)

are individual-based (Liu and Ashton 1995). The manpower invested in forest models reflects the enormous importance of forests, both from the economic (timber production and other forest services) and ecological (conservation of natural biodiversity) points of view.

In reviewing forest IBMs, Liu and Ashton (1995) distinguished the two main classes we examine: gap and growth-yield models. (Porté and Bartelink 2002 provide a general classification of forest models.) Forest IBMs that do not fall in either of these classes include the mangrove forest model KiWi (Berger and Hildenbrandt 2000) and so-called process models, which consider physiological processes such as assimilation and respiration (e.g., TREEDYN3 by Bossel 1996; FORMIND by Köhler and Huth 1998; for a critical discussion of process models see Zeide 2001).

6.7.5.1 Gap Models

The purpose of gap models is to understand long-term forest dynamics, in particular species composition and succession in relationship to environmental variables. Gap models are thus developed mainly by ecologists. Gap models consider the gaps created by death of canopy trees (Botkin, Janak, and Wallis 1972; Shugart 1984; Botkin 1993). Typical gap sizes used in early gap models are 0.01 hectare. Seedlings of the tree species of interest are then released in this gap. Each individual tree is characterized by its size, most often as trunk diameter at breast height. Empirical relationships are used to calculate the tree's height and biomass from its diameter.

Each individual tree is assumed to have a sigmoidal potential growth curve. The potential annual growth increment is reduced by "multipliers" that reflect the influence of competition and environmental factors. The multipliers range between zero and one, with one indicating no reduction in growth and zero indicating that no growth is possible. Competition is considered only vertically, representing competition for light. The heights of the trees in the gap determine the vertical profile of a leaf area index (representing the density of leaves over vertical distance), and this index determines the light absorption and, therefore, the amount of light that reaches each layer in the gap. Trees shaded by larger trees are thus reduced in their growth. Mortality often is assumed to depend on the growth rate: the tree is assumed to die when growth falls below a critical rate for a specified length of time.

The pioneering gap model JABOWA (Botkin, Janak, and Wallis 1972) and many of its descendants include no horizontal spatial effects: the position of a tree within a gap is not considered and the entire forest consists of gaps that do not interact with each other. However, spatial relationships have been introduced in more recent IBMs by merging gap models with the grid-based approach. For example, the model ZELIG (T. M. Smith and Urban 1988) uses ten-meter grid cells and assumes interaction among adjacent cells via shading and seed dispersal.

Liu and Ashton (1995) present a genealogical tree of nineteen gap models, but the basic design of growth, vertical competition, and mortality is quite similar in most. The great success of gap models probably has three main reasons. First, the model design is conceptually (and computationally) very simple; each individual is described by the same small number of equations, with species differing only in their parameter values and individuals differing only by their one state variable, diameter at breast height. Second, the sigmoidal growth equations are relatively easy to parameterize (A. Huth, personal communication). Third, gap models make important testable predictions: they mimic the species composition and dynamics of real forests so at least these outputs can be tested and validated (Shugart 1984).

6.7.5.2 Growth-yield Models

The second class of forest IBMs distinguished by Liu and Ashton (1995) are growth-yield models (e.g., Ek and Monserud 1974; Zeide 1989; Pretzsch, Biber, and Darsky 2002). Growth-yield models vary more in structure than gap models do (Liu and Ashton 1995), but there are several consistent ways that they differ from gap models. Growth-yield models are developed by foresters to help manage timber production, so they usually address much shorter time scales than gap models (e.g., one or several decades) but much larger areas. Also because of their management purpose, gap models represent the effect of biotic (e.g., stand structure and species composition) and abiotic environmental conditions (e.g., light and moisture availability) on individual growth. For example, most growth-yield IBMs (unlike gap models) represent the explicit spatial position of trees so that distance-dependent competition among trees can be simulated. The effects of environmental conditions are represented using empirical regression functions, so these models require much more data to parameterize than do gap models.

Growth-yield IBMs are highly empirical and lack adaptive individual behavior, but they nevertheless seem to capture some essential processes and structures of forests—especially, the horizontal structure of a stand—better than gap models do. Liu and Ashton (1995) therefore proposed development of hybrid IBMs that combine the advantages of gap and growth-yield models. The forest IBM SORTIE (Pacala, Canham, and Silander 1993) is such a hybrid: although a descendant of JABOWA, it considers the explicit spatial location of trees and the effect of one critical environmental condition—light availability—so SORTIE also shares characteristics of growth-yield forest IBMs (more about SORTIE is in section 11.5.2).

6.7.6   Summary and Lessons

Plant IBMs have a longer history (since at least 1964; Newnham 1964), are more numerous, and are almost certainly more widely used in applied ecology than animal IBMs (Liu and Ashton 1995). Yet we discuss example plant IBMs in only this one theme because plant IBMs are generally simpler and more similar to each other than animal IBMs are. To summarize the common characteristics of most plant IBMs, we apply the Conceptual Design Checklist of section 5.13 in the same way we applied it to animal IBMs in sections 6.2.5 and 6.3.5. Then we finally discuss new directions in botany and their implications for plant IBMs.

Emergence.   The plant IBMs are designed so that population-level outcomes of interest—age and size distributions, production rates, species diversity, spatial patterns—emerge from individual-level processes, especially competition among individuals. Emergence is used in part because the individual-level processes are believed to be important, but also (a primary concern for forest management models) because empirical models of individual growth—the most important process—are relatively easy to parameterize.

Adaptative Traits and Behavior.   Adaptive traits are almost completely absent from the plant IBMs we reviewed. The most essential process of plant IBMs is the effect of competition (and, sometimes, environment) on growth; but plants are not assumed to have behavioral responses to competition, such as choosing what parts of the plant to grow how much in which directions (to be discussed further). Instead, competition is assumed simply to impose a reduction in growth. The adaptive choice of growth direction in the IBM of Winkler and Stöcklin (2002) is an exception to this generalization.

With the lack of adaptive traits, the concepts of *fitness* and *prediction* are also unused in plant IBMs.

Interaction.   Interaction is a key concept of most plant IBMs: competitive interaction with neighbors is the most important (and, sometimes, only) biological process represented. Consequently, how interaction is represented is the characteristic distinguishing whole families of plant IBMs, including FRN, ZOI, and FON. Unlike most animal IBMs, plant IBMs represent interaction directly: each plant identifies the neighbors it interacts with and the effect of each such neighbor is represented. However, the mechanisms of interaction are highly simplified; with some exceptions, the details of how neighboring plants compete for light, moisture, nutrients, and space are ignored, and competitive interaction is instead modeled simply as a function of distance or spatial overlap.

When a plant has two or more neighbors, their effects are treated as an interaction field: the plant is affected simply by the sum of effects from all neighbors.

Sensing.   The plant IBMs we examined do not explicitly represent sensing because they do not assume that plants recognize and respond to the presence of neighbors. Instead, competitive interaction with neighbors is assumed simply to reduce the availability of resources that limit growth.

Stochasticity.   Stochasticity is used in plant IBMs in ways that are also common in animal IBMs: to initialize the model (plants are initially given random locations) and to represent processes like seed dispersal and mortality that are highly variable yet driven in part by processes (e.g., wind) too short-term or too complex to include in the IBM. Plant growth, the most important process in these IBMs, is modeled without stochasticity.

Collectives.   While many plants do form colonies or other aggregations that resemble collectives, none of the IBMs we examined represents collectives. Some grid models use a technique resembling a collective: early life stages are represented simply as characteristics of the grid cell. The cell may have variables representing the number and species of seeds, seedlings, or saplings; and assumptions modeling how these plants die and grow into the next life stage. While this aggregation of juvenile plants resembles how collectives can be explicitly represented in an IBM, it is simply a modeling technique and does not represent real collectives.

Scheduling.   The plant IBMs represent time using discrete time steps, usually at least one year and sometimes more than a decade in length. It is tempting to think that scheduling decisions are less important for plant IBMs because the processes represented in them are slow. However, scheduling is just as important when long time steps are used to represent slow processes as when short time steps represent fast processes. There are potentially important effects of how model actions are scheduled. Growth of each plant depends on the size of its neighbors, so how is the size of each plant updated with respect to its neighbors? When mortality is simulated, it is usually a function of growth rate; is growth updated before or after mortality risk is? Unfortunately, such scheduling assumptions have often been undocumented.

Observation. Spatial processes are important in all plant IBMs, so the ability to observe individual plants over space and time certainly helps test and understand the models. Models with graphical output of plant size and location over time (e.g., Pacala, Canham, and Silander 1993; Huth, Ditzer, and Bossel 1998; Köhler and Huth 1998; Savage, Sawhill, and Askenazi 2000; Berger and Hildenbrandt 2000; Rademacher et al. 2004) illustrate its value for testing assumptions about how individuals interact and understanding system dynamics. Many plant IBMs were implemented without such observer capabilities, undoubtedly due in part to the technologies available at the time the IBMs were built. However, many of these IBMs also address problems that allow them to be tested and analyzed adequately from aggregated, nonspatial outputs such as timber production rates, self-thinning relations, and frequency distributions of size or species.

Future Directions. One generalization about plant IBMs that stands out starkly from the checklist is the lack of adaptive traits—plants are rarely assumed to make decisions allowing them to respond to changes in themselves or their situation. Why are adaptive traits so rare in plant IBMs while they are so important in animal IBMs? One answer is that the adaptive behaviors most widely addressed in animal models—dispersal and movement—are not available to plants once seeds are established. Another answer lies with the kinds of problems addressed by plant IBMs, which tend to be large-scale and long-term. Apparently, these problems can be addressed with some success using IBMs that only coarsely represent the most important processes and neglect adaptive behavior.

As these IBMs were being developed, however, plant physiologists have been discovering many important and fascinating adaptive behaviors (or "responses," the term used in botany) and the detailed mechanisms—often, chemical signaling—plants use for sensing and interacting (e.g., Cosgrove et al. 2000). Whereas the IBMs we examined simply assume that proximity of other plants reduces growth, it is now known that plants can sense neighbors even before shading occurs by detecting changes in the light spectrum and can adapt to high densities of neighbors via mechanisms such as allocating more growth to stems and less to roots (Schmitt, McCormac, and Smith 1995) or adapting their above-ground shape (Umeki 1997). It is also now known that plants defend themselves against herbivores and pathogens with an array of active responses, not just latent resistance; upon attack, plants release chemicals specific to the type of damage and to their own state. For example, plants can respond to insect herbivores by releasing chemicals that attack the insects directly by interfering with feeding, growth, and ability to reproduce; and indirectly by attracting the insects' predators and parasites (Walling 2000). And in some systems, allelopathy ("chemical warfare") is considered potentially as important as passive competition in regulating spatial patterns and population dynamics.

Learning how important these individual adaptive traits are to the population dynamics of plants is obviously an important research goal for IBE, and in fact some progress has already been made. The Winkler and Stöcklin (2002) IBM (section 6.7.4) showed that adaptive behavior can be critical to the kinds of problems plant IBMs often address. Umeki (1997) is one of several studies of how the predictions of forest IBMs could be affected by adaptive behavior. Umeki assumed that individuals could adapt to neighbors that block light by growing partly sideways toward light instead of only straight up; this behavior strongly affected the predicted density and size distribution of trees.

A second, related, generalization we can make about the plant IBMs is that the full IBE theory development cycle (chapter 4) has not yet been widely applied. At the individual level, IBMs have typically assumed models for individual behavior, often parameterizing them with field observations but rarely contrasting alternative theories for individual traits to see which best explains observed higher-level patterns. At the population or community level, plant IBMs have typically addressed only one or two particular patterns, rarely a wide variety of patterns or long-term dynamics. Much more could be learned by testing alternative theories for individual plant behavior by how well they reproduce a wide variety of observed higher-level patterns. These analyses should start with simple approaches such as ZOI and FON, but eventually should also examine more mechanistic models of interaction and adaptive individual behaviors.

6.8   STRUCTURE OF COMMUNITIES AND ECOSYSTEMS

Most IBMs deal with populations of one species. This is natural because the motivation behind early IBMs, both pragmatic and paradigmatic (chapter 1), was to overcome conceptual and technical limitations of classical population models. It also seems wise not to aim too high at the beginning: we should learn how to use IBMs for populations first before tackling communities and ecosystems. Some have even argued that IBMs are unlikely to be useful for more than one species because even single-species IBMs are difficult to develop, analyze, and understand—difficulties that could multiply or even increase exponentially as more species and interspecies interactions are added.

One of the ultimate goals of IBE, however, is to understand how traits and behavior of individuals affect the structure and dynamics of ecological systems in general, including communities and ecosystems. Werner and Peacor (2003) reviewed the empirical evidence and concluded that adaptive traits of individuals can have effects on communities at least as strong as the density effects that ecologists traditionally focus on. By "communities" we refer to assemblages of several species that live in the same environment and interact with each other directly or indirectly, for example, via competition, predation, or facilitation. By "ecosystems" we refer to assemblages similar to communities, but with abiotic factors and environmental heterogeneities addressed more explicitly. (We thus do not address the abstract view of ecosystems as compartments and flows of nutrients and energy [e.g., Odum 1971], although such ecosystem elements could be included in IBMs.)

One strategy to achieve this ultimate goal of IBE is to apply relations and understanding developed with population-level IBMs into more aggregated community- and ecosystem-level models (Schmitz 2001). Unfortunately, these aggregated models (e.g., community matrix models) tend, for the sake of analytical tractability, to be extremely artificial so it is not clear how the complex dynamics emerging in population IBMs could be included even indirectly. Another strategy is to follow the general and pattern-oriented modeling guidance of chapters 2 and 3 to design community and ecosystem IBMs that are structurally realistic but still tractable. This guidance tells us that when building an IBM to study interactions among species, each species is likely to be modeled in a different, usually less detailed, way than when modeling problems that address only one species.

In this theme we present two community IBMs and one ecosystem IBM. These IBMs are quite different from each other but show that new and important insights can be gained by applying individual-based approaches directly to levels of organization higher than populations. Other community and ecosystem IBMs include the theoretical plant succession model of Bartha and Czárán (1989); individual-based forest models, which usually include several—up to more than 100—species (Liu and Ashton 1995; section 6.7.5); the grid-based model of a semiarid shrub community of Wiegand, Milton, and Wissel (1995); the grid-based savanna model of Jeltsch et al. (1996, 1997); the plant community model of T. M. Smith and Huston (1989); and the fish community model of Shin and Cury (2001).

6.8.1   Adaptive Traits in the Community IBM of Schmitz

The IBM of Schmitz (2000) addresses the structure of an early-successional old field community. The community contains twenty-five different grass and herb species; five of these species, all perennials, provide more than 90 percent of the total vegetation biomass. The dominant herbivore is a grasshopper species, and the major predators on grasshoppers are three different species of hunting spiders. Field experiments showed that the spiders exerted a strong influence on plants via direct interactions with grasshoppers, by reducing grasshopper density and by causing grasshoppers to change their foraging behavior to avoid predation. In experiments with the spiders prevented from actually killing their prey (by gluing the spiders' mouth parts), spiders still influenced plant biomass even without reducing grasshopper density.

Schmitz concluded that in this system indirect effects of the spiders on the plants arose largely from the grasshoppers changing their behavior in response to the spiders, not just from spiders reducing grasshopper abundance. Without spiders, grasshoppers preferred eating grass, probably because it is more nutritious. In the presence of spiders, grasshoppers feed more on herbs, which are structurally more complex so grasshoppers are harder for spiders to detect and capture. The effect of spiders altering grasshopper behavior could clearly be demonstrated in experiments lasting one season, but the long-term consequences of this effect on community structure and dynamics were not clear. Therefore, Schmitz developed a community IBM describing two types of plants (preferred grass and safe herbs), herbivores (grasshoppers), and predators (spiders).

Unlike most of the IBMs examined in this chapter, this community IBM was not designed or implemented in software from scratch. Instead, it was implemented using Gecko, a generic platform for IBMs developed by Booth (1997). (Gecko was in turn implemented in the Swarm software platform; see section 8.4.3.) The way individuals are represented in Gecko is similar to the ZOI and FON approach used for plant IBMs (sections 6.7.2, 6.7.3): individuals are represented as spheres that project as circles onto a plane. The radius of an individual defines its world: for two individuals to interact, their spheres must overlap. Interactions and production of resources used by the individuals take place in the plane. However, the third dimension—the volume of the spheres—is used to determine the biomass of the individuals from allometric relationships. Mobile animals move in the plane; sessile individuals stay at the location where they were released, but via growth they increase the area over which they influence, or are influenced by, other individuals.

Gecko's representation of individuals as spheres may seem awkward (to those unfamiliar with spherical cows; Harte 1988), but it is important to understand that the spheres do not represent individuals physically but their "zone of influence." This generic model of an individual makes Gecko especially useful for modeling different kinds of species and, in particular, for modeling multiple species (i.e., communities) simultaneously. To create his community IBM, Schmitz (2000) had only to translate the properties of his species and their traits into the "language" of Gecko (Figure 6.21). We can see from the following summary of the model's assumptions that it is not crude or simplistic compared with other ecological models or even other IBMs.

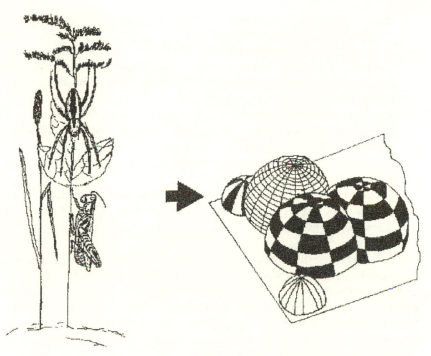

Figure 6.21 Translation of a real community into Gecko, by Schmitz (2000). The community has two plant species, a grasshopper species, and a spider species that preys on grasshoppers. In Gecko (Booth 1997), individuals (or their zone of influence) are represented as spheres on a plane. (After Schmitz 2000.)

- Population dynamics of the four species emerge from individual survival and reproduction, which depend on growth and predation (spiders eat grasshoppers; grasshoppers eat plants).
- Resource consumption and competition is represented using direct interaction: a herbivore eats a specific plant, a predator kills a specific herbivore. Individuals gain resources at a rate proportional to the area their sphere projects on the plane. Plants gain resources over their entire area; herbivores gain resources over the area where their circle overlaps that of their plant resource. Overlap of individuals of the same species represents intraspecific competition.
- Large individuals consume resources faster than smaller individuals of the same species do, including asymmetric competition (section 6.7.2).
- Growth is modeled as the difference between resource intake and metabolic demands. Inadequate resource intake (e.g., due to competition or resource depletion) can have effects on individuals ranging from reduced growth to loss of size, reduced fecundity, and death.

- Animals move following a biased random walk, modified by additional rules describing sensing and interactions.
- Herbivores select the plant within their circle that yields the highest rate of resource intake (grass, if present in the circle). Within a time step (ten per day), "feeding" means to take one "bite" from the plant, which reduces the plant's size.
- Herbivores sense predators with a size-dependent detection radius. If they sense a predator within this radius, they exhibit an adaptive behavior: moving toward a "safe" plant (herb) in hopes of avoiding the predator.
- Reproduction is asexual and individuals reproduce only if they attain sufficient size.
- A season consists of 190 days. During the rest of the year, plants are dormant as roots and seeds, herbivores exist as dormant eggs, and predators overwinter as juveniles.

While calibrating his model Schmitz made two interesting observations regarding the significance of adaptive traits. When the adaptive trait for responding to predators was taken away from herbivores, their population never persisted beyond half of the first season. Similarly, if the herbivores had the adaptive trait but no "safe" plants existed, they went extinct by the middle of the second season. These experiments use a technique discussed in chapter 9: simulating unrealistic scenarios.

To test the community IBM, it was run for one season using input corresponding to the field experiments—including, for example, the experiment with spiders' mouths glued so they cannot actually kill grasshoppers. The results of these simulations, and the general within-season pattern of community structure, matched the empirical observations. Then, after calibration, the model was applied to the problem it was designed for: studying long-term effects of the predator on community dynamics and structure. Experiments simulating ten seasons contrasted community dynamics with or without the spider predators. Without predators, abundance of the plant species preferred by herbivores declined steeply, and it took about five seasons for the community to reach a new steady state (figure 6.22a). With predators, the density of both plant types remained almost constant at a high level very close to that observed in the one-season experiment and simulations (figure 6.22b and c).

Schmitz concluded that in this system within-generation processes are sufficient to predict long-term community dynamics because seasonality removed the serial dependence in population abundance over time: the system is "reset" each season. This finding has important implications about the significance of many ecological time series, which are derived from observations taken once per year and thus *ignore* within-generation processes. Schmitz (2000, p. 482) concluded that: "Ignoring the information value of within-season interactions may lead to the wrong interpretation about community dynamics and

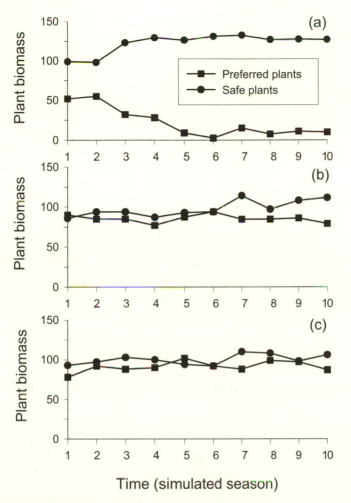

Figure 6.22 Long-term time series of biomass of preferred and safe plants in the community IBM of Schmitz (2000). Predators are either absent (a), present so that they affect herbivore behavior but without actually killing them (b), or present and preying upon the herbivores (c). (After Schmitz 2000.)

thus decrease our ability to elucidate causality." For example, analyses of an annual census time series (as performed by Schmitz on his annual time series of IBM output) may find no autocorrelation, implying the absence of density dependence in population dynamics. But such a conclusion could be misleading: strong *within-generation* density dependence may be the reason why no between-generation density dependence is detected. (Grimm and Uchmański 2002 were led to similar conclusions by another IBM; see section 6.5.2.)

The model of Schmitz (2000) is an important example, not only for what it teaches about a particular community but mainly because it illustrates a productive, individual-based way to study community dynamics. Community IBMs can use highly simplified representations of individuals, use generic platforms like Gecko to reduce further the design and implementation effort, and still produce important knowledge of how individual traits affect community dynamics. Finally, Schmitz's study illustrates an important part of IBE that we have seen few examples of in this chapter: field studies designed specifically to support model design and analysis. The elegant experiment that manipulated perceived predation risk versus actual predation rate by allowing spiders to stalk but not kill grasshoppers produced a pattern that was very important in conceptualizing and validating the model.

6.8.2   The Plant Community IBM of Pachepsky and Colleagues

The model of Pachepsky et al. (2001) was designed to explain general patterns in relative abundance distributions, which describe the distribution of individual abundance among species. Communities considered "in equilibrium" are expected to have a log-normal relative abundance distribution, with few species having either extremely high or low abundance but more species having low than high abundance. In contrast, disturbed communities considered out of equilibrium are expected to have geometric relative abundance distributions, with a few very abundant species and more rare species. Several theories based on considerations such as recruitment and metapopulation dynamics have been developed to explain these patterns; Pachepsky et al. wanted to explore how individual-level processes and variation affect the distributions.

The IBM of Pachepsky et al. describes an artificial community of plant species. Individuals have twelve physiological traits that govern how processes such as resource intake, internal allocation of resources, reproduction, and survival vary with the plant's developmental state. A species is generated by drawing a random value from the parameter probability distribution for each trait. A model run is initialized with one individual of each such artificial species, and when individuals reproduce they pass their traits on unmodified. (In some very general ways, this approach resembles the artificial evolution approach to modeling individual traits discussed in sections 6.9 and 7.5.) The model is grid-based: each plants occupies—exclusively—one grid cell, but the size of a plant's interaction neighborhood (a zone of influence; section 6.7.2) depends on its development stage. Competition for resources is represented as mediated interaction: if two plants use resources from the same grid cell, resources are divided in proportion to the plants' current resource uptake. Competition for space occurs at the seed stage: seeds are dispersed randomly within a limited distance but can only germinate on empty grid cells.

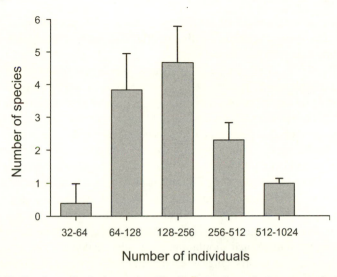

Figure 6.23 Abundance distribution produced by the Pachepsky et al. plant community IBM on a 30 × 30 grid. The values were averaged over the time, during one simulation, when fifteen species coexisted. Error bars show the standard deviation among time steps. (Modified after Pachepsky et al. 2001.)

The simulation experiments varied the size of the space between 100 and 2,500 grid cells, started with 75 randomly located individuals (and, therefore, 75 species), and ran for 50,000 time steps (1,250 generations of 40 times steps each). Typically, the number of surviving species decreased exponentially at the start of a run and stabilized after about 5,000 time steps. The number of species that persisted for long periods at the end of a run increased with system size in a manner consistent with observations from real communities that are disturbed or managed. During this "equilibrium" period, the abundance distribution was log-normal, as expected for stable communities (figure 6.23). Pacheptsky et al. reported that the distribution was, also as expected, more of the geometric form during the initial, nonequilibrium period.

To better understand their IBM, Pachepsky et al. used a technique discussed in section 9.4.4: reducing the model's complexity. They tested the importance of variability in the twelve physiological traits to the relative abundance distributions by, one at a time, assuming each trait does not vary among species. They found out that variability in only the traits affecting time to reproduction and fecundity is needed to reproduce the abundance distribution patterns of the full IBM. The species remaining during the equilibrium period had traits falling along a linear trade-off between time to reproduction and fecundity. The main conclusion of Pachepsky et al. (2001, p. 926) is thus "that the trade-off between time to reproduction and fecundity sustains the diversity in the community, and governs the form of the resulting abundance distribution." Further, observing that this trade-off is manifested at the individual level, Pachepsky et al. (p. 926) suggest that "replacing species with individuals as the fundamental ecological accounting unit" can be productive.

The Pachepsky et al. study illustrates how simple, very tractable IBMs can reproduce fundamental patterns at the community level and be used (along with clever analysis techniques such as reducing complexity) to explore the mechanisms from which the patterns emerge. But several unique aspects of how this model was designed and communicated leave interesting questions unanswered. The article was published in a journal (*Nature*) with severe space limitations, so we know unfortunately few details of the model design (we discuss strategies for this problem in section 10.3). The article also includes a differential equation version of the model that—in our opinion—was not the best use of the limited publication space (another communication issue we address: section 10.4.1). One unique characteristic of this IBM is merging the concepts of individual and species. At the start of each simulation, each individual was given unique traits that were passed on to offspring, so each initial individual essentially gives rise to a separate species. The important trade-off between time to reproduction and fecundity varies among individuals in the initial population, but with initial individuals equivalent to species it is not so clear whether the fundamental ecological unit in the IBM is the individual or the species.

This study shares a characteristic of other highly abstract IBMs designed to address questions of classical ecological theory (section 6.5.4): although these models and the research based on them provide important insights on classical theory, they leave us wondering how applicable the results are to real ecological systems. The study focused on one community-level pattern, the relative abundance distribution. Showing that the IBM reproduced this one pattern is important but not sufficient to give us high confidence that results are applicable to real systems. Perhaps the relative abundance distribution pattern would have been explained by other processes such as environmental variability or immigration if they had been in the IBM; in fact, this is such a general pattern that it seems possible that many different processes could explain it. The risk in using an abstract IBM is giving up the opportunity to build confidence in the IBM by showing that it reproduces a variety of observed patterns that capture essential behaviors of a real system and its individuals.

6.8.3   The Beech Forest Model BEFORE

Now we look at an ecosystem IBM that simulates only one species. A single-species IBM is extremely different from how many ecologists traditionally think of ecosystem models: simulating flows among pools of energy and nutrients. From the traditional ecosystem perspective, one might assume that an ecosystem IBM would include *all* species (and abiotic factors) of the system. But such an IBM would suffer from "naive realism" (section 2.1), the belief that a model needs to include everything known about a system to be "realistic." But community and ecosystem models, like all others, need to be designed for specific questions; and for some systems and some problems, the best model may describe only a few, or even only one, species. The forest IBMs described in section 6.7 are examples: they were designed to address questions about species composition in tree communities; other plants and all animals were ignored. Therefore, questions regarding animal-plant interactions or invasions of exotic herbs cannot be answered with these models. But these forest IBMs still capture much of the essence of their ecosystems simply because the most important elements of forests are, after all, trees.

The ecosystem IBM we examine here describes the spatial population dynamics of beech (*Fagus silvatica*). Nevertheless, the model describes important aspects of an entire ecosystem because in Central Europe natural beech forests would be almost entirely dominated by this one species of tree. Beech tolerates a wide range of environmental conditions and casts heavy shade while tolerating shade well itself, so it can exclude other tree species. Beech forests are ecosystems for which we can represent most of the energy and nutrients (except in soil) by modeling a single species.

Before modification by people, Central Europe was actually dominated by single-species beech forests. Now, except for a few remnants in Bohemia and the Balkans, there are no natural beech forests. Even the oldest reserves are rather small and at most 130 years old, so they still reflect their history of being managed. But what would natural beech forest ecosystems look like, and what processes would drive their spatial and temporal dynamics? What are good indicators of the naturalness of a beech forest ecosystem? How large should beech forest reserves be to allow natural processes and dynamics to emerge?

To answer these questions, the model BEFORE was developed (Neuert 1999; Neuert et al. 2001; Rademacher et al. 2001; Rademacher et al. 2004). Because this problem addresses a specific ecosystem and requires predicting system behavior under conditions never observed, a primary modeling concern was building confidence in the IBM's ability to capture essential processes and dynamics of beech forests. From chapter 3 we know that building this confidence means reproducing a variety of patterns observed in modern beech forests—especially vertical and spatial (horizontal) patterns in tree size and density. Perhaps one of the existing forest IBMs could have been modified to address the questions BEFORE was designed for, but no existing IBMs were explicitly designed to reflect the vertical and horizontal structure of a forest and its dynamics for thousands of years over very large areas.

The structure (state variables) of BEFORE was designed so that important observed patterns could emerge in the model. An especially important pattern is that natural beech forests consist of a mosaic of small patches that can be assigned to three different developmental stages. These stages are characterized by vertical structure. The "optimal" stage has a closed canopy and almost no understory, whereas the "growing-up" and "decaying" stages have understory and gaps in the canopy. This pattern can only emerge if a model has both a small spatial resolution and explicit representation of different canopy layers. BEFORE is therefore grid-based in all three dimensions. Horizontal space is divided into cells approximating the crown area of a very large beech (about 0.02 hectares); this cell size is much smaller than the typical area (0.1–2 hectares) of the developmental stage patches observed in beech forests. Vertically, beech are grouped into four height classes (figure 1.2): seedlings, juvenile trees, and lower- and upper-canopy trees.

The seedling and juvenile height classes are not represented as individuals but instead simply as the fraction of cell area covered by plants of that class. The lower- and upper-canopy trees are modeled individually, but the position of the trees within the cell is not considered. Up to eight lower-canopy individuals may be present in a cell, and the crown size of each individual is fixed at one-eighth of the cell area. Similarly, up to eight upper-canopy individuals can exist in a cell, but the crown size of each can grow from one-eighth up to eight-eighths of cell area. However, the total crown area of all trees in the upper canopy of one cell cannot exceed the cell area. Competition occurs if the upper canopy is completely covered by tree crown; in this case trees with a larger crown size can grow at the expense of trees with a smaller one. This asymmetric competition can even lead to the death of smaller individuals.

The processes driving dynamics of the model forest are growth and mortality. Growth is driven by the light reaching individual trees and mortality is driven by growth and by storms that cause windfall. (Because these forests are dominated by large beech trees, recruitment of seedlings was assumed to be a far less important process than light competition and was represented in a simple, stochastic way.) The light reaching trees of the three lower height classes depends on the cover in the higher height classes in both the same cell and in neighboring cells; the positive effect of indirect and oblique light from canopy gaps in neighbor cells is taken into account. The relative amount of light reaching each height class determines its growth and mortality rates.

BEFORE includes one environmental process—windfall—because it appears to have strong effects on the ecosystem. Simulated storms have different wind directions and three different strengths: "normal" storms occur in 89 percent of time steps (which represent fifteen years), 10 percent of time steps include "strong" events, and "extreme" events occur in 1 percent of time steps. Storm effects are modeled at the individual level: each tree has a certain probability of being wind-thrown. This probability is sharply increased, however, if there are gaps in the upper canopy of either upwind or downwind cells; upwind gaps increase the wind a tree is exposed to and downwind gaps reduce the support provided by neighboring trees. Wind-thrown trees are assumed to damage the upper and lower canopy of the neighbor cells into which they fall; the damage affects a row of up to three cells. The IBM's scheduling of wind damage

(a)

Figure 6.24 Damage in the upper canopy (see figure 1.2) caused by (a) "normal" and (b) "extreme" storm events in the beech forest model BEFORE (Rademacher et al. 2001). The figure shows only 15 × 15 cells (4.7 hectares) of a 54 × 54 cell (64 hectares) model forest. Prior to the storm, the forest had a typical mosaic of small patches (averaging 0.3 hectares) in different developmental stages. The damage ranges from opening the entire cell as a gap (*black*) to no damage at all (*white*). The storm wind was from left to right. (From Rademacher et al. 2004.)

was carefully designed so that natural patterns such as aisles of wind-damaged trees can be reproduced (figure 6.24): when a simulated storm is executed, the model action simulating damage in each cell is executed on cells in order from upstream to downstream. In this way, the risk of wind damage to trees in a cell is affected by damage occurring upwind in the same storm; a gap opened in one cell increases the probability of windfall in the downwind neighbor cells, making chains of damage possible.

Overall, we see that the patterns and problems that BEFORE was designed to address lead to a hybrid model design. The model is grid-based in horizontal

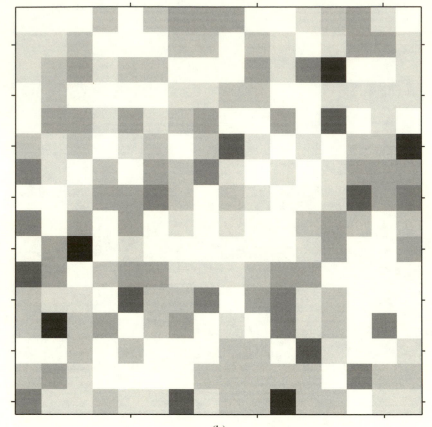

(b)

Figure 6.24 (*continued*)

and vertical dimensions; the beech trees are represented partly in an aggregated way (seedling and juvenile stages) and partly as individual-based (lower- and upper-canopy trees). The individuals are described only coarsely: lower-canopy trees mainly by their age (they can survive in a kind of "dormant" stage for a number of years but not forever), and upper-canopy trees by their age and crown size. Most individual-based forest models (section 6.7.5) describe individuals in more detail, but for the purpose of BEFORE—understanding large-scale and long-term spatiotemporal dynamics of natural beech forest ecosystems—more detail was unnecessary.

BEFORE succeeded in reproducing important patterns observed in real beech forests: the presence of three developmental stages (optimal, growing-up, and decaying), their mosaic spatial pattern, and the sequence and average duration of the stages. The fact that these stages and patterns were not imposed by the model rules but emerged from the IBM's structures and mechanisms provides confidence that BEFORE captures essential characteristics of the beech forest ecosystem. But BEFORE was designed and calibrated with these patterns in mind, so there still is a risk that wrong model structures or parameter values were chosen because they cause these patterns to emerge.

To develop more confidence that a model is structurally realistic, we can test its ability to make independent predictions of system properties not used to design or calibrate the model (section 9.9). BEFORE can make independent predictions about individual trees, because the patterns used to design and calibrate the model are about developmental stages (the development stage of a cell is assigned according to the total cover in the different height classes, not the state of any particular tree). Rademacher et al. (2001) asked two questions regarding individual trees: how are very old, large trees ("giants") distributed? and what is the age structure of canopy trees? Predictions from BEFORE matched— qualitatively—observations from remnant natural forests and reserves. First, "giants" are present all the time, and 80 percent of giants have another within forty meters. The presence and distribution of giants turned out to be a much more reliable indicator of the naturalness of a beech forest than the spatial percentages of the three developmental stages, which were cyclical and can have extreme values after extreme storm events. Second, the average age difference of two neighboring canopy trees is about sixty years. Thus, even apparently homogeneous canopies are very heterogeneous in age.

Given all these indications that BEFORE captures essential structures and processes of the beech forest ecosystem, the model should be useful not only for the questions it was designed for but also for new questions. One such new question concerns the amount and distribution of woody debris. In the original model, dead trees were ignored and the stem volume of canopy trees was not calculated. To adopt BEFORE to questions about woody debris (Rademacher and Winter 2003), rules were added to describe the production and fate of dead wood. Canopy trees of different age and crown size were assigned stem volumes. This modified model also produced predictions—both "dependent" and independent—that matched observed patterns; one example is that the total amount of living and dead wood is constant over time. The model then could be used to investigate strategies for increasing the dead wood in managed forests, for example, by creating "dead wood islands" where no wood is removed.

It is interesting to compare BEFORE to the earlier model of Wissel (1992a) that also attempted to explain spatiotemporal dynamics of natural beech forests. This model was designed around only one pattern, the spatial mosaic of developmental stages. Patterns in vertical structure were ignored. The Wissel model was able to reproduce the mosaic pattern, but foresters were skeptical of it. The model was not rich enough in structure to allow those familiar with real beech forests to look at it from different angles or to test predictions that could make the case that it was structurally realistic. In contrast, BEFORE has more

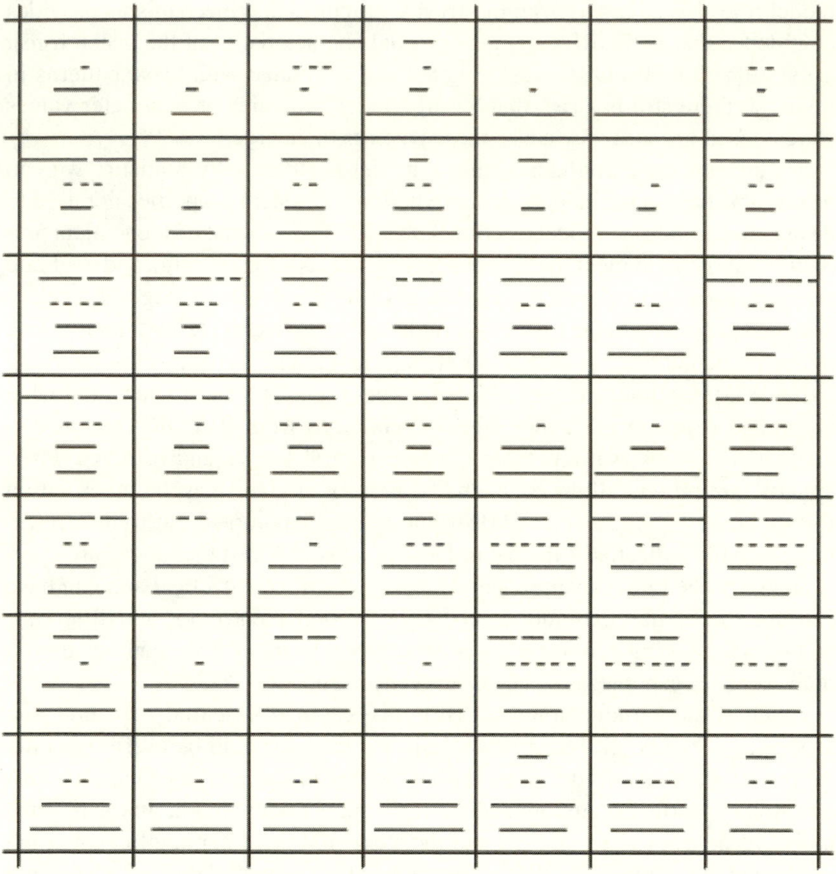

Figure 6.25 Graphical representation of the forest structure in the model BEFORE (see figure 1.2). Each cell represents an area of about 15 × 15 meters (the maximum canopy area of a single beech tree). Within each cell, the cover in the four tree height classes is indicated by four rows of line segments. In the upper canopy (the top row of line segments in each cell) and in the lower canopy (second row of lines), zero through eight trees may be present. Upper-canopy trees can have crown sizes between one-eighth and eight-eighths of the cell size, but the sum of sizes over all upper-canopy trees cannot exceed the cell size. Lower-canopy trees all have a crown size of one-eighth of cell size. For the two lower height classes (the lower two lines in each cell), the lines represent percentage cover. (From Neuert 1999.)

structure (figure 6.25) because it was designed to reproduce several observed patterns simultaneously.

It is also interesting to contrast BEFORE with the forest IBMs discussed in section 6.7.5. Because BEFORE was designed for ecosystem questions, it uses a relatively coarse description of trees and the forest while simulating large areas and long times. BEFORE does not use, for example, detailed models of growth rates and how they vary with environmental conditions. In fact, no processes in BEFORE are modeled in a completely mechanistic way. Instead, all the traits of trees are partially stochastic and based on "if-then" rules developed from empirical experience: if a tree (or cell) is in a certain state, then the probability of event X is Y percent (section 7.4). ("If there is a normal storm and all upwind and downwind neighbor cells have full canopy cover, then the probability of the cell's upper-canopy tree being wind-thrown is 0.01.") BEFORE is clearly not suited for many questions that other forest IBMs are used for, such as planning harvests or predicting what kind of forest would occur in different soils or moisture conditions.

6.8.4 Summary and Lessons

The three IBMs examined in this theme were chosen mainly to illustrate one point: that by using modeling practices similar to those we recommend in chapters 2 and 3, it is quite possible to develop community and ecosystem IBMs that are simple and tractable enough to learn from. By focusing on specific problems and patterns, and simplifying appropriately, we can build these IBMs almost as easily as we can population IBMs. In Schmitz's (2000) study, the community was simplified into four representative species and modeled using a platform, Gecko, that provides a generic template for multispecies IBMs. Pachepsky et al. (2001) represented plant communities in general using an abstract IBM in which species differences were simplified to variability in twelve basic traits, and then these differences were even further simplified via model analysis. The developers of the BEFORE model recognized that important dynamics of a whole ecosystem could be represented by modeling only its dominant species, beech. Despite these simplifications, the IBMs all provided new insights on real systems or ecological theory.

These examples also show that we can use IBMs to link patterns of biodiversity to traits of individuals. The model of Pachepsky et al. was designed to look at a classic problem of biodiversity: what traits of separate species allow them to persist together in a community (see also T. M. Smith and Huston 1989)? Schmitz's study looked at how an adaptive trait of grasshoppers (avoiding risky foraging behavior) affects their persistence and relative abundance within the community—an important reminder that adaptive behavior can have strong effects on communities and ecosystems, not just individuals and populations. The BEFORE model looked at diversity in forest structure, studying the environmental processes and individual traits that cause the diversity of forest stages observed in natural forests.

The abstract model of community diversity by Pachepsky et al. provides an interesting contrast with the other two IBMs in how we think of "general" research. The Pachepsky et al. IBM was designed to explain one particular pattern—a community's relative abundance distribution. Because it simulated an artificial system, this one pattern was the only way to show how "valid" the IBM was. In contrast, both the community IBM of Schmitz and the BEFORE model were built around a number of patterns observed in real systems, in some cases patterns developed from field experiments designed specifically to support modeling. As a result, these IBMs could be parameterized with field data and analyzed to build confidence that they capture essential characteristics of the real systems. Consequently, we feel we learned something we can believe, if only about the particular systems that were modeled. When abstract models are used to address "general" problems, though, we often find ourselves wondering what real systems, if any, the results might apply to.

6.9 ARTIFICIALLY EVOLVED TRAITS

In this theme we examine a "high-tech" approach to the central problem of IBE, finding adaptive traits that explain observed population-level behaviors. With this technique, models of the individual behaviors that cause a particular system behavior to emerge are artificially evolved within an IBM. The technique does not model evolution of the species but uses computer algorithms that mimic evolutionary processes to calibrate traits so they reproduce particular population behaviors.

Artificial evolution is often used to create artificial, digital worlds of evolving agents or species. (Adami 2002 provides an overview of artificial ecosystems; Tesfatsion 2002 discusses artificial economies.) In these digital worlds, agents typically compete with each other in an unstructured environment and the systems never settle down into a stable state: new traits continually appear because they exploit weaknesses of previous traits, only to be displaced by even newer traits. However, here we look at a completely different application: using artificial evolution as a computational technique to solve a specific modeling problem, the calibration of individual traits to solve real-world fitness problems in IBMs. Mitchell and Taylor (1999) provide an excellent overview of artificial evolution and its application to problems in biology and ecology, including problems of IBE; Mullon, Cury, and Penven (2002) provide an interesting application of artificial evolution to traits of the anchovy.

All the IBMs in this theme were developed at the University of Bergen, Norway. While the three studies differ in the problems they address, they share a general technique for artificial evolution of adaptive traits. Huse, Strand, and Giske (1999) describe and test this "individual-based, neural network, genetic algorithm" (ING) technique. One of the most interesting aspects of the ING technique is that it uses a highly simplified representation of how decision making actually occurs at its lowest levels within organisms: as an interaction of genes and networks of neurons, driven by sensed information about the individual's environment and internal state. A second interesting aspect is that different traits for different behaviors (e.g., movement, energy allocation, and reproduction timing) can be evolved simultaneously in such a way that the traits combine to produce realistic emergent behaviors.

The ING technique represents an adaptive trait as an artificial neural net (ANN). An ANN is a software representation of a simple network of neurons, commonly used to model complex decisions from multiple inputs. As input, an ANN takes information that the individual senses from the environment or from itself; each ANN can accept several kinds of such input. As output, it produces the decision that the trait models, usually as one or several variables standardized to range between zero and one. For example, Huse and Giske (1998) represented traits (discussed later) using an ANN with three "layers", each layer being a set of nodes. Nodes are analogous to neurons: a node accepts input values from senses or other nodes, transforms the inputs, and if the sum of transformed inputs is high enough, the node "fires" by producing an output. Thus, node j of a layer that receives n inputs can be represented as:

$$F_j = \sum_{i=1}^{n} \frac{1}{1 + \exp\left(-W_{ij} I_i - B_j\right)}. \tag{6.5}$$

F_j is the output variable that ranges between zero and one, I_i the input information, W_{ij} a weighting factor for each input, and B_j the node's so-called bias. The sigmoid function used to calculate F_j means that nodes respond nonlinearly to their inputs. The three layers used by Huse, Strand, and Giske (1999) include one that uses sensed information as input (with one node per type of information), a second "hidden" layer, and a final output layer. Each node in the hidden and output layers receives, as a separate input, the output F of each node in the previous layer. The output layer has one node for each of the ANN's outputs. For example, if the ANN produces only a yes-no decision, its output layer has only one node; if F for that node is close to 1.0, the decision is yes; otherwise the decision is no.

ANNs are capable of producing quite complex and sophisticated decisions from their multiple inputs, but only after they are "trained" by calibrating their parameters W and B. This training is an extremely complex problem because the number of parameters is large: each layer in an ANN has one B parameter for each of its nodes, but the number of its W parameters is the number of nodes times the number of inputs. Huse, Strand, and Giske (1999) modeled a trait using a three-layer ANN with six sensed inputs, a five-node hidden layer, and four output variables—and, therefore, seventy-one parameters. And, of course, the individual- and population-level consequences of changing any particular parameter are completely unpredictable. The ING concept thus uses artificial evolution, via genetic algorithms (GAs) to do this training of the ANNs: the genes in these algorithms are simply the values of W and B for each of the ANN's nodes.

GAs were developed by one of the pioneers of CAS, John Holland (1975), and are now a common technique for solving extremely difficult problems that can be represented in simulation models. Without going into detail (extensive literature and software is available for GAs), a GA represents a potential solution to the problem as an artificial gene: a string of values analogous to a string of DNA. The simulation model is used to impose "natural selection" (actually, digital selection): only genes producing good solutions in the simulation model survive. These surviving genes then are reproduced, but with processes such as recombination and mutation to continually produce new strings. Typically, a GA has hundreds or thousands of individuals, each with its own gene, competing and reproducing. Thus, GAs coarsely approximate the genetic and selection processes that drive biological evolution.

The ING technique for finding adaptive traits in IBMs can be summarized in the following steps (see also section 7.5.4). The following subsections illustrate the technique in more detail.

1. Completely define the fitness problem faced by individuals by developing a full-life-cycle IBM that is complete except for the adaptive traits that are to be artificially evolved. In the subsequent steps, the adaptive traits are evolved so they provide individuals with behaviors allowing them to survive and reproduce in the IBM's virtual environment.

2. In the model individuals, represent each of the adaptive traits to be evolved as an ANN. The ANN has inputs for each kind of sensed information that are assumed to drive the trait, and its output is the decision that the trait represents. The problem now is to "train" the ANNs: find parameter values that produce realistic decisions.

3. Couple the IBM with GA software that conducts the artificial evolution of ANN parameter values. The ANN parameter sets are treated as artificial genes. At the start of a "training run" of the IBM, individuals have ANNs with random parameter values. Then, as the training run proceeds, selection sorts out the successful parameter sets. Individuals that have parameter sets producing behavior good enough for them to survive and reproduce in the IBM then create offspring. The offspring inherit their parents' ANN parameter values, altered by crossover and mutation. The genes producing poor ANN decisions disappear from the GA because the individuals containing them die before reproducing.

4. After sufficient generations of artificial evolution, the ANN parameters have evolved into highly successful adaptive traits. The artificial evolution may result in all individuals having similar parameter values so there is essentially one best trait; or there may be a variety of traits that are quite different but each successful when competing with the others.

5. The modeler can analyze the IBM to determine whether the evolved ANNs produce behavior matching the behavior of the real system. If not, there

could be several problems: the IBM may not adequately represent the fitness problem faced by individuals (e.g., the distributions of food and mortality risks over space and time) or the ANNs may not represent the decision traits—inputs, outputs, or level of complexity—well enough.

6. When the evolved ANN parameter sets are considered adequate, they are then implemented as fixed traits in the IBM. Now the IBM can be used as any other to solve ecological problems.

Now we take a look at three IBMs that used the ING technique. The third example modified the technique to address a well-known limitation of ANNs.

6.9.1 The Huse and Giske Model of Horizontal Movement in Marine Fish

Huse and Giske (1998) modeled horizontal movement of fish such as herring or capelin in an environment resembling the Barents Sea. From day to day, horizontal movement is important for finding and following concentrations of prey while avoiding conditions where predation risk is high; but horizontal movement must also include annual long-distance migration to areas suitable for spawning. If the IBM's individuals are not successful at finding food while avoiding excess risk *and* migrating to suitable spawning areas, they do not reproduce. This study examined whether the ING approach could reproduce all these movement behaviors. The study also assumed that two different movement traits—reactive and predictive—might be needed to reproduce both the short-term and migration movements.

"Reactive movement control" is the trait for how the fish move in response to local, short-term variation in prey and predation risk. Reactive movement was represented with an ANN having four variables as input: the individual's current growth rate, the predation risk it senses, the temperature it senses in its current cell, and the temperature of the cell it was in the previous day. "Predictive movement control" is the trait for how the fish move seasonally, also represented by an ANN. The predictive movement ANN has as input the date, the fish's age and size, the current temperature, and the fish's current location. Huse and Giske point out environmental cues (e.g., day length) and physiological mechanisms (e.g., ability to sense geomagnetic fields) justifying the assumption that fish can sense date and location. The fish also have a separate simple trait they use each day to decide whether to navigate via predictive or reactive movement control.

The ANNs produce four outputs: yes or no values for moving east, west, north, or south. Together, these four outputs determine if and where the fish moves. For example, if only the outputs for north and west are yes, the fish moves to northwest; if only the outputs for north and south are yes, the fish does not move.

Huse and Giske used artificial evolution in a GA to find parameter values for the reactive and predictive movement ANNs and for the trait that decides when the fish use reactive or predictive movement. After 300 generations, evolution in the GA had slowed and the individuals made highly successful movement decisions. The traits evolved by the GA were quite heterogeneous: apparently there were several quite different yet successful traits.

Huse and Giske conducted several analyses to learn more about the traits they developed via artificial evolution. They investigated the importance of reactive and predictive movement traits: do individuals need a trait for basing decisions on predicted seasonal changes, or is it sufficient to have just the ability to react to current and recent conditions (as many behavioral models and IBMs have assumed)? Huse and Giske found that their individuals used reactive movement most of their lives, but without predictive movement they were unable to evolve the migrations needed for successful spawning. Both kinds of movement were necessary for the population to persist.

A second analysis tested how well the evolved traits perform when the environment was altered in one important way. During the artificial evolution to train the ANNs, the simulated environment included food depletion: high densities of fish drove prey densities down over time. What would happen if food depletion was turned off, *without retraining* the ANNs? Huse and Giske predicted that with food depletion off, food concentrations would stay high so local densities of fish should be higher *if* the movement trait produced emergent density dependence realistically. The simulation experiment confirmed the prediction, finding fish densities averaging nearly twice as high with food depletion off.

6.9.2   The Model of Strand and Colleagues: Vertical Movement, Energy Allocation, and Spawning

Strand, Huse, and Giske (2002) modeled a fish called Müller's pearlside, in a Norwegian fjord environment. The study was designed to develop traits for three different behaviors: vertical movement (whether the fish moves up or down each time step), energy allocation (whether fish use excess energy for growth or for fat reserves, with this allocation decision made once for each month), and spawning (whether the fish spawns during the month, a decision made once for each of the seven possible spawning months). To examine the effects of variability in environmental conditions that affect juvenile survival, the traits were evolved twice: once with a constant juvenile survival probability and once with survival probability that varied among and within years.

The trait for vertical movement was represented with an ANN, the output of which is the depth the fish occupies (figure 6.26). Inputs are the visual range of the fish (a function of light level), prey density, temperature, how full the

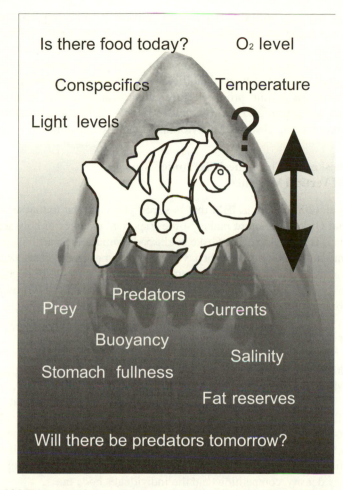

Figure 6.26 This cartoon demonstrates the environmental factors, both internal and external, that influence a fish's decision whether to move up or down at any moment. A trait for this decision was created via artificial evolution by Strand, Huse, and Giske (2002). (From Strand 2003.)

fish's stomach is, and what the fish's current fat reserves are. Because prey (zooplankton) density at the study site is dominated by tidal advection, prey availability was assumed unaffected by consumption by the modeled fish: there is no competition among the individuals.

The traits for energy allocation and spawning were not represented as ANNs, but instead as simple strings of zero or one values, one per month: for example, if the spawning trait's value for June is one, then the fish spawns in June. All the parameters for the vertical movement ANN, and for the energy allocation and spawning traits, were evolved in a GA. In this model, the GA converged until there was very little variation among individuals: apparently, one best set of traits was found instead of there being several successful traits for the same behavior.

Strand et al. compared the behavior and population dynamics produced by traits evolved with and without variability in juvenile survival probability. These two different sets of traits produced different vertical movement behavior and population age structures. However, neither set was consistently better than the other at reproducing observed behavior and population dynamics.

6.9.3   Giske and Colleagues' "Hedonic" Model of Vertical Movement

Giske et al. (2003) built an IBM that is superficially similar to that of Strand, Huse, and Giske (2002) but makes some important changes in how artificial evolution is implemented. The Giske et al. model simulates an artificial marine organism and its one adaptive behavior: vertical movement. This organism lives in a one-dimensional environment: temperature varies with depth, light level varies over a diel cycle, and the distribution of food organisms moves vertically during the diel cycle. Mortality risks resemble those from a sight-feeding predator with risk dilution: risk increases with light level and decreases with density of the modeled individuals.

The artificially evolved trait determines whether the individual moves up a meter, down a meter, or stays put. For each of these three possible destinations, the individuals can sense four variables: food concentration, light level, temperature, and density of individuals. Individuals that survive and attain sufficient size and energy reserves through one life cycle (which is artificially short—only two days, with seventy-five time steps per day) reproduce and pass their traits on to the next generation. Growth and energy reserves are modeled using simple feeding and bioenergetic algorithms that depend on temperature, light level, prey density, competition, and the individual's body mass.

The important difference between the model of Giske et al. and the previous two models is an attempt to overcome a well-known limitation of ANNs: the difficulty of understanding the trait once it has been evolved. Once the ANNs in an IBM have been "trained," it is relatively easy to analyze how closely the IBM's emergent behaviors match behaviors observed in real populations. However, it is not trivial to describe how an ANN itself works or how its output varies with all the different inputs—so it is very hard to understand exactly what the individuals are doing. ANNs are "black boxes" because the only way to describe their behavior would be to conduct experiments similar to sensitivity analysis (section 9.6).

Giske et al. (2003) developed a new approach they call "hedonic modeling" because it describes an individual's response to sensed information in terms of emotions like fear and attraction. The most important characteristic of this approach is representing a trait not with an ANN but instead with a set of equations that link the individual's response to the trait's inputs. First, each input sensed from the environment (food concentration, temperature, etc.) is modified by the individual's internal state to produce a "hedonic tone" for that input. For example, the hedonic tone for temperature, H_T, has the form:

$$H_T = M_T \, S_1^{m_{1T}} \, S_2^{m_{2T}} \, T.$$

S_1 and S_2 are two internal state variables, stomach fullness and body mass, both scaled so values are between 1.0 and 2.0; T is the sensed temperature; and M_T, m_{1T}, and m_{2T} are coefficients. Similar equations determine a "hedonic tone" for each of the externally sensed variables: food, light, temperature, and density of individuals. The individual's net attraction or avoidance of a decision alternative (e.g., moving up a meter) is determined by adding the hedonic tone values for all four externally sensed variables. The coefficients for these hedonic tone equations are analogous to the ANN parameters in the previous two models: they determine the direction and strength of the individual's response to sensed information, and they are calibrated using artificial evolution.

After completing the artificial evolution, Giske et al. illustrated how their approach facilitates understanding of traits by identifying three individuals and examining the vertical movement trait of each. While the traits are still complex, by just looking at their coefficient values we can describe the traits using terms such as how attraction to light and to other individuals changes with body mass. The three individuals analyzed in this way were quite different in some hedonic tone coefficients, indicating the presence of multiple successful traits in the population.

6.9.4   Summary and Lessons

In this theme we have explored ways to develop highly sophisticated adaptive traits by borrowing approaches—ANNs and GAs—widely used in artificial complex adaptive systems and to solve complex engineering problems. Even though these techniques have some "black box" characteristics, a considerable amount of biological knowledge is needed to use them. First, the modeler must decide exactly what adaptive traits are important for the population and problem being modeled: what decisions of the individuals need to be modeled? Before the parameters for an ANN (or a set of hedonic equations) are artificially evolved, the modeler must also decide what external and internal variables are sensed by the individual and used by the trait: to what conditions does the trait allow the individual to adapt? Finally, the rest of the IBM must be designed to represent the environment and physiological context in which the adaptive trait must work.

One compelling feature of the ING technique is its resemblance, however simplified, to real mechanisms that organisms use to make adaptive decisions:

signals sensed from external and internal conditions and then processed by algorithms that have been shaped by genetic evolution under the influence of natural selection. Equally compelling is the ability to evolve jointly several different traits for different decisions so they act together to produce realistic emergent behavior.

Even though the ING technique resembles physiological and evolutionary mechanisms, we must remember that traits developed via artificial evolution are *empirical*. Artificially evolved traits are "trained" only to solve the problems posed to them in the IBM during their evolution. However, the simple experiment of Huse, Strand, and Giske (1999), testing how well a trait evolved in one environment (with food depletion) performed in a different environment (without food depletion), indicates that there is much to learn about how general evolved traits can be. Can we evolve traits that perform well under a very wide variety of conditions, including many not present during "training"? Can even these simplistic representations of neural processes produce behavior resembling that produced by predictive, fitness-seeking theory (section 7.5.3)?

We briefly touched on another of the many intriguing facets of the Bergen models and experiments: the possibility of evolving several different traits that succeed simultaneously for the same individual behavior in the same population. Both Huse, Strand, and Giske (1999) and Giske et al. (2003) found a variety of "genes" in their evolved traits: multiple sets of parameters, sometimes producing quite different behavior, that survived the artificial evolution in competition with each other. Evolutionary ecologists expect traits to be more diverse and frequency-dependent for decisions driven by intraspecies competition. Is it a coincidence that the one model *without* competition for resources (Strand, Huse, and Giske 2002) was the only one of the Bergen models to converge to one set of parameter values?

6.10   SUMMARY AND CONCLUSIONS

The primary purpose of summarizing all these IBMs and the research done with them is simply to illustrate conclusively that there is much to be learned from IBE. We clearly have learned much already, even before the use of IBMs has gelled into a cohesive approach to ecology. Each of the studies in this chapter is interesting and exciting and produced new insights that could not have been produced by other approaches; and there are many more such studies we could have included.

Another main purpose of this chapter is to illustrate points made throughout the rest of the book. In fact, three of the main points from earlier chapters especially stand out from the example studies.

First, when model design is guided by patterns observed in a real system (or a generic class of real systems, such as forests), good things happen. Models are testable throughout the entire modeling cycle. Even early versions can be tested qualitatively, and the final version can often even make testable independent predictions. Pattern-oriented modeling leads to models that are rich enough in structure and process to be testable in many ways, yet still simple enough to be understood.

Second, when the IBE theory cycle (chapter 4) is followed, ecologists can learn important things about their individuals and systems and maybe even about individuals and systems in general. When the cycle is not followed, mainly by not testing alternative theories of individual behavior, we learn less about the detailed mechanics of how systems and individuals are interrelated. But the examples also include many studies that did not focus on testing theory for adaptive individual behavior yet were very useful and productive.

Third, the examples illustrate the need for a standardized, comprehensive way to think about and describe IBMs and, therefore, the value of the concepts and terminology developed in chapter 5. Sometimes these concepts help us succinctly describe some part of an IBM; other times they help us think about important design questions that the modeler did not think about—or document—enough. From the highest levels of an IBM (e.g., what outcomes are emergent or imposed? exactly what assumptions about fitness are the basis of how decisions are modeled?) to the lowest (exactly how are model events scheduled?), there are assumptions that do not fit into traditional modeling frameworks. Yet we must think about and describe these assumptions because they are of utmost importance for understanding and reproducing IBMs.

However, the example studies also illustrate the challenges to individual-based modeling discussed in chapter 1. Someone wanting only to criticize IBMs could draw sweeping, negative generalizations from the examples: IBMs have not been *coherent* as an approach to ecology. They have not followed common research agendas, and their design is not based on common principles or theory. Consequently, it is difficult and ineffective to compare different IBMs, to relate their results to each other, and to see a general, overarching framework. IBE has also been inefficient, with little reuse of formulation methods and software. The models and their software are complex and detailed, but not fully communicated and therefore not reproducible; and the software could easily be full of mistakes.

These generalizations are indeed true—as generalizations. But they certainly are not inherent limitations of IBMs and IBE. How do we know that? First, the studies examined in this chapter include clear exceptions to all these generalizations, so we know they are not insurmountable. More important, most of this book is dedicated to avoiding these kinds of problems. We have already seen in parts 1 and 2 how we can bring coherence to the use of IBMs by using appropriate modeling strategies, developing theory, and using a new conceptual framework. But we still need to address the more mundane issues of making

IBE more efficient and effective: How do we formulate and analyze IBMs in detail? How do we make software efficient, reliable, and useful? Finally, how do we communicate our work so its value and credibility is conveyed? In part 3 of this book we roll up our sleeves and get busy learning how to deal with these issues that, experience shows, we actually spend most of our time on.

PART 3

The Engine Room

Chapter Seven

Formulating Individual-based Models

Technical skill is mastery of complexity, while creativity is mastery of simplicity.

—Sir Erik Christopher Zeeman, 1977

7.1 INTRODUCTION

This chapter begins our descent into the "engine room" of individual-based modeling. According to the statement by biomathematician E. C. Zeeman, part 2 of this book focused on the creative phases of individual-based ecology: approaches for reproducing and explaining ecosystem complexity using relatively simple concepts and theories. Now, in part 3 we address the technical skills needed to master the complexity we create when we implement an IBM, execute it, and start to conduct science with it. A major theme of part 3 is that many of the technical skills needed for IBE are different from those typically used in classical modeling. The individual-based ecologist needs familiarity with discrete mathematics and software engineering, the ability to design the experiments that provide understanding of complex digital worlds, and a thorough understanding of the natural history and autecology of the organisms being modeled.

The implementation of an IBM typically begins with preparing a written *formulation*: a fully detailed description of the model as it is to be implemented in software. The formulation (called a *specification* in software engineering) captures the conceptual design developed from the problem the model addresses (chapter 2), the patterns used to define the modeled system's essence (chapter 3), theories of adaptive traits (chapter 4), and individual-based modeling concepts (chapter 5); then it fills in the details needed to translate the IBM into executable software. Writing a detailed formulation makes the whole research cycle more productive: it helps us think through the model design thoroughly, and have the design reviewed, before starting to invest time in software development; documents how and why we made design decisions; tells the project's programmer (if there is one) exactly what to do; and provides the reproducibility we need to publish our work and have it accepted as credible.

Formulating a model's detailed design is never an isolated activity. We start modeling by developing a conceptual design and some of its specifics in our head, but—as we all know—trying to write our ideas down so they are complete and unambiguous can be painful and frustrating. Often we are happy with the model in our head, but when we start writing it down, we discover inconsistencies and gaps. Writing the model's software inevitably identifies more ambiguities and ideas that do not seem quite so smart when we try to actually implement them; and once we start using and analyzing the model, we continue to revise and improve its formulation. Yet when it is time to publish work based on IBMs, it is especially important (and challenging) to provide a thorough description of the model. This chapter is not intended as a comprehensive guide to the detailed design of IBMs (such a guide would be a big project by itself) but as a way to link the conceptual design process of part 2 with the following phases of the IBE modeling and research cycle discussed in chapters 8–10. We provide formulation techniques that help modelers prepare for and manage the challenges of software development, analysis, and communication.

We start by outlining the kinds of information that need to be included in a typical formulation. Then we discuss techniques for three specific problems: formulating spatial elements, probabilistic rules, and decision-making traits. The remaining sections recommend general ways to limit an IBM's uncertainty and make its parameterization, analysis, and communication easier and less controversial; ways of formulating an IBM that lead to a smooth, easy transition from conceptual design to written description to software; mathematical styles appropriate for IBMs; and techniques sometimes needed for coping with extremely high numbers of individuals.

7.2   CONTENTS OF AN IBM FORMULATION

In chapter 5 we discussed many essential characteristics of IBMs that cannot be described with just equations and parameter values. To describe these characteristics, an IBM's formulation must contain more kinds of information than is usually needed to define a classical model. Here we list the kinds of information that are essential, or very helpful, to include in the detailed formulation of an IBM.

We must emphasize that the formulation needs to describe not just *what* is in the IBM but *why*. Ecologists using IBMs constantly combat the misconception that IBMs are inherently ad hoc; the purpose of much of this book is to show that we can often have a strong biological basis for our modeling decisions. Instead of simply writing down what assumption, equation, or parameter value was used, write down why it was chosen—even if the best justification is "our field observations X, Y, and Z indicate that this assumption captures essential elements of the process" or "the literature indicates that there are no good assumptions so we chose this simple one."

An IBM's formulation needs the following elements (see also section 10.3):

1. *Statement of model purpose.* A clear statement of the problem that the IBM is intended to solve, which can serve as a compass for the rest of the formulation (section 2.3). It can help the modeler and reviewers understand why some processes need to be included and others do not.
2. *Major structural assumptions.* Description of the kinds of entity represented (habitat units, individuals, etc.) and their state variables, the time step, the spatial extent and resolution, and the environmental variables or processes that drive the model; and why these were chosen.
3. *Conceptual design.* Essentially, the checklist from section 5.13. It is especially important to define early what model outcomes emerge from what traits of individuals.
4. *Overview of submodels.* What major processes are included and why (section 7.6). These include submodels for environmental processes as well as traits of the individuals.
5. *Observer plan.* A description of what model results need to be observed, why, and how (section 8.6.5). This plan determines how each kind of output should be summarized statistically and written to files or graphical displays.
6. *Schedule.* A complete description of how submodels are grouped into actions and the order in which actions are executed (section 5.10). The schedule must include everything in the IBM, including habitat simulations and observer actions (outputs).
7. *Initialization.* How the habitat and individuals are created and placed in space (and, sometimes, time) at the start of a model run, and why.
8. *Submodel details.* The complete description of each part of the IBM. The description includes assumptions and equations and justification for their selection; parameter values; and the literature, data, and methods used to test and calibrate the submodel. While these details are absolutely essential for reproducibility, they are still details compared with the preceding elements. It may be best to isolate the submodel details near the end of the document so they do not interfere with the reader's ability to develop first a clear understanding of the conceptual design.
9. *Input data.* A description of the model's input data, such as data used to represent environmental conditions. Data sources and field methods may need description, as may any methods used to process or transform data.

7.3 FORMULATING AN IBM'S SPATIAL ELEMENTS

One of the major structural assumptions of most IBMs is how space is represented. Most IBMs are spatial because most questions of IBE require modeling the spatially heterogeneous environment that individuals occupy and interact with locally. Basically, representing space is very simple: we must decide on the extent of the area (or volume) to represent, define a coordinate system for this area, and assign coordinates to each spatial object. Then, the fundamental design decision is whether to represent space as discrete units (cells) or continuously. A related decision is whether to represent explicitly the spatial extent of individuals.

7.3.1 Discrete Space

Discrete-space models represent space as a collection of discrete cells, with spatial characteristics varying among, but not within, the cells. Many such models are called "grid-based" because they use a grid of uniform, usually square, cells. (Other names in use are "interacting particle system" and "lattice gas," but these names reflect their roots in physics and should be avoided in ecology.) A general description and examples of grid-based plant models are in section 6.7.4; for overviews of the approach, see Hogeweg (1988), Ermentrout and Edelstein-Keshet (1993), Czárán (1998), Wissel (2000), and DeAngelis, Mooij, and Basset (2003). Some discrete-space models, though, use irregular cells designed to match the geometry of the natural environment; examples include the stream fish models of M. E. Clark and Rose (1997), Van Winkle et al. (1998), and Railsback and Harvey (2002; section 6.4.2).

The discrete-space approach is widely used and successful for several reasons. The basic assumption that spatial variation and processes are negligible within small areas (within cells) yet important over larger areas (among cells) can greatly simplify models. The approach also has computational advantages, especially for identifying an individual's neighbors: for uniform grids it is very easy to identify a cell's neighboring cells by incrementing or decrementing the cell's grid coordinates. Another advantage is the wide availability of grid-based spatial data from remote sensing, Global Positioning System (GPS), and Geographic Information System (GIS) technologies. The lynx dispersal model of Schadt (2002; section 6.4.1) is an example of how satellite data can be made into the spatial element of a grid-based IBM. With GPS or other surveying technologies and GIS, it is now easy to generate grid data even for very small spaces and grids.

One of the most important potential advantages of discrete space is the potential to give model behaviors to the spatial cells. In some IBMs (e.g., the lynx model of Schadt 2002) cells have no behavior and only provide environmental information to the individuals. In others (e.g., the stream fish models just cited), cells model the depletion and regeneration of resources such as food. Some IBMs even represent the individuals as cell behaviors for some or all of their life cycle; the BEFORE forest IBM (section 6.8.3) tracks seedlings and juvenile trees only as a cell characteristic—the fraction of cell area covered by the trees.

Models that represent *all* of the population's behavior as traits of the grid cells fall within a general category called cellular automaton models (von Neuman and Burks 1966; Wolfram 2002). These models characterize each cell by an ecological state, for example, the presence of a species, the amount of biomass, or the size of a local subpopulation. This state then changes every time step according to "transition rules" that consider the state of the cell itself and its neighbor cells. In ecological models, these rules often are probabilistic (section 7.4).

A limitation of the grid-based approach is that the spatial resolution (grain; or cell size) of the model is not just a model parameter that can easily be changed but a design decision that affects the design of all other parts of the model and even parameter values (section 2.3.3). Therefore, it is essential that the formulation of any grid-based IBM explain why its particular grain was chosen. Of particular concern is showing that it is reasonable to assume that spatial variation and effects can be ignored over areas less than the cell size.

The discrete-space approach is especially suited for IBMs that do not explicitly represent the spatial extent of individuals. Many animal IBMs, especially, represent the individuals' locations but not how much space they occupy, because animals move over much greater spaces than they occupy. (An exception is IBMs of animals assumed to always maintain a territory; the territory size can be considered the individual's spatial extent.) Discrete-space IBMs that do represent the spatial extent of individuals run the risk of cell size artifacts: model outcomes such as the density of individuals can be affected in nonlinear ways by the cell size. For example, if adults each occupy a 1×1 meter square and juveniles occupy a 0.5×0.5 meter square, a cell of 1.9×1.9 meters can hold one adult and five juveniles; but a slightly bigger grid of 2.1×2.1 meters could hold four adults and no juveniles.

To avoid such artifacts of cell size, grid-based IBMs of sessile organisms (especially plants; see sections 6.7.4, 6.8.3) or territorial animals (e.g., Tyre, Possingham, and Lindenmayer 2001) often simply assume that the grid size is equal to the spatial extent of one individual. This assumption can greatly simplify the IBM, but also introduces another limitation: individuals grow, so there is no single grid size that matches the size of all individuals or even one individual over its life-span. A solution that is acceptable in some plant models is to use the size of adults as the grid size and ignore or greatly simplify the early life stages of individuals (section 6.7.4). Or, as in the Winkler and Stöcklin model (section 6.7.4), clusters of grid cells can be used to represent individuals; in this way, the spatial extent of individuals can vary at the price of making the model more complex conceptually and computationally.

7.3.2 Continuous Space

In continuous-space models, the position of an object is described by continuous variables. Whereas an individual's location in a grid-based model might be described as "within the 10×10 meter grid cell that is centered at $x = 250$, $y = 190$," in a continuous-space model the location might be described as "at $x = 245.734$, $y = 193.806$." Continuous space has obvious advantages for problems strongly affected by the exact spatial arrangement of individuals; examples include fish schooling (section 6.2), growth of microbial colonies (Kreft, Booth, and Wimpenny 2000; Kreft et al. 2001), and competition among plants (section 6.7). Continuous space also allows modeling the spatial extent of individuals of very different sizes without having to worry about how the individual's size corresponds to grid size.

One difficulty with continuous space is representing resource dynamics. In a discrete-space model, it is typically assumed that an individual consumes the resources within its cell; and depletion and regeneration of the resources is modeled at the cell level. A continuous-space model cannot simply assume that an individual uses the resources at its location: its location is a point, and the amount of resources available at a point is of course zero. The usual way to overcome this problem is to assume individuals occupy—or consume resources over, or interact with their environment over—a zone surrounding their location. Examples of this approach are "zone of influence" plant models (section 6.7.2) and Gecko (section 6.8.1). The area occupied or influenced can easily vary among individuals or increase over time as an individual grows.

However, modeling spatial variation in resource availability, as well as dynamics of spatial resources, remains a problem with continuous-space models. For example, there is no simple way to represent environmental data (e.g., from GIS) in a continuous space, or to model how resources consumed by an individual moving through the space are regenerated. These limitations can be overcome by using hybrid approaches (section 7.3.3).

Several computational problems are often encountered with continuous-space IBMs. These IBMs are particularly useful for representing direct, spatially dependent interactions among individuals, but doing so requires that each individual know which other individuals are its neighbors. Computing which individuals have overlapping zones of influence, or which individuals are "seen" by another individual that has a specific zone of vision, can be done simply by comparing all pairs of individuals but computation time increases with the square of population size. For models large enough for this simple comparison to be a problem, the problem can be overcome with sophisticated software techniques (Hildenbrandt 2003; section 6.7.3) or, again, by using hybrid approaches. Another problem is computing the areas of overlap when more than two individuals have overlapping "zones of influence." Computational approaches to this problem are illustrated by Wyszomirski (1983) and Weiner et al. (2001).

7.3.3 Hybrid Approaches

Many of the limitations of both discrete and continuous representations of space can potentially be overcome by not sticking rigidly to either. Some ways that spatial approaches and resolutions can be hybridized are:

- Representing the environment as discrete cells while tracking individuals in continuous space. This approach allows use of grid-based spatial data and allows resource dynamics to be modeled at the cell level. Berger and Hildenbrandt (2000), for example, use a discrete bitmap of salinity data as the environmental component of their otherwise continuous-space forest model.
- In discrete-space models, using one spatial resolution for environmental processes and another resolution for individuals. Consider an IBM of animals that move rapidly through a complex habitat; it may be best to model the complex habitat with a small grain size (e.g., plant production on ten-meter grids that vary in soil type, moisture, slope, and aspect) and then aggregate the habitat variables to a larger grain size more appropriate for the animals (e.g., total food availability on one-hectare grids).
- Using different spatial resolutions for different behaviors of individuals. The trout IBM of Railsback and Harvey (2002; section 6.4.2) represents food competition within each cell, but assumes an individual can move each time step to any cell within a radius that increases with trout size. Trout can use hiding habitat many meters from where they feed, so hiding habitat was represented in each cell as the distance to the nearest hiding spot, whether that spot is within or beyond the cell. The continuous-space Gecko-based community model of Schmitz (2000; section 6.8.1) uses one radius to represent resource intake by herbivores and another radius for predator detection.
- Modeling different life stages differently. One example discussed previously (section 7.3.1) is the BEFORE beech forest model, which uses a cellular automaton approach for juveniles and treats adults as full individuals. In the extreme, completely different models with different spatial (and temporal) resolutions can be used for different life stages (e.g., the salmon IBM of Railsback and Jackson 2000).

7.4 FORMULATING LOGICAL AND PROBABILISTIC RULES

Because IBMs are discrete event simulators (section 8.3.1), they contain rules for how model objects change state upon certain conditions. For example, in the fish school model of Huth and Wissel (1992; section 6.2.2), once per time step each fish changes its swimming direction in response to the spatial configuration of its neighors. If a neighbor is too close, the fish turns to avoid collision; if neighbors are close but not too close, the fish turns to swim parallel to them; or if the nearest neighbor is far away the fish turns to swim toward it. We can see from this example that *logical* ("IF-THEN") rules are a natural way to model the behavior of individuals—and other processes such as habitat dynamics—in IBMs. The "IF" part of these rules can include several logical conditions combined by Boolean algebra using the operators AND, OR, and NOT. (Usually, no more than two or three conditions are combined in one rule, to avoid logic errors.) But logical rules are not always adequate by themselves: often we know from observations that the process we are modeling does not rigidly follow a simple set of such rules. The rules may capture important trends but often there is variation in process outcomes not captured by a purely logical approach.

We also often want to base an IBM's rules on empirical information about the system we are modeling. When the available information includes data of sufficient types and quantities, we can use it to formulate *probabilistic* rules. If real individuals have been observed to exhibit a certain behavior (e.g., dispersal in the marmot model described in section 6.3.2) on X percent of their opportunities, then a simple probabilistic model rule is that on each opportunity the model individuals have a probability of exhibiting the behavior equal to X. This simple probabilistic approach has limitations, too. If used as a trait for individual behavior, the behavior is not adaptive—the individual makes the same decision with the same probability no matter what conditions it is in. Also, the data are often too sparse or uncertain to estimate the probability value with confidence. (The use of stochasticity in general is discussed in section 5.8.)

Combining the logical and probabilistic approaches overcomes many of each approach's limitations. Combined rules take the form "IF condition X occurs, THEN event Y occurs with probability Z." Most important, adding logical conditions to a probabilistic trait for individual behavior can make the behavior adaptive: it gives individuals the ability to behave differently in different conditions. The adaptive trait for woodhoopoe dispersal (section 6.3.1) assumes birds have a probability of dispersing from their natal territory that increases with their age and decreases with their social rank: "IF age = 3 and rank = 2, THEN execute dispersal behavior with probability of 0.3." A random number between 0 and 1 is then drawn; if it is less than 0.3, the dispersal behavior is executed.

The other major advantage of combined logical and probabilistic rules is that they can take advantage of many kinds of "soft" information. Often, people familiar with the system and its individuals may know, or suspect, what outcomes are possible, and which are more likely, under various conditions. Experts may also have a feeling for how predictable the outcomes are. Often there are also small or uncertain sets of field observations—for example, only a few observations of rare but important events. These kinds of information can be very valuable, sometimes the most important information we have about a system. Such information cannot be boiled down into probabilities or fixed rules reliably but can be used to formulate logical rules that include probabilities to represent uncertainty. The probability values can (and should) be treated as parameters and the effects of their uncertainty on model results examined (sections 9.6–9.7).

There is, in fact, an extensive literature on modeling the behavior of systems and individuals by using logical and probabilistic rules derived from the knowledge of experts and sparse data. In the 1980s and 1990s, this "Expert Systems" technology was developed as a kind of artificial intelligence. The Expert Systems literature (e.g., Cowell et al. 1999; Jackson 1999) covers such topics as how to extract useful information from experts and data, Bayesian statistical methods for calibrating probability parameters, and software tools. This approach has also been developed specifically as a way to formulate IBMs: MOAB (Carter and Finn 1999) is a platform for assembling and executing spatial IBMs of terrestrial animals with traits formulated entirely from logical rules.

7.5 FORMULATING ADAPTIVE TRAITS

Adaptive traits—models of how individuals make decisions in response to other individuals, the environment, or changes in themselves—are often the "guts" of an IBM. Chapter 4 discusses developing theory for adaptive traits, and much of chapter 5 provides a conceptual basis for the theory. Here, we provide some approaches for formulating decision-making traits in detail once the theoretical and conceptual approach has been selected. First we describe a framework— four distinct steps to address—for designing an adaptive trait. Then we briefly describe four very different approaches for evaluating decisions alternatives.

7.5.1 A Framework for Modeling Decisions

It is useful to think of an adaptive trait as involving four distinct steps. Designing and describing each of these steps in a decision-making trait separately should help avoid hidden or implicit assumptions and make assumptions that are reasonable and effective.

1. Specify when the decision is executed. The modeler must determine how often, or under what conditions, the individuals execute their decision-making trait and have the opportunity to change the behavior prescribed by the trait. Often, decision traits are simply scheduled to be executed every time step so individuals are continually able to reconsider and change their decisions. However, it is not unusual for IBMs to include some decisions that are only occasionally executed, being triggered by some internal or external event. For

256

example, Bernstein, Kacelnik, and Krebs (1988) modeled habitat selection by assuming that predators move only when prey capture rate falls below a threshold: one decision (whether to move) is executed every time step, but a second decision (where to move to) is executed only when triggered by the first decision. An IBM representing competition among mobile animals might assume animals decide whether to move only when a new competitor enters their territory. Some decisions are even irreversible, so can only be made once; examples are the selection of dispersal destinations by marmots and coyotes in IBMs depicted in section 6.3.

2. Identify alternatives. The range of alternatives can be either discrete (with the individual choosing among a limited number of specific alternatives) or continuous (choosing where to be on a continuous gradient). This step is trivial in the case of decisions for which the same alternatives are always considered (e.g., "Should I feed or should I hide?" "How much of my daily energy intake should I expend on growth, reproduction, or reserves?" "In which direction should I extend my roots?"). For many decisions, however, identifying alternatives is an important process by itself and can be an important constraint on the decision. How many potential mates does an individual consider before choosing one? How many habitat patches does an individual consider when deciding where to forage? Are these patches defined as being within a circle around the individual's current location, or within a square, or along a line in a random direction? The assumptions that define which alternatives are considered can be as important as the rest of the decision-making formulation.

Real organisms make many *contingent* decisions: two or more decisions that are highly dependent on each other. One example is a contingent choice of habitat and activity: animals decide which habitat cell to occupy, but this choice depends on whether they decide to feed or hide from predators. Habitat that is good for feeding may not be good for hiding, and vice versa. At the same time, the activity choice between feeding and hiding depends on what habitat is available: hiding may be a better choice if there is currently no good feeding habitat available, or the presence of good feeding habitat may make hiding undesirable. Because the range of alternative activities is sufficiently limited (only feeding or hiding), this contingent decision can be modeled by assuming individuals identify the combinations of activity and habitat as distinct alternatives: feeding in habitat cell A is one alternative, hiding in habitat cell A is a second alternative, feeding in habitat cell B is a third alternative, and so on.

3. Evaluate alternatives. If the alternatives are discrete, then this step involves rating the relative value or desirability of each alternative. For continuous alternatives, the evaluation typically involves determining the best point on a continuous gradient. As we discuss extensively in chapter 5, modeling how individuals evaluate alternatives to make adaptive decisions involves many assumptions about what information the individuals sense and how the individuals use the information. General approaches are discussed in sections 7.5.2–7.5.5.

4. Select an alternative. Even after an individual has identified and evaluated its alternatives, there is a final step of selecting an alternative that the individual then implements. Often (but not always), IBMs simply assume that individuals choose the alternative that was ranked highest in the evaluation step. The individual chooses what it perceives to be the best alternative, but the optimality of this choice is bounded by the individuals' limited ability to identify and evaluate alternatives.

7.5.2 Probabilistic and Logical Rules

The probabilistic and logical approaches discussed in section 7.4 are often used to model adaptive traits. These rules are used to determine the probability of the individual choosing each alternative; then a random number is drawn to determine which alternative is actually chosen. These approaches are best suited for selecting among a fixed set of discrete alternatives.

7.5.3 Direct Fitness-seeking

Another common approach for modeling decision making is what we call direct fitness-seeking (section 5.3): the individuals evaluate a *fitness measure*—an indicator of their expected success in passing genes on to future generations—for each alternative and then select the alternative providing the best value of the fitness measure. In many early IBMs and behavior models, this approach was implemented by simply assuming individuals select the alternative providing highest growth or lowest mortality risks. The *state-based, predictive* approach to fitness-seeking theory (Railsback et al. 1999; Railsback and Harvey 2002; similar approaches have been proposed or used by Tyler and Rose 1994; Giske, Huse, and Fiksen 1998; Thorpe et al. 1998; Grand 1999; and Stephens et al. 2002b) represents how fitness depends on both growth and mortality; and on the individual's life history state and energy reserves; consequently, this approach provides a more general adaptive ability. The elements of this approach are all discussed in chapter 5; here we summarize the steps in developing state-based, predictive fitness measures. An example application to habitat selection is described in section 6.4.2.

1. Identify one or several appropriate fitness elements (section 5.4): fitness "goals" such as future survival or reaching reproductive status that are most important for the individual at its current life stage.
2. Identify a time horizon appropriate for the fitness elements: the future period over which expected fitness will be evaluated. The time horizon

can be a fixed future date (e.g., the beginning of the next reproductive season; the end of a winter period of reduced activity) or may be a constant number of days after the current date. The end of the current life history stage may be a useful time horizon (e.g., Grand 1999).

3. Identify internal state variables and environmental variables that depend on the decision alternatives and affect the fitness element. These are variables that affect the individual's survival, growth, and the like, and that also change with the individual's decision.
4. Develop appropriate models for how the individual predicts the variables identified in step 3 over the time horizon (section 5.5).
5. Develop the full fitness measure: a model of how the current and predicted future values of the variables identified in step 3 directly affect the fitness element(s) over the time horizon.
6. If appropriate, repeat the process to develop different fitness measures for different life stages.

7.5.4 Artificially Evolved Traits

Instead of carefully formulating direct fitness-seeking traits, we can empirically develop traits that mimic the fitness-seeking behavior of real organisms. In this approach, model individuals are given generic traits that can be "trained" to produce successful adaptive decisions. The training is accomplished using an artificial evolution (or "evolutionary computation") technique. Mitchell and Taylor (1999) provide an overview of artificial evolution and its applications to biology and ecology. Huse and Giske (1998; see also Strand, Huse, and Giske 1999; Strand, Huse, and Giske 2002; Giske et al. 2003) have applied these techniques to the problem of modeling behavior in IBMs; their approach is described in section 6.9.

Artificially evolved traits have great promise for modeling complex behaviors for which direct fitness-seeking theory is unavailable. Another potential advantage of this approach is its ability to develop several different but related traits at the same time (e.g., how the individual feeds and how it allocates its energy intake). However, the approach has limitations and trade-offs. First, even though the traits are "trained" to convey fitness, we must remember that the artificially evolved traits are highly empirical. Unlike traits based on direct fitness-seeking theory, we cannot automatically expect artificially evolved traits to solve problems they were not exposed to during their artificial evolution. Second, using artificial evolution requires a full-life-cycle IBM that simulates how the traits of interest affect the individual's ability to survive and reproduce. Third, the approach is fairly simple (at least conceptually) for traits with clear, direct effects on individual fitness, but could be very challenging for traits having only indirect effects on fitness. Finally, of course, artificial evolution requires additional software and computational effort. However, the general approach of modeling behavior via artificial neural networks that are "evolved" using genetic algorithms is well established. There are dozens of books on evolutionary computation (Mitchell 1998 is a popular introduction) and dozens of software packages for adding artificial evolution to models (lists of these packages are currently available at websites such as www.aic.nrl.navy.mil/galist/ and www.geneticprogramming.com).

7.5.5 Decision Heuristics

Decision heuristics have been promoted—most conspicuously by the Center for Adaptive Behavior and Cognition, Max Planck Institute for Human Development, Berlin (Gigerenzer and Todd 1999)—as a biologically realistic model of adaptive decision making in situations where information, time, or cognitive power are very limited. Decision heuristics can be thought of as very simple rules for making decisions that are *usually good*—but rarely optimal and possibly bad. Heuristics generally use minimal information or computation. Another characteristic of some heuristic approaches is minimizing the number of alternatives considered. For example, "satisficing" approaches do not closely follow the four-step decision process described in section 7.5.1; instead, individuals are assumed to identify repeatedly one more alternative, evaluate it, and then decide whether to accept that alternative as "good enough."

Some heuristics for human or animal decisions analyzed by Gigerenzer and Todd (1999) are:

- When selecting among alternatives, choose one that you recognize. When choosing a stock to invest money in, for example, go through a list of stocks and select the first one you recognize. (This heuristic works for the stock selection problem because stocks of bigger, well-known companies *generally* outperform stocks of smaller, more obscure companies.)
- When selecting among alternatives that have different values for several different "cues" (e.g., variables that might affect fitness), (a) identify the one cue that is most important, then (b) select the alternative that has the best value of that one cue.
- To make a good choice among many alternatives, evaluate a small number of the alternatives and then pick the next alternative that has a higher value than any of the previous. (This heuristic is well known from mate selection theory.)

The first of these three example heuristics solves the problem of evaluating alternatives (step 3 of the decision process; section 7.5.1). However, the second two only address step 4—how to select among alternatives that have been evaluated; the often-difficult problem of how to evaluate the alternatives is not addressed.

The agenda of the Center for Adaptive Behavior and Cognition is to develop a "toolbox" of heuristics from which the right rule for a particular decision and environmental situation can be selected (Gigerenzer and Todd 1999). Then, adaptive traits could be represented as a process of first selecting the right heuristic and then applying it to evaluate or select among alternatives. So far, we are unaware of any attempts to use this approach in IBMs. Heuristics seem promising for the kind of trait that Gigerenzer and Todd advocate them for: decisions that must be made with little information or an uncertain number of alternatives. The mate choice problem (Andersson 1994) is a good example of such a decision. However, many traits in IBMs represent decisions that very strongly affect an individual's fitness—often determining whether the individual eats, or is eaten—and have been made over and over for many generations in environments rich with information and cues. Such decisions do not seem to be the kind of problem that heuristics have been designed for. Still, until heuristic traits are further developed and tested we do not know what kind of adaptive decisions they could model well.

7.6 CONTROLLING UNCERTAINTY

One of the key themes throughout this book is finding the right level of complexity for an IBM: an overly simple model cannot solve the problem it is intended for, while excess complexity produces uncertainties that make an IBM hard to test, understand, and learn from. Chapters 2–5 focus on finding the right level of complexity at the conceptual modeling level: determining what structures, variables, processes, emergent processes, and so on should or should not be included in an IBM. In this section we discuss controlling complexity and uncertainties in an IBM's detailed design as it is formulated.

Once the processes that must be included in an IBM have been chosen, efforts to reduce uncertainty then focus on reducing the uncertainty in each process. Key to doing so is organizing the model (and, subsequently, its software) in a way that lets each process be designed and tested independently. Each major process can be treated as a *submodel* that has its own distinct inputs and outputs. The formulation (including design, parameterization, testing, and validation) of each submodel becomes a separate task. For example, one part of a plant model is a set of equations and parameters that calculates growth for a time step from the solar radiation. Treating this process as a submodel to be developed, parameterized, and tested separately is much easier than testing and calibrating growth as part of the whole IBM. To do so, we develop a *test code*: a little bit of simple software that implements the submodel. Developing a test code helps identify and resolve any ambiguities in the submodel's formulation, and then the code becomes an important piece of documentation (section 8.6.4): it is a complete, independent description of the submodel.

Here we present three techniques that are useful for reducing uncertainty in the submodels that represent an IBM's major processes.

7.6.1 Keep Submodels Simple

Often it is necessary to capture only the most basic dynamics of a process to represent adequately its effects in an IBM. In making the inevitable decisions of how much detail to include in a submodel, keep in mind the benefits of minimizing complexity and the number of parameters. IBMs often can produce realistic dynamics with highly simplified or partial representation of key processes. For example, many IBMs use very simple representations of resource dynamics, such as how the availability of food for animals or light for plants varies with season, time, or consumption by individuals. These IBMs can still reproduce many resource-dependent dynamics when they include the processes controlling the ability of individuals to obtain the resources. An IBM that represents how an individual animal's ability to obtain food varies with habitat type and competition may capture many important food-related dynamics even if it ignores how food production varies over space and time.

7.6.2 Consider Borrowing Existing Submodels

Sometimes a process can be represented in an IBM by using an established, existing model. Using an existing model allows the IBM to borrow the existing model's credibility, while avoiding the need to make up a new and potentially controversial submodel and its parameters. (It also shows that the IBM developer is familiar with the literature on the process in question.) Useful existing models are especially likely to be available for environmental or habitat processes because mechanistic models are widely used in environmental chemistry, meteorology, hydrology, and other fields. The trout model described in section 6.4.2, for example, uses (1) a popular river hydraulic model to simulate depth and velocity in habitat cells, (2) a modification of an existing model of how fish food intake depends on hydraulic conditions, (3) a widely used bioenergetics model to predict fish growth from food intake and temperature, and (4) an existing model of the probability of trout eggs being destroyed by high flows as they incubate in river gravel. Of course, it is important to avoiding misusing existing models. Common misuses include extrapolating an empirical model beyond the conditions used to parameterize it and applying models and their parameters to temporal or spatial scales for which they were not designed.

7.6.3 Design Submodels Carefully and Thoroughly

The development of every submodel in an IBM should be treated as an independent modeling exercise that includes documentation, literature review, parameter estimation, and testing. The primary reason to conduct all these steps is of course to find the best possible way to represent the submodel's process, with the fewest uncertainties. An important secondary reason is to convince reviewers and clients that uncertainties in each part of the IBM were thoroughly considered and reduced as much as practical. In an influential article on how potential clients should decide if an IBM is useful, Bart (1995) emphasized the importance of submodels that are fully described, tested, and shown to produce realistic predictions. Therefore, the following stages of submodel development should be described in the IBM's formulation document.

A good first step in designing a submodel is searching the literature for models that could be adapted or seeking conceptual models, parameter values, or even field observations that could be useful in designing the submodel. Even if little of use is found, it is worthwhile to document the literature review and to indicate why no existing approaches were used.

After a submodel is formulated, it should be implemented in a test code so it can be calibrated, tested, and analyzed by itself. Spreadsheet software is often useful for test coding because many of us are familiar with spreadsheets, their graphing capabilities are useful, and they are a convenient platform for using the test codes to test the full IBM's software (section 8.5.1). Modelers have also developed test codes in other convenient platforms such as MathCad, MatLab, and S-PLUS.

Next, the submodel can be parameterized to the extent possible. In some cases parameter values will be available directly from the literature, and in other cases parameters can be fitted to available information using formal techniques (e.g., Hilborn and Mangel 1997). Often, however, it is necessary to simply "guestimate" parameter values for simple submodels, selecting values that produce believable results. Some parameters cannot be estimated with any certainty until the entire IBM is assembled; such parameters should be identified as among those needing calibration (section 9.8) and given an estimated value.

Finally, the submodel should be analyzed to develop a thorough understanding of its behavior under all conditions that could possibly be encountered during simulations. It is important to explore how the submodel behaves over all possible ranges of inputs and to document (often, graphically) that it produces realistic, believable results. Submodel errors or unrealistic behaviors will cause much more trouble if they are found later instead of sooner.

7.7   USING OBJECT-ORIENTED DESIGN AND DESCRIPTION

In chapter 5 we show how IBMs are best understood using concepts such as individuals, adaptive traits, interactions, actions and schedules, and observation. In chapter 8 we discuss *object-oriented* software platforms that implement IBMs using very analogous concepts: objects with methods that define their behavior, messages for interaction among objects, schedules of actions, and observation tools. Clearly, following the same object-oriented style in an IBM's formulation can provide a smoother, more natural link between conceptual design and software (as partly illustrated by the coyote model description in section 6.3.3). Using a consistent style of description from conceptual design to detailed formulation to software makes each of these phases easier, makes it easier to track each idea all the way through the design and implementation phases, and helps make sure that the model formulation is complete—that every assumption used in the software is also documented in the text.

An object-oriented model formulation can be developed by, first, understanding the object-oriented software platform in which the IBM will be implemented (section 8.4), and then describing the detailed model design in a way that can be translated directly into the platform. Software engineers have developed somewhat formal techniques for doing so, and it is helpful to at least examine techniques such as Unified Modeling Language and Object Modeling Technique (described in many software books; figures 7.1 and 7.2 are examples of the kinds of diagrams used in these techniques). (See example applications of such techniques to IBMs by Laval [1995, 1996].) However, less formal approaches are often suitable. Instead of the traditional approach of listing equations and parameters, an object-oriented model formulation can completely describe an IBM's detailed design by describing the following kinds of model components (defined in more detail in section 8.3). It should be clear how each part of this object-oriented description fits into the formulation organization recommended in section 7.2.

Classes and *instance variables* describe the model's structure. Each class defines a different kind of entity in the IBM (the individuals, habitat units, etc.). A class's instance variables define the state variables of its objects: age, sex, weight, location, and the like.

Methods and *parameters* define all the IBM's submodels, including the traits of its individuals. Each class has a number of methods that define the behaviors that objects of the class can execute.

Messages are the ways that objects tell other objects either to execute a behavior or to provide some information. Messages typically define such important model characteristics as how individuals conduct interactions and sensing.

Actions and *schedules* define an IBM's model of time. They determine which behaviors of which objects are executed in which order, and define the IBM's temporal resolution.

Observer tools and *observer actions* define how data are collected from the IBM and reported to the modeler. The tools describe what information is reported and how (e.g., summary statistics written to output files, spatial data reported graphically), and the observer actions define how observations are scheduled with respect to other model actions.

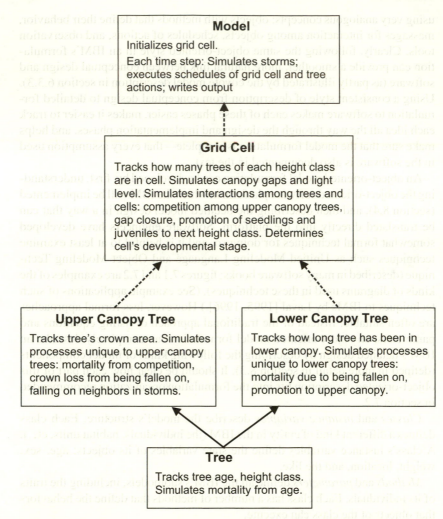

Figure 7.1 An informal, hypothetical, class responsibility diagram for the beech forest IBM described in section 6.8.3. Each box represents a class of entity in the IBM and describes the class's responsibilities: the variables and processes it represents. Dotted arrows indicate *ownership* relations: the model "owns" (creates and controls) the grid cells; the grid cells "own" the trees. Solid arrows indicate *class hierarchy* relations: the Tree class is the superclass of the Upper-Canopy Tree and Lower-Canopy Tree classes. (See also section 8.6.4.)

7.8 USING MECHANISTIC AND DISCRETE MATHEMATICS

Two issues of mathematical style can help formulate IBMs clearly and avoid errors. First, the mechanistic nature of an IBM can be expressed and reinforced by using variables, equations, and rules that have clear meanings and consistent

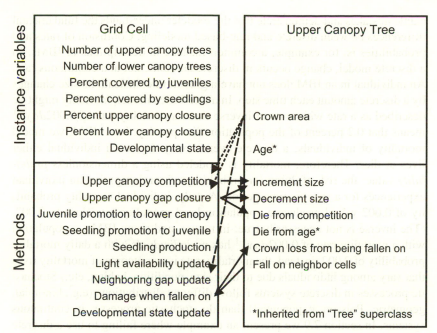

Figure 7.2 A hypothetical class relationship diagram for Grid Cell and Upper-Canopy Tree classes of the beech forest IBM described in section 6.8.3. "Instance variables" are variables for which each object of the class (each grid cell or each tree) has its own value. "Methods" are processes simulated by the class. (For classes representing individuals, instance variables correspond to the individual's state variables; and methods correspond to individual traits.) Dotted arrows indicate *information flow* among classes: methods of one class that use instance variables of another class as input. Solid arrows indicate *control flow* among classes: methods of one class that execute methods belonging to another class. (See also section 8.6.4.)

units. By "clear meaning" we mean that variables represent actual, measurable quantities; and equations and rules represent specific, real processes. By "consistent units" we mean that each variable in an equation has explicit units (e.g., joules of energy per gram of body weight), and the units on one side of the equation match those on the other side. Engineers and physicists are well trained to use these kinds of mechanistic equations, but ecologists are not always—instead, our models often rely on empirical relationships (e.g., statistical models) in which parameters lack clear and measurable units. Using the engineering style whenever possible can help keep the model design focused on real processes, make the model easier to understand, provide the ability to check the units consistency of equations (a safeguard against errors), help make it clear what parts of the model are mechanistic representations of real quantities and processes, and make it possible to parameterize and test as much of the model as possible against data.

The second issue of mathematical style is using discrete mathematics. IBMs are discrete models, so using the mathematics of discrete events can help avoid subtle mistakes and demonstrate that the modeler understands the fundamental differences between discrete and rate-based modeling. Confusion of rates and probabilities is, for example, a common mistake in descriptions of IBMs. In a discrete model, change occurs in discrete amounts, not at a continuous rate. An individual in an IBM does not have a growth rate; instead, its size changes by a discrete amount each time step. In a rate-based model, mortality might be described as a rate with units of inverse time: a mortality rate of $0.002 day^{-1}$ means that 0.2 percent of the population dies per day. In an IBM, we model mortality of individuals, a discrete event: every day, each individual either lives or dies. Therefore, mortality is modeled using a dimensionless *probability* value: the risk of dying (or probability of surviving) that an individual experiences for a specific time period. A daily, individual mortality probability of 0.002 would produce a population-level mortality *rate* close to 0.002. (The inverse is not necessarily true: in nature, and many IBMs, a population with a mortality rate of $0.002 day^{-1}$ has no *individuals* with a daily mortality probability of 0.002; instead, the mortality rate is an outcome of mortality risks that vary among individuals due to their size, condition, habitat, etc.) Stochastic processes in discrete systems follow different distributions (e.g., binomial, geometric, Poisson, exponential) than do stochastic processes in continuous systems. In section 7.9 we present an example where failing to use a discrete distribution could cause serious errors.

Ecologists may find it useful to browse a few textbooks on discrete mathematics. The discrete-event simulation literature (section 8.3.1) also provides background on this topic, with content more likely to be directly applicable to IBMs.

7.9   DESIGNING SUPERINDIVIDUALS

The concept of "superindividuals" is used to make simulation of very large populations computationally feasible. A superindividual is a model object that is generally treated as an individual but represents many (N) individuals (figure 7.3). Huse and Giske (1998), for example, used superindividuals to make a large-scale marine fish IBM computationally feasible; their IBM represented a population of 15,000,000,000 fish by using 15,000 superindividuals with $N = 1,000,000$ individuals represented by each superindividual. This technique has been used especially for organisms like fish that have high reproductive and mortality rates, so a population having a moderate number of adults periodically has extremely high numbers of juveniles.

Usually, a superindividual has state variables (e.g., size, location) that apply to all the individuals it represents; all its N individuals are therefore assumed to be identical. The behaviors of superindividuals are generally treated as individual behaviors, but with some obvious exceptions: when a superindividual eats,

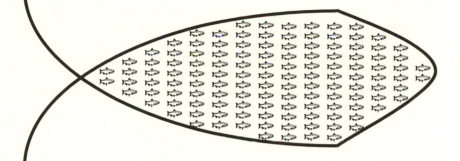

Figure 7.3 A superindividual.

for example, it consumes N times the food of one individual. IBMs that use superindividuals clearly have some parallels to age- or size-structured models—treating the population as a collection of groups, within which individuals are identical—but they still meet the four criteria for IBMs set out in section 1.5. While some of the variability among individuals is obviously lost, this loss can be minor (discussed later). Most important, superindividuals can still exhibit the adaptive behaviors and local interactions that give rise to emergent population dynamics in IBMs.

Several approaches for modeling superindividuals have been explored by Rose, Christensen, and DeAngelis (1993) and Scheffer et al. (1995). These articles mainly address how the relation between the number of superindividuals and N is handled as population size decreases due to mortality. There are three general approaches to this problem, each with advantages and disadvantages.

1. Assume that mortality reduces N of each superindividual while the number of superindividuals remains constant. This approach has the advantage of retaining more variability; as mortality proceeds and N decreases, the model becomes more and more truly individual-based.
2. Assume that mortality reduces N, but then combine superindividuals as needed to keep N relatively constant. When a superindividual's N falls below half its original value (e.g.), it is combined with another to create a new superindividual with N near the original value. A disadvantage is that variability is lost as superindividuals are combined. However, maintaining a relatively constant N can avoid spatial resolution problems: the most appropriate resolution can depend on the value of N (discussed later). The computational advantage of this approach is that the number of superindividuals decreases over time.

3. Assume that an entire superindividual lives or dies together: when mortality occurs, the whole superindividual dies. This approach has the advantages of the second approach: N is constant and the number of superindividuals decreases over time. It does not have the first approach's advantage of retaining more of the original variability, but this limitation can be overcome by using more superindividuals with lower N—if it is computationally feasible. This approach requires fewest assumptions about what superindividuals do.

There is no simple algorithm or theory for figuring out how many superindividuals are needed to retain an acceptable fraction of the population's variability among individuals. Instead, we must experiment with the IBM to understand the effects of using superindividuals of various designs. The value of N should be a parameter that is varied to explore its effects. However, in discrete-space IBMs it may not be trivial to vary N without introducing spatial resolution artifacts. Spatial elements of IBMs using superindividuals must be designed with great care to avoid errors due to inappropriate spatial resolution, and the most appropriate spatial resolution may vary with N. A cell with enough food to support 20 individuals may be excellent habitat for a superindividual with $N = 10$ but very poor for a superindividual with $N = 100$. In general, it is safest to design the spatial resolution and N so that the number of individuals a cell can support is much greater than N.

Superindividuals may be especially appropriate for IBMs that include relatively few adults, treated as individuals, and many juveniles that are represented using superindividuals. If such models are spatial, superindividuals can be designed so that the spatial resolution appropriate for adult individuals is also suitable for the juvenile superindividuals.

A risk in using superindividuals is the temptation to mix rate-based approaches into IBMs, which are fundamentally discrete (section 7.8). Modeling mortality provides an example: it is tempting to apply a mortality *probability* to a superindividual as a mortality *rate*. If an individual's daily probability of mortality via predation (P) is assumed to be 0.05 (resulting in 79 percent mortality per month, not unrealistic for juveniles of some species) and N is 1,000, it is tempting to simply multiply N by the mortality probability P as if it were a rate, determining that 50 individuals die so N becomes 950. Problems obviously set in as N gets lower and $N \times P$ is no longer an integer: rounding up exaggerates mortality and rounding down underestimates mortality. The solution is to stick with discrete math. The number of individuals dying each time step can be drawn randomly from a binomial distribution, which models the *integer* number of times an event (death) occurs in a specified number of trials (N individuals), with P as the probability of the event occurring each trial.

7.10   SUMMARY AND CONCLUSIONS

This chapter describes techniques for designing the detailed formulation of an IBM in a way that facilitates the following phases of the individual-based modeling cycle. Many of the techniques are ways to design and describe an IBM in a style appropriate for complex, discrete, object-oriented simulators. Primary concerns are reducing the real and perceived uncertainty of the IBM and providing a consistent and smooth flow of ideas from the IBM's conceptual design through the formulation and on to its software.

Several of the techniques we recommend in this chapter are fairly radical departures from the way ecological models have traditionally been formulated. One is treating each of an IBM's major processes as a separate model and using many kinds of information to develop, justify, test, and parameterize the submodel. Primary among the kinds of information useful for developing submodels is the autecology and natural history of the organisms being modeled, another way to keep our models grounded in biological reality. Object-oriented model design and description is common in many disciplines and very natural for IBMs, although still rare in ecology. Discrete mathematics and mechanistic equations with consistent units should be used to formulate IBMs, whereas classical models typically use rate-based math or statistical parameters that lack physical or biological meaning.

One of our objectives in this chapter is to make clear the links between an IBM's conceptual design (the subject of parts 1 and 2 of this book) and its translation into a complete, working model that is then used to solve problems (the subject of part 3). Even more important than the techniques we present is the basic idea that the way we formulate a model—translate a conceptual model into a complete written description—largely determines the efficiency and productivity of our modeling project. We also hope that readers understand how the formulation's style can facilitate its translation into software. This translation of formulation into working software is of course the next step in developing an IBM and the topic of our next chapter.

Chapter Eight

Software for Individual-based Models

Early in the development of a scientific field scientists typically construct their own experimental equipment: grinding their own lenses, wiring-up their own particle detectors, even building their own computers. Researchers in new fields have to be adept engineers, machinists, and electricians in addition to being scientists. Once a field begins to mature, collaborations between scientists and engineers lead to the development of standardized, reliable equipment (e.g., commercially produced microscopes or centrifuges), thereby allowing scientists to focus on research rather than on tool building. The use of standardized scientific apparatus is not only a convenience: it allows one to "divide through" by the common equipment, thereby aiding the production of repeatable, comparable research results.

> —*Nelson Minar, Roger Burkhart, Chris Langton,*
> *and Manor Askenazi, 1996*

8.1 INTRODUCTION

This is by far the most difficult chapter in the book, certainly for the authors and probably also for the readers. Developing the software for an IBM is a major step, yet this topic easily takes on a negative tone. For an unfortunate number of early IBMs, the modeling cycle was not propelled forward by the software development phase; instead, the cycle ground to a halt. Many IBMs were implemented in home-made software that contributed to common, yet avoidable, problems: (1) far too much of the researcher's time and budget was spent on programming instead of on science; (2) much time and money was wasted because errors were found *after*, instead of before, a model was put into use; (3) the IBMs lacked both credibility and usefulness because important parts were unobservable and untestable; and (4) models soon fell into disuse because—each being totally different—they were difficult to communicate, understand, or share (Axelrod 1997; Grimm, Wyszomirski, et al. 1999; Minar et al. 1996; Lorek and Sonnenschein 1999).

For new individual-based modelers, we feel compelled to provide plenty of advice so these problems can be avoided, but then we risk scaring beginners away. Developing software for a large IBM is indeed a major job, so we cover software design and testing extensively. Yet we hope beginners do not miss another key message: that there are tools available now that can make software development much easier, even trivial for some IBMs.

Ecologists who are already experienced simulation modelers are also likely to find parts of this chapter frustrating. They will hear the message "You could be doing a better job!" but find only general information on what approaches and technologies would help them do better. Software tools and technologies are themselves complex and adaptive, and different technologies are best for different IBMs. We cannot make this a software engineering book, and if we did it would likely be out of date by the time you read it. So please accept that our aim can only be to help you know what kinds of tools you might need and where to look for help, but we cannot provide a cookbook for software development.

Experienced IBM developers may also not like the many parts of this chapter that discourage software practices that are widespread in ecology. We made a tough decision: where practices that are standard in ecology differ sharply from standard practices of software engineers that specialize in agent-based and discrete-event simulation, we went with the software engineers. Nothing in this chapter is considered new, controversial, or excessive in the field of simulation software engineering (which has, by the way, been around considerably longer than IBMs have).

The underlying reason why this chapter is difficult is that ecology as a discipline still has a naive attitude toward software. Universities rarely provide ecology students with a strong background in software design or even (in many programs) modern programming skills; instead, students follow the "do it yourself" approach that Minar et al. (1996) describe in our chapter motto as characterizing an immature science. This attitude is frustrating because we do not need to develop new or expensive technologies, only to adopt and adapt software and modeling approaches that are already cheap and widely used—just as we have adopted other technologies such as geographic information systems, statistical software, remote sensing, and telemetry. Talented software developers take great pride in never programming anything they do not absolutely have to; instead, like any good scientist, they search out proven tools and methods.

How can we keep software from making IBE more difficult or controversial than other model-based approaches to ecology? In our vision of IBM software heaven, modelers could describe their IBM on paper using some kind of language that (1) people can understand intuitively, (2) is widely used throughout ecology, (3) provides "shorthand" conventions that minimize the effort to describe the IBM rigorously and completely, and (4) can be converted directly into an executable simulator without the possibility of programming errors. After converting the model description into an executable simulator, the modelers then could turn the simulator into a *simulation laboratory* by attaching experimentation tools: probes to collect data; displays to show results visually; and controls that automatically generate, execute, and interpret the kinds of analysis experiments we discuss in chapter 9. What would such a language look like? Possibilities include graphical languages in which the modeler draws a "picture" of the model, which is then turned into an executable code; menu-driven systems in which a model is described by selecting menu options; and simplified programming languages that use statements resembling human language.

But what do we do until this vision of software heaven is realized? (This question is, of course, the topic of this entire chapter.) First is to be aware that many elements of the vision are already real. There already are graphical languages, menu-driven systems, and simplified programming languages that can take much of the work out of IBM software. Currently these languages or platforms are an essential, but only partial, solution. For most IBMs, the most valuable approach we can follow now, both to facilitate our own research and to promote IBE in general, is to use one of the "framework" platforms (section 8.4.2) that are, essentially, programming languages specifically for agent-based simulation. These frameworks are designed to be the standardized scientific apparatus that Minar et al. (in the chapter motto) say will catalyze progress. And we should consider it an important part of our job to support development of these tools, not only financially but by contributing software and helping other users. Mainly, we need to overcome the temptation—very strong in an immature field—to start over from scratch every time. The more we share (and contribute to!) common software tools, the more rapidly these tools will improve toward our ideal vision.

The history of statistical and geographic software should be an inspiration: as more people used the same tools, the tools rapidly improved and standardized; and it therefore became far easier to do and communicate our work. Now, in a journal publication we can describe an elaborate statistical procedure by simply saying "we used PROC GLM in SAS." Sharing and contributing to platforms specifically for agent-based modeling will help us get to the same point.

Our goal in this chapter is to help ecologists prepare for and manage the inevitable challenges (and pleasures) of bringing an IBM to life in well-designed software. A key message is that software *design* and *engineering* must be taken seriously; programming skills are not enough and, in fact, are not always even necessary for the ecologist. But individual-based modelers do not have to become experts in software engineering; we try to show that ecologists can be more productive by knowing the *right* stuff, not necessarily more stuff, about software. Therefore, we focus on familiarizing readers with the tools and techniques that are available and equipping readers to participate in the growing world of agent-based simulation software. Especially, we try to help ecologists estimate the level of effort needed for software, a crucial step in planning an IBM project. Another important goal is to help research program managers and proposal reviewers better understand the resources—plans, software tools, interdisciplinary collaboration—needed to keep software development from sinking an IBE project.

We begin by explaining why software is more important for IBMs than for conventional models. Next, we briefly describe a handful of software engineering concepts that are particularly important for developing IBMs and communicating with those who design software for agent-based modeling. We then present a number of specific strategies and techniques for efficiently designing and implementing an IBM's software. Two of these strategies—selecting an appropriate platform and testing software—are so important that we discuss them in separate sections.

8.2   THE IMPORTANCE OF SOFTWARE DESIGN FOR IBMS

Modelers often think of a model's software as simply being an "implementation" of a model that also exists as a written "formulation" (discussed in chapter 7). However, this concept of the software being only a computer-executable version of the model does not work for IBMs because the software must do much more than just implement the model. John Holland, a pioneer of both computer science and simulation of complex adaptive systems (CAS) illustrates the issue using the analogy of a flight simulator program (Holland 1995, p. 157):

> To be useful, the flight simulator must successfully mimic the real plane under the full range of events that can occur. Solid theories of aerodynamics and control, a natural cockpit-like interface, and superb programming are vital ingredients of an acceptable flight simulator. Given this complex mix, how is one to validate the resulting simulator? Even relatively simple programs have subtle bugs, and flight simulator programs are far from simple.
>
> Enter the experienced pilot. The pilot "takes the simulator out" for a series of test flights.... If the simulator performs as the pilot expects, we have a reality check; if not, back to the drawing board....
>
> This means of attaining a reality check sets a goal for simulations that mimic real systems. Individuals experienced with true *cas* should be able to observe familiar results when executing familiar actions in the simulator. This puts a requirement not only on the programming, but also on the interface provided.

The flight simulator analogy illustrates the first of several reasons why software is a bigger concern for IBMs than for other kinds of models: an IBM's software must provide a laboratory for experimenting on the model. We need to think of ourselves sitting in front of a display that gives us a thorough feel for the complex simulations going on in our IBM and how closely they resemble the real system we are studying. And we can only understand and learn from IBMs by conducting simulation experiments to figure out how results arise from what processes and traits (the subject of chapter 9). A computer program can implement an IBM's formulation perfectly, but we can still learn nothing until it also allows us to "see" and conduct experiments on the virtual ecosystem.

The greater complexity of IBMs is a second reason their software is more important. Many IBMs simulate a variety of processes for one or several kinds of organisms, plus habitat dynamics. The software must manage and collect data from large numbers of individuals that are variable and continually changing. As a consequence of this complexity, designing and writing the software is more work, with more potential for error.

We must also pay more attention to software because errors are more difficult to detect in IBMs. By their nature, IBMs produce complex and novel outcomes, making the consequences of programming mistakes difficult to distinguish from useful results (or from errors in the formulation). "Taking the simulator out" to see if it looks and feels right is a necessary, but by no means sufficient, testing procedure. In most IBMs there are essentially infinite states and control pathways the software can attain, so IBMs often have errors that are manifested only in rare situations or in a few individuals. And because we are simulating complex adaptive systems, we cannot assume that rare or small software errors are unimportant. IBMs, like the real systems they represent, can be sensitive to rare events and sometimes can tip sharply from one state to another when a threshold is crossed (e.g., Huse, Railsback, and Fernö 2002b; Lammens, van Nes, and Mooij 2002).

Finally, communicating an IBM's software to its "clients" is much more important than for classical models. Clients may be a student's advisers, a research program's sponsors, the readers of a journal article, or resource agency staff that might use the IBM to make management decisions. These people know, even if only intuitively, that any seemingly small detail of how an IBM is implemented can affect its results—so they cannot understand and trust an IBM unless they can understand and trust its software. A key job of the software development process, and the software itself, is therefore to promote the clients' understanding and trust.

8.3   SOFTWARE TERMINOLOGY AND CONCEPTS

In this section we briefly introduce a few terms and concepts applicable to software for IBMs. One purpose of this section is to encourage ecologists to think about software in more ways than the traditional perspective of a computer program that reads in data, executes an algorithm, and then prints out results. The second purpose is to equip ecologists with the tools to participate in the large community of scientists and engineers building and using software for agent-based models. Participating in this community is one of the most essential keys to success with IBMs. Each of the following subsections introduces a term or concept that is important to understand in thinking and communicating about software for IBMs.

8.3.1 Discrete-event Simulation

IBMs are of a class of model known as discrete-event simulators. This term denotes that instead of representing processes as occurring at continuous rates, processes are modeled as discrete events happening independently at specific times. Events can be represented using a wide variety of nonlinear and complex algorithms. Discrete-event simulation has been used in many fields for many decades and has produced a large body of established theory and software (e.g., Fishman 2001; Zeigler, Praehofer, and Kim 2000). Much of the software technology now available for IBMs is simply an extension of preceding discrete-event simulation technology. Whenever we need an algorithm that is not specifically biological, we should look in the discrete-event-simulation literature (software and books) before making one up.

8.3.2 Software Platforms

By software platform we refer to the programming language or environment used to convert the model into executable code and then run it. Software platforms used in ecological modeling range from procedural programming languages (e.g., Basic, C, FORTRAN) to high-level environments in which specific kinds of models can be built and executed with little programming (e.g., RAMAS for risk assessment using matrix population models; SAS for statistical models; StarLogo for simple IBMs). Platforms for IBMs are discussed in section 8.4.

8.3.3 Observability

Observability refers to the ability, provided by software tools, to see what is going on in an IBM, including environmental conditions, individual behaviors and interactions, and spatial patterns. Techniques for this job are discussed in section 8.6.5.

8.3.4 Object-oriented Programming

Object-oriented programming (OOP) is a software design paradigm that is the conventional approach for implementing agent-based models. Here, we very briefly describe OOP and its advantages for IBMs. We strongly encourage anyone building an IBM to become familiar with OOP, perhaps by reading one of the many books on the topic (e.g., Gilbert and McCarty 1998; Weisfeld and McCarty 2000; NeXT 1993 provides an excellent introduction for beginners and nonprogrammers and can be freely downloaded from the Internet). Prominent languages for OOP include C++, Delphi, Java, Objective-C, and Visual Basic.

Many of us learned to program using the procedural programming paradigm, in which program statements are executed in the order they are written except when execution is controlled by loops (FOR, WHILE), logical statements (IF . . . THEN . . . ELSE), and subroutine calls. In procedural programming, data are typically stored in arrays. For example, the population of individuals in an IBM could be represented in an array, with a row for each individual and a column for each variable representing the individual's state. In contrast, OOP represents program code and data in discrete objects. An IBM might include a collection of objects that each represent an individual, another collection of objects representing the habitat units, one object that controls the individuals and habitat units, and other objects collecting data and providing graphical output. The programmer writes *classes* that each contain the code for one type of object. (The word "class" refers both to a type of object and to the software that implements the objects; objects are also called "instances" of the class.) All the objects of a class share the class's code, but each has its own data describing its state, in *instance variables*. A bird IBM, for example, would include a class Bird that codes all the traits of the bird individuals. The Bird class also defines instance variables for sex, age, size, and location. When the model is executed, the Bird class is used to create many objects that each represent an individual bird. Each such bird object has its own value for the instance variables, thereby defining the bird's unique sex, age, size, and location.

A class's code is broken into separate *methods*, each method coding one particular thing that its objects do. Methods are analogous to subroutines in procedural programming languages, except that in OOP *all* code is broken into methods. Objects communicate with each other (and among their own methods) by sending *messages* that tell a specific object to execute a specific method. Messages serve two purposes: causing objects to execute their methods, and transferring information among objects.

There are many reasons why the OOP paradigm has become the standard approach for discrete-event simulation and agent-based models. Following are some of OOP's important advantages for IBMs.

Code That Resembles the Model.   The primary advantage of using OOP for IBMs, in our opinion, is that it makes the code resemble the system being modeled more closely. IBMs are designed to represent individual organisms, habitat patches, and other discrete entities; and the specific ways these entities interact and communicate with each other. In OOP, these discrete entities are represented as separate objects, and the modeler explicitly decides what each object knows about each other object, and how the objects communicate and interact. Conventional programming approaches can be used to code the same kinds of communication and interaction among individuals, but it is much more natural to use OOP.

The great benefit of code that more closely resembles the model is that much less abstraction is required to convert an IBM from its written description into working code. The more closely the code structure resembles the real system being modeled, the less translation is required between the modeling concepts and the code design. A population of some species can be considered a collection of unique individuals with a common genetic heritage; in an OOP code, this population is treated as a collection of unique objects of a common class, not as rows of numbers in an array. The resemblance between code and modeled system makes it easier to implement natural processes like interaction among individuals and makes the code easier to understand. (In section 7.7 we encourage formulating an IBM in the same object-oriented paradigm, to reduce abstraction between the formulation and the system being modeled as well as between formulation and software.)

Metaphor.   Metaphor is a way of understanding something by comparing it to something else. OOP promotes metaphor as a way to make code easier to design and understand. Metaphor makes the code seem less abstracted so we can think about it using everyday concepts and terminology instead of having to think about what is actually happening within the computer.

Metaphor is fundamental to the basic concepts of OOP presented previously. Calling a small piece of computer memory an "object" and the code for the object's behavior a "class" is metaphor, describing the computer operations using simple, everyday concepts. OOP programmers routinely talk about whether an object "knows" some information (has its own variable representing the information) or has to "ask" another object for it (send a message to the other object, which returns the information in response to the message). One object "tells" another object to do something (by sending it a message that executes a method).

Metaphor specific to a model is also encouraged by OOP. During design of an IBM, the modelers and programmers may think about, for example, whether a "rabbit" (a code object of the class representing rabbits in a rabbit population IBM) "knows" the food availability (has a variable for food availability) or whether the rabbits can "get" (or "sense") the food availability (send a message, receiving the food availability value in return) from "their habitat" (the habitat patch object pointed to by a rabbit's variable representing its current patch). This kind of metaphor increases the abstraction in the description of the code, so *less* abstraction is needed between the system being modeled and the code description. As a result, the modeler can think about and document the model and its code using concepts and terminology that

are familiar to ecologists, while still providing a rigorous description of the
code.

Hierarchical Organization.   Building an OOP code requires the modeler
and programmer to make a number of explicit decisions about how the code is
organized, including:

- The code's class hierarchy (also known as its *constituency*). What classes
  are needed, and which should be subclasses of other classes.
- Which classes should contain the code for each part of the model.
- In each class, how many methods there are and what function each has.
- What state variables each class has—or the inverse decision. Which class
  should each model variable be in.
- What objects get what information from what other objects.

Making and implementing these decisions can lead to a well-organized, hierarchical code design. (Tools like Unified Modeling Language and Object
Modeling Technique are simply ways to document these decisions; see section 7.7.) For example, all the code for what the individuals in an IBM do is in
one place, and all the code for what the habitat does is in another place, so the
two are not confused. Within a class, we generally write one method for each
major equation or assumption in the model. This organization makes it easy to
find and modify the code for each assumption.

Flexibility in Process Control.   OOP makes it easy to use great flexibility
in process control; it is simple and natural for any object to pass execution
control to any other object. This makes it easy to program natural processes
like interactions among individuals.

Protection of Code and Data.   The OOP approach isolates both data and
code to make them less subject to unintended alteration. Data are encapsulated
and protected within objects instead of contained in public variables or arrays.
Code is encapsulated in several ways. Each class has its own completely separate code; code for one class has no effect on code of other classes. Within
each class, code is further encapsulated in separate methods. The variables
within each method are local (except specially declared exceptions), so ten
methods can each have a variable called aParameter and these variables have
no effect on each other (just as ten classes can each have a method called
dailyUpdate, each doing completely different things). Having each important
process or equation in a model encapsulated in a single method avoids the
risk that altering one part of the code unexpectedly affects some other part.
Likewise, having each individual's state variables encapsulated in a separate
object is much safer than representing the individuals in an array where state
variables can be accessed and modified by many parts of the code. Experienced programmers know that this kind of protection often saves a lot of
trouble.

8.3.5   Causality

The concept of causality refers to *how* a simulation model reaches the states
and results that it does. Understanding the causality of a model result is usually
as important, or more so, than the result by itself. However, figuring out how an
IBM's results arose is often a challenge. The causality of a model result can be a
combination of the software's algorithms, the input data driving the model, and
the initial conditions; often, just changing model input or parameter values can
completely change the order in which various algorithms are executed. One of
the potential benefits of using IBMs instead of studying real ecosystems is that
causality can eventually be figured out: given adequate software tools, we can
observe and experiment as needed to determine how results arose (chapter 9).

8.3.6   Software Evolution and Maintenance

Because model development and use follows a cycle of testing, revision, and
experimentation, software development is never a one-time job. The software
is routinely changed as an IBM is developed and tested. Even after an IBM is in
full use, its code requires occasional maintenance such as providing additional
kinds of output needed for new applications; fixing bugs that were previously
undetected; adapting the code to new versions of the software platform or computer operating system; improving execution speed when we need to simulate
bigger systems; and, of course, updating the documentation to reflect all such
changes. In addition, good models are often revised, extended, and adapted to
new problems long after they were originally implemented.

   At the same time, we very often need to go back and reproduce simulations
made in an older version of the model. A typical example is when comments
on a journal submission require reproducing simulations that were conducted
many months previously. If the code or input files have been modified in undocumented ways, or cannot be identified, or do not run on the new computer we
are now using, we cannot reproduce our own results—a serious credibility
problem. We need not only to plan for future maintenance and evolution of the
software, but also to manage software change in a way that lets us reproduce
old versions.

8.4   SOFTWARE PLATFORMS

Selecting an appropriate software platform is one of the most important steps
in IBM software development because the right platform can make the entire
process much more efficient and likely to succeed. Selecting the right platform
is how we take advantage of all the work that other people have done for us: for

any kind of IBM, there are now platforms that provide tools we need and greatly reduce the programming effort.

Unfortunately, we cannot simply recommend the best platform because different platforms are most appropriate for different IBMs and because new platforms regularly appear. Instead, we provide some criteria for selecting a platform and review the types of platforms now available.

8.4.1 Criteria for Selecting a Platform

The following criteria are useful to consider in selecting a platform for an IBM. No platform provides all the capabilities listed here; however, most IBMs will need most of these capabilities. Whatever necessary capabilities the platform does not provide will have to be developed from scratch.

Support for Individual-based Simulation. A platform is much more useful if it not only *allows* individual-based simulation but also actively *supports* IBMs with built-in designs and code. How much of the IBM's formulation can be implemented using the platform's capabilities instead by writing new code? Does the platform provide facilities for creating and managing collections of model objects, scheduling model events, storing spatial information, and performing spatial functions (e.g., tracking locations, calculating distances between objects, identifying nearest neighbors), and generating random number sequences from a variety of distributions? The more IBM support a platform provides, the less new code must be written, debugged, tested, and documented.

Code Analysis Capabilities. How does the platform help the modeler understand, debug, and test the software? What tools are available to help understand causality as a model executes? Competent debugging facilities are essential (except perhaps for high-level modeling environments). Code profiling tools, which tell how much execution time is spent in specific parts of the code (section 8.7.4), are very useful for understanding how the code executes during a simulation. Graphical programming environments (described later) illustrate the software design visually, which can help modelers understand and communicate the model and its software.

Observer Capabilities. What tools does the platform provide for observing the IBM as, and after, it executes? Does the platform provide graphic user interfaces (GUIs) allowing behavior of model individuals to be observed? How deeply into the model does the platform allow you to probe—for example, is it easy to identify a single individual and track its calculations and behavior? Is it easy to obtain summary output of any desired kind (e.g., to get the weekly mean, minimum, maximum, and standard deviation of individual weight, broken out by species and age class)?

Links to Other Software. Often an IBM can be greatly enhanced by linking its software to other programs. A simple kind of link is writing output files so they are easily imported into spreadsheet or statistical software to analyze results, or designing input files so they can be generated by a geographical information system (GIS). The ability to link directly an IBM's software so it uses other packages *as it executes* can also be very desirable; for example, why write code for spatial or statistical functions when these functions could be provided by linking to a GIS or statistical package? The Swarm platform (section 8.4.3) has a facility making it easy to save, in the HDF5 database format, the state of all the individuals at any time during a simulation. This database can then be analyzed directly with the R statistical language or used to restart the simulation in its saved state.

Tools for Model Analysis. In chapter 9 we discuss approaches for analyzing an IBM once it is implemented, and some analyses can be at least partly automated. What analysis tools does the platform provide? Can multiple model runs be created and executed automatically to compare scenarios or analyze sensitivities? Are tools provided for Monte Carlo analysis or parameter fitting?

Ease of Use. How easy is it to learn to use the platform, and how productively can models be implemented? How complete are training classes or materials, documentation, and user support? Unfortunately, there is usually a trade-off between ease of use and the platform's flexibility and generality: more complex IBMs are likely to need platforms that require more effort to use.

Cost-effectiveness. The cost of modeling platforms varies widely, but it is important to consider cost-effectiveness: would a platform's cost be offset by increases in productivity and usability of the IBM? Many platforms cost little or nothing, but a very expensive high-level platform could be cost-effective if it saves months of effort for writing and testing code and allows models to be analyzed rapidly.

User Community. Is the software platform widely used for individual-based or agent-based simulation? Does it have an active user community? Many ecologists participate in the user communities of geographical information systems and statistical packages by subscribing to email lists, attending user conferences, and reading and contributing to platform-oriented publications. Some of the platforms discussed here for IBMs have user communities focused on simulation modeling in general, or specifically on agent-based simulation. Our experience (primarily with Swarm) has been that the user community can provide the following very important benefits: first, a forum for interacting with scientists working on agent-based simulation in a wide variety of fields, not just ecology, a cross-fertilization that is extremely valuable, especially because much of the pioneering work on complex systems and agent-based simulation takes place in fields other than ecology; second, user support such as help in fixing bugs and finding the best way to design software; third, shared code, from useful, reusable classes to complete models that can be adapted to new purposes; fourth, feedback from users to the platform's developers on how the platform can best be improved. But the most important advantage of using the same platform as many other agent-based modelers is promoting the standardization of tools that we need for this new scientific approach to mature (section 8.1).

Execution Speed.   Execution speed is rarely the factor that limits how quickly good science is conducted with an IBM, and other characteristics of a platform are usually much more important. However, platforms can vary greatly in how fast they execute the same model, and for some large IBMs the difference can be important. Unfortunately, reliable data comparing the speed of different platforms are rarely available, and the speed of each platform may be highly dependent on subtle characteristics of its design. Software engineers or experienced modelers may be able to provide useful advice on the relative speed of alternative platforms.

8.4.2   Types of Platforms

Software platforms likely to be considered for IBMs generally fall into the following categories. We offer a few comments on the relative merits of platforms typical of each category, although each platform is different so our comments should be treated as generalizations likely to have exceptions.

Procedural Programming Languages.   Many IBMs have been implemented using conventional, procedural programming languages like C and FORTRAN. However, these languages meet few of the preceding criteria for good platforms. Procedural programming languages provide no direct support for IBMs, so the entire model must typically be coded "from scratch." Especially when used with traditional techniques such as storing state variables for individuals in arrays, these languages are relatively clumsy and risky for individual-based simulation. Observer tools are especially lacking, although external graphics libraries may be useful to some extent for viewing simulations as they occur. Compared with alternatives such as agent-based simulation frameworks, procedural programming languages have many disadvantages and few benefits other than familiarity.

Object-oriented Programming Languages.   As discussed previously, the OOP paradigm has important advantages, being more naturally suited to IBMs. For popular OOP languages like C++ and Java there are many library classes with potentially useful tools and observer capabilities. Like other programming languages, however, these platforms provide little direct support for IBMs so they require the code to be written mainly from scratch; they do not provide reusable software designs that make different IBMs easier to compare; and they lack tools for observing and experimenting with an IBM.

General High-level Modeling Environments.   This category includes platforms that typically provide (1) a simplified programming language, (2) code for a number of common modeling tasks, and (3) graphical output. An example many modelers are familiar with is MATLAB, a platform for matrix mathematics and modeling. Of more interest for IBMs are high-level platforms for object-oriented modeling such as MODSIM and Simscript. Compared with OOP languages, these platforms offer a simplified yet highly flexible programming language, support for discrete-event simulation, and graphical display capabilities. However, we are not aware of any general high-level environments that provide capabilities specifically for agent-based simulation.

Graphical Modeling Environments. These platforms allow models to be coded using graphical symbols, with a simplified script for coding details that cannot be depicted graphically. Stella is one of the oldest and most popular of these platforms. Many of the graphical modeling environments were designed for rate-based modeling and are not suited for IBMs. There have been a few, however, designed for discrete-event simulation and possibly useful for some IBMs. These platforms typically provide observer tools like GUIs and at least a few provide model analysis capabilities such as tools for Monte Carlo analysis. We are not aware of any IBMs that have been implemented using one of the commercial graphical modeling environments.

Agent-based Modeling Frameworks and Libraries. A *framework* can be thought of as a standardized, general design for implementing a class of models. A *code library* is usually a collection of reusable object-oriented code classes. Typically, libraries are used as building blocks as a programmer constructs a model's software; in contrast, a framework provides the overall model structure with the programmer writing new code to fill in the details for a particular model. A framework can provide a consistent set of software design concepts, conventions, and tools; these not only make it easier to implement models but also to organize and communicate each model's software, compare different models, and share code and techniques. While a framework is a set of software concepts, it is typically implemented as a code library, a set of reusable OOP classes with which the programmer customizes the framework to a specific model. An example of a library developed for implementing IBMs and grid-based models in C++ is EcoSim (Lorek and Sonnenschein 1999); Swarm and RePast (section 8.4.3) are frameworks and libraries for implementing all kinds of agent-based models and IBMs.

There has recently been some effort toward adding graphical modeling capabilities to these agent-based modeling libraries and frameworks. Tools are in development that allow modelers to at least outline a Swarm or RePast program (defining the types of objects, the schedule, etc.) graphically, while detailed behaviors must still be encoded using a programming language. A good example is RePast's "SimBuilder" tool.

High-level Agent-based Modeling Environments. A small but growing number of packages allow specific kinds of IBM to be implemented very easily. Well-established examples are AgentSheets, EcoBeaker, NetLogo, and StarLogo. More experimental and less commercial products include the MOBIDYC platform, which provides built-in "primitives" for individual behaviors (Ginot, Page, and Souissi 2002, which also provides a review of IBM

platforms). These platforms allow modelers to use menus or simplified programming commands to customize agents and their environment in a world
that is otherwise highly structured. Most of these environments allow users to
write at least limited code for custom behaviors. In addition to greatly reducing
the work of designing and implementing software, these environments greatly
reduce the effort to document and communicate a model and its software.
The relatively few lines of code (or set of menu choices) needed to implement an IBM, along with the platform's standard documentation, can describe
the model completely enough to make the model reproducible. Some of these
environments (e.g., NetLogo) have active user communities.

Although these high-level environments are less flexible than other platforms,
they can be used to create virtually infinite kinds of agents and environments
and to conduct meaningful simulation studies. StarLogo is designed for use
by precollege students, yet Camazine et al. (2001) developed StarLogo models for several of the self-organizing biological systems they explore, and An
(2001) used a StarLogo simulation to develop a fundamentally new understanding of an important medical problem. NetLogo is an extension of StarLogo
intended for higher-level applications. These platforms can be excellent for
abstract simulations exploring fundamental ecological concepts, for many
IBMs of real systems, and for rapid prototyping of model concepts. Given
their advantages, we recommend modelers thoroughly explore and consider
the high-level environments, especially for prototypes and relatively simple
IBMs.

8.4.3 Swarm and Related Frameworks

Swarm and several related platforms fall in the "agent-based modeling frameworks and libraries" category, but we discuss them in more detail because
they currently are the platforms most widely used specifically for agent-based
modeling and IBMs. We discuss Swarm in particular because it is the most
prominent and because we have more experience with it.

The Swarm project at the Santa Fe Institute pioneered frameworks for agent-
based simulation (Minar et al. 1996). The goal of the project was to establish
something like a programming language specifically for agent-based models.
The product of this project, Swarm, includes a powerful yet general framework, a code library that implements the framework (and provides a variety of
other useful tools), and a community of users. Currently, Swarm is maintained
by the nonprofit Swarm Development Group (www.swarm.org). The general
Swarm approach has also been implemented and modified in several other
software libraries; currently the RePast project at the University of Chicago
(http://repast.sourceforge.net) is doing so most actively. RePast is also widely
used, and most of what we say about Swarm applies equally to RePast.

The overall concept of Swarm is that modelers use a normal programming language to define the behaviors of the specific entities in their model—so individuals, habitat units, and the like can have any characteristics the modeler wants—while Swarm's code provides common functions, such as representing space, managing collections of model objects, organizing and scheduling actions, controlling execution, and taking observations. A key element of the Swarm framework is the concept of a "swarm." A swarm is a software object that includes a collection of objects and a schedule of actions that the objects conduct. A simple IBM implemented in Swarm would include a "model swarm" containing the individuals, their habitat, and a schedule of actions that the individuals perform; and an "observer swarm" containing the model swarm, the observer objects (animation windows, graphs, output files), and a schedule integrating model actions with observer actions. However, any hierarchy of swarms is possible. "Individuals" in one swarm could each be a lower-level swarm. Higher-level swarms can perform such functions as running multiple-simulation experiments, or coordinating multiple swarms that each model different life stages of a species at different temporal and spatial scales. For example, we have modeled salmon using one Swarm program that includes a nonspatial swarm operating at a monthly time step to represent adults in the ocean; and, at a daily time step, a branched one-dimensional swarm for upstream spawning migration, multiple higher-resolution swarms for spawning and egg incubation at various spawning grounds, and another large-scale swarm for migration to the ocean.

Some important characteristics of Swarm and similar frameworks are:

- Providing reusable software *designs*, in addition to reusable code. This characteristic makes frameworks especially valuable for users, like most ecologists, who are not experienced software designers. These reusable software designs do not limit in any way the design or capabilities of the IBM itself.
- Providing a common software organization and terminology. Key aspects of an IBM's software can be described concisely by, for example, listing what swarms are included, what kinds of agents are in each swarm, and what kinds of schedules are used and what actions are on each schedule. Modelers familiar with the framework can rapidly understand each others' code.
- Being designed as a laboratory for experimenting on IBMs, not just implementing them. Swarm provides a number of sophisticated tools for collecting data on an IBM and performing manipulation experiments (some illustrated in figure 8.1). These include GUIs and a very powerful and unique tool: "probes" that allow users to select any model object (including simulated individuals), then observe and even manipulate the object in the midst of a simulation.

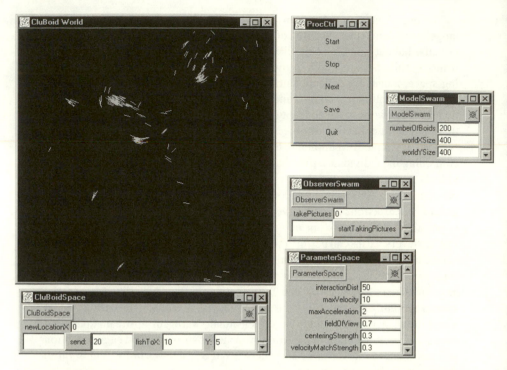

Figure 8.1 Swarm graphical interfaces for observing and experimenting with the "CluBoids" model of Huse, Railsback, and Fernö 2002 (see also section 6.2.3). The animation window ("CluBoid World") shows the location, direction, and speed of each CluBoid: the displayed line segments point in the direction the CluBoid is moving and segment length is proportional to speed. The control panel (*upper right*) allows users to stop and restart execution, or (by clicking "Next") to execute just one time step. The "ModelSwarm" window is a Swarm probe to several model parameters—it allows the user to change the number of CluBoids and the size of the model space before simulation begins. Similarly, the "ObserverSwarm" window is a probe controlling an additional observer tool: whenever the user clicks the "startTakingPictures" button, the software starts writing a "picture" of the animation window to a graphics file after every time step. These pictures can later be made into a movie of the simulation for use in presentations or web sites. The "ParameterSpace" window provides probes to parameters for the CluBoids themselves: at any time, parameter values can be changed via this window. Finally, the "CluBoidSpace" probe allows the user, at any time, to execute a method that sends a selected number of the CluBoids to a specified location in the space (e.g., "send: 20 fishToX: 10 Y: 5").

- Having users code the specific behaviors of the entities in their model using an object-oriented programming language (Java or Objective-C for all the current implementations of Swarm). However, Swarm provides library code that reduces the effort needed to program behaviors.
- Promoting user communities that share ideas and code. One of the greatest benefits of Swarm, in our experience, is that it was developed and is used by very clever people who put a great deal of thought into the very modeling

and software issues that individual-based ecologists must deal with. Swarm users interact via email lists and annual conferences, and routinely share software, ideas, and programming help.

- Being extendible downward into specialized frameworks. Users can write reusable code classes that extend the framework to provide more tools for specific kinds of IBM (Bruun 2001). For example, Swarm users have written additional libraries to interact with geographical information systems and to support ecological IBMs.

8.4.4 Summary for Software Platforms

Selection of an appropriate platform is one of the most important decisions in the process of designing and implementing software for an IBM. Unfortunately, when modelers have not investigated alternative platforms their default choice has very often been to write code entirely "from scratch." This approach maximizes the amount of code that must be written and tested and makes the modeler fully responsible for the software's design, instead of taking advantage of existing designs and code.

Why do so many scientists (not just ecologists) tend to write models in languages like Basic, C++, or FORTRAN when so many other platforms would provide more capabilities for less work? One reason is the energy required to learn a new platform. The effort to learn a new platform may seem even greater if it includes identifying and selecting among a number of alternatives, when none are established as a standard. And we cannot go to the bookstore and find a shelf full of books on agent-based simulation platforms or (usually) take a class on them from our computer science department. However, there are introductory and reference guides for platforms like EcoBeaker, RePast, StarLogo, and Swarm; there are users of these platforms at many universities; and a small but growing number of schools do indeed provide classes on them. A second reason seems to be that many scientists see programming as part of the creative process of modeling and fear that using a specialized platform would take away this opportunity to think about their model in a very low-level way. In fact, the platforms we discuss do *not* reduce the modeler's need to think about their IBM: everything that is unique about a particular IBM must still be programmed from scratch. But the platforms let modelers focus just on what is unique about their model while spending much less effort on the common, nonunique aspects of its software.

The high-level simulation environments like AgentSheets, EcoBeaker, and NetLogo have great advantages as long as the platform is compatible with the IBM's formulation. These platforms should be considered for any IBM. In fact, the potential to use these platforms is yet another good reason why simplifying an IBM "to the threshold of pain" (section 2.3) may result in learning more instead of less: if you can squeeze your IBM into a high-level simulation environment, you can be busy doing ecology with the model much sooner. For more complex IBMs, frameworks and libraries like Swarm can greatly reduce the model design and programming effort while providing essential observer tools and imposing no limitations on the kind of model that can be implemented. As important as these programming benefits are, the nonprogramming benefits of widely used platforms—ease of describing and communicating models, standardized software designs, diverse and talented user communities—are at least equally important.

8.5   SOFTWARE TESTING

A thorough, aggressive testing program is, like selecting an appropriate platform, one of the most important strategies for successful and efficient software development. Minimizing the amount of code we have to write and test is a primary consideration in selecting a platform, but many IBMs will still require a significant programming effort. This section focuses on effective testing of the code we must write.

Unfortunately, testing (also often referred to as verification—that is, verifying that the software faithfully implements the model's formulation) is a critical area of software engineering that few ecologists have experience in. A familiar sign of the naive modeler is the attitude that as long as the code was written carefully and compiles without error, and the output looks reasonable, there are probably not any important problems. This attitude is never productive and often leads to disaster. Early in the code development and testing process, errors tend to be major, widespread, and easy to find—for example, a key equation coded incorrectly so it always produces the wrong result. Later, errors tend to be subtle, affect results only under special circumstances, and be difficult to detect. It is important to remember that there is practically an infinite variety of states an IBM's code can attain during execution; errors that occur only rarely (e.g., only in a few individuals, under special circumstances) are common but require special techniques and determination to find. And it is not safe to assume that seemingly minor or rare errors have negligible effects on model results.

The only productive attitude for IBM developers to have toward software testing is that *errors are inevitable and must be searched out*. We discuss testing early in this chapter to emphasize that testing needs to occur throughout software development. Goals of the testing process are to: (1) find errors as early as possible—to minimize the time and money wasted using erroneous code; (2) search comprehensively for errors throughout the code and throughout the software development process—because many errors only occur under special circumstances and because each change to the model and software can introduce new errors; and (3) document the testing process—to help make sure it is efficient yet complete and to provide credibility with software-savvy reviewers and clients.

Like other kinds of model analysis (chapter 9), software testing should be treated as a research process with a clear experimental strategy. Testers should design experiments, predict the outcomes of those experiments, collect data from the IBM software, compare observed outcomes to predicted to identify discrepancies that are likely due to errors, and finally explain differences between predicted and observed outcomes. In its later stages this testing process is linked to the cycle of analyzing the model itself (chapter 9): we look for unexpected results, then attempt to determine whether they are due to software errors or model formulation problems, or whether they are valid and interesting results. In testing the software, however, we must search very deeply into the model, not just examine general results. Testing can be time-consuming but can also be a fun and creative kind of detective work.

8.5.1   Testing Methods

A hierarchical approach to software testing is efficient because it finds the major, obvious errors early in code development while comprehensively testing code as it is completed. The following hierarchy of code testing methods is the *minimum* necessary to provide reasonable assurance that an IBM's software is ready to be used. (Even very simple models implemented in high-level platforms should undergo these tests, except probably the systematic tests against an independent implementation; but the tests will be easy.) Our experience has been that the first three testing methods—code reviews, spot checks, and pattern tests— typically find most of the errors that have widespread effects on results, but the final systematic tests invariably find additional errors even in IBMs of modest complexity. We also discuss when in the modeling cycle these testing methods are appropriate.

Code Reviews.   Code reviews are the first line of defense against major programming errors and poor software design, so they need to take place throughout the modeling cycle. Whenever a programmer finishes drafting each part of the IBM's code, it is immediately reviewed by someone else. If someone other than the modeler writes the code, then the modeler conducts these reviews. If the code is written by the same person writing the model formulation, then the reviewer needs to be someone else familiar with (or willing to get familiar with) the software platform and the IBM. The reviewer directly compares the code to the model's written formulation to look for mistakes but also thinks about the overall design (see also section 8.7).

Visual Tests.   These tests are exactly the "test flights" John Holland describes in the quote in section 8.2. Observing behaviors from the GUIs is an easy and absolutely essential way to test for errors in coding (and, just as important, in the model formulation and input data; Grimm 2002). In every new model we have implemented in Swarm, there has been at least one very important error, missed in careful code reviews, that was immediately detected from the GUIs.

Visual tests are conducted simply by running the model and observing its behaviors. Modelers should spend time playing with the model, running it under a variety of conditions, and looking for anything unexpected. Because visual tests are so easy and so important, they should be conducted every time a model's software is modified; in fact, they should be conducted every time *any* change is made to the model, including use of new input data or parameter values. Modelers should develop the habit of carefully scrutinizing an IBM's execution visually before using the results for any kind of analysis.

A good example of visual testing is from a model of juvenile salmon migrating down a large river (an early version of the model of Anderson 2002). Salmon movement was modeled as resulting from the river's two-dimensional velocity field plus a component due to random swimming. As soon as we executed the movement code, we noticed from a GUI that the salmon tended to drift toward the inside of a river bend, which (as any canoeist knows) is unexpected. This tendency was clearly caused by underprediction of movement in the Y (north) dimension when the Y component of velocity was low, and underprediction of movement in the X (east) dimension when the X component was low. Consequently, we immediately diagnosed the error: the code was truncating instead of rounding off the fishes' location coordinates. This error had strong effects on model results but was extremely unlikely to have been detected except by the GUI. Such experiences are the rule, not the exception, in coding IBMs.

Spot Checks.   Spot checks verify a few selected model calculations by comparing results to those calculated by hand. For classical models, spot checks alone can be sufficient to show that major parts of a model are coded correctly; for testing an IBM, however, they are of important but limited usefulness. Spot checks are most important as an early test for major and widespread errors, especially in how equations are coded. Spot checks can be conducted for each piece of code as it is implemented or modified.

Extreme Input Tests.   Whenever visual tests and spot checks are conducted, they should include "extreme" values of parameters and input data, which tend to expose subtle implementation problems. Examples of extreme input include setting parameters to very high and low values and to zero (because zeroes often cause funny things to happen), and testing input data sets (e.g., for weather) containing extreme values. Extreme input tests should be part of testing and calibrating each submodel (section 7.6.3), not just testing the final software. Our natural tendency is to avoid exposing our code to risky situations that could illuminate its faults, until we remember that the sooner we find the faults the better off we are.

Systematic Tests against Independent Implementations.   Comparing output from two completely independent implementations is generally considered the only way to test code for a complex model with acceptable completeness. (Even this approach is not completely reliable because different programmers tend to make the same mistakes; see Knight and Leveson 1986.) This sounds like doubling the programming effort, but compromise approaches can be very effective with little additional programming.

Systematically testing key parts of an IBM's code against a separate implementation is essential because this is the only way to test the code over a sufficiently wide range of model states. For example, many IBMs have processes that depend on different factors under different conditions. Growth of a plant might be limited sometimes by temperature and sometimes by light availability; an error in how temperature affects plant growth will not be apparent when growth is limited by light. In many IBMs there are also many different pathways through the code—different orders in which different parts of the code are executed. Software testing needs to sample these multiple pathways with reasonable thoroughness.

When software reliability is of utmost importance, it is common for two (or even more) teams to program the entire model independently and then compare both intermediate and final model results. However, this effort is not always justified for ecological models. The following process is an efficient way to conduct systematic code tests for most IBMs; it can test all the separate submodels but not the full IBM. While the process is relatively easy and inexpensive, our experience is that it typically finds a small number of sometimes-important errors, even after the other testing methods have been completed.

1. Use the test code developed during the model formulation process (section 7.6.3) as the independent implementation of each submodel. We typically use spreadsheet software for test codes, with columns for all the inputs, intermediate calculations, and final results for the submodel. For example, the feeding and growth test spreadsheet for a fish model starts with columns for the fish and habitat variables driving growth (fish size; water temperature, depth, and velocity). Calculation of growth requires calculation of a number of intermediate results like food intake rate, the temperature effect on metabolism, and the total metabolic rate; these intermediate results are each coded in separate columns. Each spreadsheet row models an individual fish.

2. Program the IBM software to write output files reporting the input variables, key intermediate results, and final result for the submodel being tested. For the fish growth example, the code would write one line of output for each fish on each simulated day, and this line would report the fish size; temperature, depth, and velocity; food intake rate; metabolism temperature factor; total metabolic rate; and daily growth.

3. Run the IBM using test input that forces the model over a wide range of conditions, including extremes. For the fish growth formulation, the model might be run over twenty days in which temperature varies over

the full range a fish might ever experience (including 0 degree), using a wide variety of fish sizes and depths and velocities. It is extremely important (and easy) to generate a wide variety of test cases; we typically examine tens of thousands of results to test one submodel thoroughly.

4. Compare both intermediate and final results from the IBM software and the test code. The IBM output file is imported into the spreadsheet that contains the independent implementation. The spreadsheet is then used to reproduce the IBM's calculations and to calculate the difference between results calculated by the IBM and calculated by the spreadsheet. For the fish growth example, spreadsheet columns would be added to calculate the percent difference between the IBM software and the spreadsheet in calculated values of intermediate (food intake, temperature effect, total metabolic rate) and final (growth) results. Differences that cannot be attributed to computer rounding error (e.g., differences with absolute value greater than 0.001 percent) are then identified.

5. Finally, undertake the detective work to explain the differences found between implementations. Often, differences due to an error are small (only 0.01 of 1 percent or less between the two implementations), or occur in a small minority of cases. However, such differences do indicate that there are potentially important errors to be identified and corrected. Most errors can be explained quickly, but occasionally it requires a substantial investigation to figure out an especially subtle error. Occasionally, errors are found in the software platform itself. One Swarm user, when independently checking results, even discovered errors caused by the computer chip.

When in the modeling cycle should this intensive level of testing be conducted? The answer is: before any significant investment (of time, money, credibility) is made in analyzing or using the IBM's results. Early prototyping and model design (e.g., implementing simple "null" models; see section 2.2) may not require this level of software testing, but it is a mistake to postpone comprehensive software testing until an IBM is "finished." Remember that the purpose of these tests is to save time and effort by finding mistakes as early as possible, so they should be completed before the cycle of model analysis, testing, revision, and parameterization (chapter 9) is entered in earnest. Parts of the software that have been tested comprehensively need not be retested unless they have been altered. However, when an IBM reaches the stage of the modeling cycle where we test and revise alternative versions of key submodels (as in theory development; chapter 4), the submodel code will need to be thoroughly retested each time it is significantly revised. The keys to keeping retesting from becoming painful (or, worse, neglected) are automation and documentation.

8.5.2   Automation and Documentation of Testing

Automation and documentation of software testing go hand in hand: both help make testing (and, especially, retesting as an IBM is revised) efficient and reproducible. Automating parts of the testing process instead of conducting them entirely by hand may take more effort initially but will save substantial work as parts of the software are retested during the modeling cycle. By "automation" we do not necessarily mean creating a giant program that does all of the software testing for you. Instead, there are simple yet effective tricks to make specific code tests easier and more reproducible. Examples include:

- Archiving special input data sets used for testing.
- Providing permanent code that creates optional debugging output files.
- Creating and archiving spreadsheets (or similar programs) that test the debugging output as discussed in section 8.5.1.
- Creating special programs that "test drive" the IBM's submodels by making them execute over wide ranges of input.

Any little technique that makes it easier to repeat and reproduce software tests should be considered, especially for testing code that is likely to be revised.

Automation can also make it easier to document software testing. Most of the value of testing software is lost if the tests are not documented. Documentation should include logging the kinds of tests that were conducted, on which pieces and versions of the code (and which versions of the formulation they implement), using exactly which parameter values and input data, on what dates, and by whom. Test records, such as the IBM outputs that were tested, methods used to automate the tests, and the code providing an independent test implementation, should be archived. This information can, for example, all be put in a spreadsheet file; each time a new version of the IBM is tested, the spreadsheet can be copied and updated.

One reason to document code tests is to make testing more efficient. Tests are often repeated to determine whether some change had the expected effect (e.g., did an apparent problem go away after fixing what the programmer *hoped* was its cause?). Often, such tests must be repeated in exactly the same way, and without adequate documentation it is too easy to get confused by whether differences in test results are due to the code change being tested or due to differences in how the test runs were performed. Documentation also helps the modeler keep track of what code has and has not been tested, making it less likely that code is unnecessarily retested or accidentally used without testing.

The second essential reason to document software tests is to record the methods and software used for testing so they can be reused or improved in the future. Documentation is also, of course, very important for providing assurance to the model's clients that software quality was given due attention.

8.6 MOVING SOFTWARE DEVELOPMENT FORWARD

The right platform and a good testing program alone are not sufficient for avoiding all the common traps that ecologists can fall into while developing an IBM's software. In this section we describe several additional strategies for keeping software development from bogging down.

8.6.1 Keep Model Formulation and Software Development Separate

Model formulation and software development are part of the same modeling cycle: changes in formulation are inevitably identified during software development, and model testing and analysis lead to revisions and software changes. However, formulating an IBM and producing its software need to be treated as two separate jobs. The first reason for this separation is to make sure both jobs are done well. When *formulating* an IBM, a modeler needs to be focused on biology and the modeling concepts presented in chapters 2–5; and when *implementing* an IBM, a programmer needs to be focused on the software engineering issues discussed in this chapter. Especially, the modeler needs to avoid making model design decisions simply because they are computationally convenient instead of being biologically justified.

A second concern is the utmost importance of having a written description of the model that accurately and completely matches the software. Especially when modelers write their own software, there is a strong temptation to explore formulation ideas by writing them directly in the code instead of designing them in a written document and in test code (discussed in chapter 7). It then becomes very difficult to keep the written description accurate; after the modeler moves on to other tasks and forgets what changes were made, no one knows exactly what the code is supposed to be doing. When the same person writes both the model and its software, that person must be able to switch back and forth between these two roles without mixing them.

8.6.2 Collaborate with Software Professionals

The question of whether ecologists building IBMs should collaborate with software professionals can be a difficult one. (The term "software professional" is vague, potentially ranging from self-taught programmers to those with extensive training and experience in engineering or computer science. For an IBM project, professionals should at least have expertise in designing and implementing object-oriented software for complex models.) Collaboration can take many forms, but here we focus on the value of a programmer and modeler working together to produce software for a particular IBM. There are many advantages to this kind of collaboration, especially for complex IBMs or research programs developing a series of IBMs:

- A programmer's expertise can greatly improve the software's reliability and usability while reducing the time needed to produce it.
- Instead of spending their time learning software skills and writing code, the ecologists have much more time to focus on formulating, testing, and conducting ecology with the IBM—tasks that are also time-consuming (chapter 9).
- Modelers are more likely to focus on ecology, not programming, while formulating the model.
- At least two people are involved in code development, which has many benefits. These include sharing ideas, forcing the code to be clear and well organized, identifying ambiguities and mistakes in the model formulation quickly, and keeping the project from collapsing if one key person leaves. "More than two eyeballs on the code" is a rule software developers live by.
- Funding agencies often encourage interdisciplinary collaboration, especially in fields like IBE where the collaboration seems natural and productive. IBE projects are likely to offer opportunities for research in software engineering or computer science as well as ecology.

On the other hand, ecological modelers often do produce their own software, for valid reasons. (In section 8.8 we mention some not-so-valid reasons why ecologists have chosen not to collaborate with software professionals.) In the absence of generous funding there may be no alternative. When an IBM can be implemented in a high-level platforms with little or no "from scratch" programming, there may be little need to work with a software professional. And collaborating with a programmer can be frustrating if the programmer does not have enough time to keep up with the modeler.

Our experience has been that software for IBMs of at least moderate complexity has been completed successfully by ecologists alone, but the probability and degree of success increases with the level of support from software professionals. Successful implementation of many IBMs requires software skills well beyond those of typical ecologists, which has certainly been a limitation to the success of IBE. This is in no way a criticism of ecologists—instead, we are saying that ecology is progressing to the point where we require specialists to help build our tools. Learning the skills and tools to design and build IBM software cost-effectively may not be a good use of an ecologist's time or a project's resources. (On the other hand, it may be very worthwhile for an ecologist dedicated to individual-based approaches to invest in developing software skills—as long as the ecologist really learns modern software design and development skills, not just how to program.) At several universities, departments (not yet in ecology, in the United States, as far as we know) have hired software staff to support researchers that use agent-based models by helping design and code models, producing software tools that expand what researchers can do themselves, and enhancing grant proposals by demonstrating the ability to conduct research cost-effectively via interdisciplinary collaboration.

We have worked successfully with programmers in two ways. The first way is hiring a programmer to write the first prototype software for a model and, at the same time, teach us how to use the platform. Often, one or two weeks of intense collaboration is enough to get a project off the ground and make the ecologist fairly competent with a new platform.

However, for sustained high productivity on a major IBM project or program, integrating the following software development cycle into the overall model cycle has worked well. Note that in this cycle *the ecologist developing the IBM remains fully in control of, and responsible for, the software*. The modeler must understand the software platform well enough to read the code and check it, but need not be a competent programmer.

First, the modeler designs the first draft or prototype of an IBM by attempting to write out the draft formulation in full detail. The goal (unlikely to be met completely) is to specify the draft model so thoroughly it can be implemented unambiguously by a programmer. In addition to specifying the model's formulation, the modeler also develops an observer plan that identifies the model outcomes that need to be observed and how.

Next, the programmer implements the draft model from the formulation. This typically involves collaboration between programmer and modeler in designing the software's structure and organization, user interfaces, and so on. This step also inevitably involves the programmer identifying ambiguities and errors in the model formulation; the modeler corrects them. Interactions between modeler and programmer are frequent as unexpected decisions must be made—for example, the modeler reconsiders parts of the model after working out the implementation details, or new observability needs are identified. The products of this step are not only draft software but also a more complete and well-considered model formulation.

As the code is drafted, the modeler reviews it thoroughly with the objectives discussed in sections 8.5.1 and 8.7.1. And, as the code is produced and reviewed, the modeler (not the programmer) prepares the software's written documentation (see section 8.6.4). This job reinforces the modeler's familiarity with the code and lets the modeler design the software's input and output files and other user interfaces.

After one or several cycles of code review and revision indicate that obvious mistakes have been found, cycles of testing and revision of both formulation and software can begin. The modeler should be the person with primary responsibility for designing and conducting code tests, because the modeler is ultimately responsible for the IBM and because the code should be tested by someone other than the person who wrote it. Finally, as discussed in section 8.7.5, software maintenance and evolution requires collaboration between modeler and programmer even after a model code is put into use.

This development cycle may seem cumbersome at first, but if the participants are dedicated to the collaboration, it can be highly productive while assuring that the modeler gets software with the necessary capabilities and quality.

8.6.3 Design Software to Resemble the System Being Modeled

The ecological system being modeled should serve as the primary metaphor in software design. When an IBM is implemented in an object-oriented platform, this practice reduces the conceptual differences between the software, the model, and the system being simulated. In deciding what parts of a model should be coded in what classes, and what variables should be stored in what objects, we continually think about what happens in the real world. Consider an IBM of prey and predator fish in a lake. Somewhere in the software the locations of the fish must be stored. Metaphorically it makes no sense that a lake would "know" the locations of the fish in it; it makes more sense to assume the fish themselves know where they are in the lake. So instead of the lake object storing locations of all the fish in it, we design the code so the fish objects each store their own location. Considering an environmental variable like temperature, it makes no sense for temperature to be a fish variable: fish have no control over temperature. Instead, it makes sense that temperature is modeled by the lake, and if fish need to "know" the temperature they would "sense" it from the lake—fish objects would send a message to the lake object that returns the current temperature value.

(However, compromises are common and metaphors are sometimes ambiguous. Sometimes we *do* have a habitat object keep track of the location of the organisms in it. That way, a fish can find out from the lake how many other fish are nearby instead of having to ask each other fish where it is.)

8.6.4 Make Multiple Representations of the Model and Its Software

There are many different ways to represent a model and its software, even beyond the written formulation and computer code that normally are the two most important model descriptions. Using a variety of representations, in human and computer languages and graphics, has benefits such as directing more thought to the code's design before coding begins, helping organize the coding process, and making it easier to fix and revise code. Multiple representations of a model and its code also are essential for communicating the model among research team members and to the IBM's clients. The importance of multiple representations is obvious to anyone who has experienced the frustration of attempting to use, test, modify, or review an incompletely documented model.

The following ways of representing a model and its code are all common and well worth considering, especially for more complex IBMs.

Detailed Model Formulation.   Chapter 7 addresses why and how we prepare a written formulation, which attempts to describe completely a model in human language.

Test Code for Submodels.   During development of an IBM's formulation, simple test codes are prepared to test the submodels (section 7.6.1). These submodels are important independent descriptions of the submodels that can also be used in software testing (section 8.5.1) and to communicate the model. If someone wants to understand how we model some particular process, giving them a spreadsheet or similar implementation of the process is a powerful way to explain the formulation.

Flow Charts and Class Hierarchy and Entity Relationship Diagrams.   Flow charts are widely used to design and describe the general sequence of events in a computer model. They can represent an entire simple model or separate parts of a more complex model. Or one chart can represent the high-level processes and show only major submodels, while other more detailed flow charts show what happens within each submodel.

In OOP several types of diagram are often used to show what kinds of objects are in a model and their relationships to each other. A class hierarchy diagram shows what classes are subclasses and superclasses of each other, information that is helpful for users that just need to understand the model but essential for anyone working with the code. Entity relationship diagrams show what the various kinds of objects in a model do and how they interact. As we discuss in section 7.7 (see figures 7.1 and 7.2), diagram-based object-oriented design techniques such as Unified Modeling Language and Object Modeling Technique can be a natural way to link how we think of the ecosystem we are modeling (e.g., as depicted in influence diagrams; section 2.3), the IBM's formulation, and its software.

In general, these kinds of diagrams are useful for developing and communicating an IBM's overall software design, but they are not well suited to describing an IBM in full detail. Diagrams detailed enough to represent all characteristics of most IBMs would be harder to understand than verbal descriptions.

Working Computer Code.   The working code of the full IBM is the most important representation of the model because it is what produces output for analysis. But well-written code (section 8.7) in a platform like Swarm that provides "shorthand" for many modeling functions can also be a surprisingly clear and concise, as well as thorough, description of an IBM.

Software Documentation.   Written documentation of the working software is well worth the effort of producing it, for any IBM. Software documentation can describe how to install and execute the code, how to prepare input files, the exact meaning of output, and where in the code each model equation or assumption is represented. Software documentation supports model users and facilitates code maintenance and revisions. Even if only one or two people use and modify the code, maintaining up-to-date documentation helps avoid mistakes and wasted time. Experienced programmers can remember the times they thought "I'll remember how to do this" (e.g., how to un-do some tentative change in the code) and then found themselves, a few months or even just days later, laboriously reading source code, spending much more time trying to figure out what they did previously than it would have taken to document the change.

In documenting the software (as in documenting the model formulation) it is important to write down not just *how* the code was designed but *why*. Often a piece of code is designed, perhaps after much trial and error, in a way that works well but seems counterintuitive. If the reasons for the design are not documented, the programmer is likely to review the code months later, see that it seems counterintuitive, and then waste considerable effort trying to "fix" it.

There are several software packages to automate partially the production of software documentation; doxygen and Javadoc are popular examples. These packages do not automatically write up a description of your software but do things like produce a nicely formatted document from comments placed at the start of each key method. At a minimum they can be useful for creating a list of the classes and methods that need to be documented.

Model and Software Revision Records. These records (discussed in section 8.7.5) maintain a useful history of the modeling cycle, documenting what changes were made, when, why, and by whom.

8.6.5 Implement Observability and Analysis Tools Early

When we think of both software development and IBE as cycles driven by simulation experiments, it becomes clear that we need good tools for observing and analyzing the IBM from the start of software development (Grimm 2002). Preparing an observer plan as an IBM's objectives and formulation are determined is a way to ensure that observability receives the attention it deserves in software design. An observer plan identifies all the model outputs needed for three purposes: testing the software, testing and understanding the model's formulation, and conducting the modeling experiments and ecological research the model is intended for. The plan then addresses, for each such output, the observability issues discussed here.

First, analyzing an IBM typically requires looking at results from a variety of views, not all of which can be anticipated before analysis starts. For example, the modeler may realize during analysis that summary data on the weight of individuals need to be broken out by sex as well as age, or that the individuals using each habitat type need to be examined separately. Designing software from the start so that the modeler can choose new output views (relatively easy in some platforms) will save time and avoid frustration.

A second important observer design issue is selecting spatial and temporal resolutions for each output. To avoid overwhelming amounts of output, graphics or file output can be updated at intervals greater than the time step of the IBM: a model with a daily time step could print output once per week. Graphical outputs can use multiple spatial resolutions; an example is code we designed for an IBM of queen conch (a large, snail-like, marine invertebrate). Because daily movement of individual conch is very small compared with the extent of the simulated space, we provided one display showing conch *density* in the model's large grid cells but allowed users to zoom in and observe the individual conch *locations* in selected cells. (This was easy in Swarm.)

Another major step in designing observer capability is determining *how* to observe each kind of model result. The following three kinds of observer tools are almost always valuable.

Summary Statistics. Typical summary outputs include the mean size of individuals, broken out by variables such as species and age; the distribution of habitat area over one or several variables (e.g., how much habitat area there is at each of ten levels of food availability); and distributions of individuals over habitat types or conditions (e.g., mean density of individuals in habitat of each level of food availability). These outputs can be displayed graphically as the IBM executes, but analysis almost always requires postprocessing this kind of output: putting IBM output files into other software for graphical and statistical analysis.

Tracing Individuals. Tracing the state of selected individuals can be useful for software testing and for understanding how and why individuals exhibit the behaviors they do. This kind of observation requires reporting the individuals' state variables (size, location, etc.), the habitat conditions they experience, and enough of their internal variables to understand behavior. One approach is to have the software write this output for all individuals, but the resulting quantity of output may make model execution and analysis difficult. Swarm's "probes" allow the user to select model individuals (e.g., via mouse clicks on a GUI) and output selected variables from those individuals.

GUIs. GUIs, especially animation windows, are essential for observing patterns over space and time, because patterns are much more easily interpreted visually. For spatial models, GUIs showing habitat and individuals over space and time (usually a map of habitat with individuals overlain, updated every time step; however, the "space" on which individuals are displayed can be in dimensions other than geographical ones) are a rich source of information on the model. In addition to benefits already discussed, GUIs are often the only effective way to observe interactions among individuals, and between individuals and their habitat; detect emergent behaviors; and identify unusual or "outlier" individuals that may be especially interesting or important. Another absolutely essential benefit of well-designed GUIs is promoting understanding and belief in a model to its clients. Clients are often wary of a complex "black box" model, but when they see that an IBM's behavior is observable and realistic, their interest and belief grows rapidly.

In chapter 9 we discuss analysis of IBMs to test, understand, and learn from them. Many analyses (e.g., exploring uncertainties and robustness) require execution of many model runs. Software tools that automate such model analyses—generating the parameter or input values, executing the simulation (often without the graphics for additional speed), and recording or even analyzing results—can be very worthwhile. It is best to include such tools from the start of software development because they will be useful for even preliminary analyses of the IBM.

8.7   IMPORTANT IMPLEMENTATION TECHNIQUES

This section describes some techniques that we have learned from software professionals and found very useful in actually programming IBMs once the software has been designed. In this implementation phase, the focus remains on making the software useful for science while avoiding opportunities for undetected errors.

8.7.1   Obtain Critical Reviews of the Code

It should not be surprising to scientists that peer review is a common and highly valued practice in software development. Peer review of code provides the same kinds of benefit that review of journal articles provides. First, review is an essential part of the code testing process (section 8.5.1). Second, review is often a source of valuable ideas for improving a code. Finally, and perhaps most important, programmers who know their work will be reviewed are much more likely to write code that is well organized and self-documenting (section 8.7.2). Without reviews, the temptation to cut corners in coding style, documentation, and testing is a major risk to producing reasonably mistake-free code. Code "clean" enough to be reviewed is also essential to the model's credibility: one of the most frequent criticisms of IBMs is that models are fully defined only by their code, and the code is unavailable or unreadable.

8.7.2   Use Defensive Programming Practices

"Defensive programming" is a software term meaning to program with testability and reliability as the primary concern. We learned these practices from talented and experienced programmers, but they are not always taught (and are sometimes contradicted) in the kinds of introductory programming training that ecologists are likely to have experienced. Most of these practices have the goal of making a model's code easy to read and understand, or "self-documenting." Self-documenting code increases the ease and quality of many parts of the software development process: debugging, code reviews, implementation of formulation changes, code publication, and code reuse and sharing.

1. Use simple and clear logic. A favorite saying of experienced programmers is that "code should be written for people, not computers." Novice programmers sometimes take pride in writing "elegant" code that has as few statements and variables as possible, so computer resources and execution times are reduced. However, the speed of accomplishing IBM-based research is much more likely to be limited by the time it takes people to review, test, and debug a code than it is by machine execution time, and these human processes are more rapid and less painful if the code is written clearly and simply. Instead of designing a code for execution speed from the start, experienced developers first write a code so it is easy for people to review and test. If, and only if, execution time is found to be a problem after the code is put into use, steps can be taken to improve execution time (section 8.7.4).

2. Use descriptive names for classes, methods, and variables; and sentence-like code statements. Modern programming languages and platforms allow names of variables, methods, and classes to be long and descriptive. Using names that convey useful information (a variable's meaning, resolution, units, etc., such as a variable named habitatDailyMeanTemperatureC) may take a bit more typing, but the information it contains can make it much easier to understand and check the code. Well-designed names can even allow code statements to read like sentences. For example, even readers unfamiliar with the programming language (Objective-C) are likely to figure out that the purpose of the following statement is to initialize the weight of a model deer using a random sample drawn from a normal distribution.

```
[aNewDeer setWeightTo:
  [normalDistribution
    getSampleWithMean: deerInitialWeightMean
      withVariance: deerInitialWeightVariance] ];
```

Programming students are sometimes taught to use comment statements generously to explain their code's function. However, excessive comments can be a problem when they take the place of, or distract from, code statements that explain their own function. The reader may accidentally "check" the code by following the logic in the comment statements instead of in the executable statements.

3. Defend against run-time errors. Even after a model code has been tested extensively, erroneous results can result from run-time errors. Especially in complex codes, run-time errors can be common yet difficult to detect unless a specific effort is made to check for them. Causes of run-time errors include uninitialized variables or parameters, division by zero, truncation in integer arithmetic, variable overflows and underflows, and invalid or corrupt input.

In fact, a reviewer pointed out to us that the preceding code for initializing a deer's weight will cause run-time errors. Drawing initial weights randomly from a normal distribution, no matter how small the variance, will eventually produce a new deer that has a negative weight or is bigger than its mother. So we always follow the random draw with an "if" statement to make sure the initial weight is within a reasonable range.

The best protection against run-time errors is defensive programming techniques such as:

- Liberal use of code that checks for error conditions during execution, for example, by making sure the denominator is non-zero before completing a division.
- Avoiding unnecessary use of public, global, and pointer variables.
- Code that checks for uninitialized variables and invalid or missing input data.
- Knowing how the software platform handles conditions like divide-by-zero and variable overflows (which do not always cause execution to stop in some programming languages!).
- Using double-precision (or even larger) floating point variables when overflows and underflows are a risk.

Even though techniques like check statements may slow the model's execution, the delay is negligible compared to the cost of errors that are found late or not at all.

8.7.3   Select a Good Pseudorandom Number Generator

Stochastic processes in IBMs are simulated using "pseudorandom" numbers produced by random number generator software. Modelers need to be aware that random number generators vary widely in quality, and poor generators can induce important biases or artifacts in simulation results (Fishman 1973; Ripley 1987; Wilson 2000; Gentle 2003). Unfortunately, many software platforms that can be used for simulation models (programming languages, spreadsheets, etc.) are likely to have built-in generators of poor or unknown quality. (For an interesting diversion, try to find out what generator is built into your favorite spreadsheet, statistical software, or programming language.) Also unfortunately, the performance of a generator may depend on a computer's hardware. These problems are well known among software engineers and simulation modelers, so modelers that do not address random number quality in their IBM's software are likely to have their work criticized. Modelers need to at least know and document which generator their platform uses, and replace it if it is substandard. The effects of random number generators on model results are easily explored by trying several generators. One benefit of a good software platform is providing quality pseudorandom number generators.

8.7.4   Reduce Execution Time—If Necessary

Throughout this chapter we state our belief that the progress of an IBE project is far more likely to be limited by how well the software is designed and tested than by how quickly the code executes. To facilitate the modeling cycle, software should be designed initially to facilitate its review and testing, without undue regard for execution speed. Many IBMs execute sufficiently fast with no further work to reduce execution time. However, execution speed is likely to be a significant issue for IBMs with large numbers of objects (many individuals or many habitat units) or whose individuals perform many complex calculations. Often the number of computations increases more than linearly with the number of objects in the model—for example, if an individual has to interact with neighbouring individuals (Hildenbrandt 2003).

The following are among the software engineering techniques that can speed up execution of an IBM's code, once the code and the model have been tested and are ready for use. These techniques are listed in approximate order of increasing difficulty and risk of introducing errors. All these techniques should be used cautiously, for example, by running standard test simulations to verify that model results are not affected.

1. Use a faster computer or more computers. Buying a faster processor (or dual processors) can be the easiest, safest, and most cost-effective way to improve performance. The cost of a new desktop computer is often a bargain compared to the time and risks involved in trying to speed up the software. Also, doing research with an IBM always involves simulation experiments that require many model runs (chapter 9). Simply executing different runs on different processors (in different computers or in a cluster) is often a simple, very effective way to produce results faster.

2. Reduce graphics and file output. For some IBMs, the GUIs can significantly increase execution time. A version of the software that bypasses the GUIs can be created and used once the model has been thoroughly tested. Platforms that specifically support nongraphics (or "batch") execution modes have the advantage of letting graphics be turned off without otherwise touching the code. Our experience has been that models with individuals conducting numerous complex calculations benefit little from nongraphics modes because the graphics updates account for little of the total execution time. Other IBMs that have individuals doing few calculations over many time steps can be speeded up significantly by turning off graphic displays. Writing unnecessarily large amounts of output to files can also slow execution, sometimes considerably.

3. Avoid slow algorithms. The computer simulation literature is filled with algorithms for common tasks (for sorting, random number distributions, and just about anything that more than one person might use). These "numerical recipes" are thoroughly tested for speed as well as reliability and certainly will perform better than home-made algorithms (yet another reason to never program anything we do not have to!).

4. Profile to focus code improvements. "Profilers" are software that reports data on how a code executes, allowing the programmer to identify the parts of the code that use the most time. Profilers are available for common programming languages and so can be used with code library platforms. Often, a significant portion of the total execution time is spent in a few small parts of the code. Once identified by the profiler, these parts of the code can be speeded up using techniques such as those discussed next.

5. Avoid slow math operations. Mathematical operations that use Taylor series (e.g., logarithms, exponentials, powers) are much slower than other operations and sometimes can be avoided. For example, the statement `cellArea = length * length` executes much faster than using the power operator ("^"): `cellArea = length^2`.

6. Reduce method calls. In OOP languages, calling one method from another (or "messaging") is relatively slow. Heavily used parts of the code can often be made faster by combining methods to reduce messaging. This approach especially requires care because it can make a program harder to understand and more vulnerable to errors during future modifications.

7. Reduce creation of new objects. Creating new objects is also slow. OOP codes can be full of small objects that are created, then dropped, as needed. Speed can be increased by reusing such objects instead of dropping and recreating them—as long as it is clear that there are no artifacts left from the previous use. Bigger objects with several variables should not be reused due to the risk and time involved in preparing them for reuse.

8. Bound decision processes. In some IBMs, much of the computation is used in evaluating decision alternatives, for example calculating the fitness each individual would expect if it made a decision in each of several alternative ways. If clearly bad alternatives can be eliminated with a few quick calculations, considerable speed-up can be obtained. This technique must be used with great caution to avoid causing individuals to make bad choices in some circumstances—even rare bad choices can have significant effects on IBM results. Remember that with this technique the actual model, not just its software, is being altered.

8.7.5 Accommodate Software Evolution and Maintenance

Section 8.3.6 describes some of the ways we must expect an IBM's software to change during and even after a research project, and the serious problems that result if we do not take steps to accommodate and manage change. Primary among these steps are documentation of the software and its revision history (section 8.6.4). "Version control" software is widely used in software development to automate partially the documentation and management of code changes. Each code file is checked into a repository and must be checked out again for editing. The version control software keeps track of what changes were made and when, documenting the history of changes and allowing them to be reversed.

Version control software by itself is not a practical way to document revision history. Keeping a simple log of changes—what was changed, by whom, and why—in both the model formulation document and the software documentation makes it much easier to recreate history.

Another important technique is periodically creating official release versions of the model when major changes are completed, and whenever the model is used for purposes (e.g., publications, management decisions) that may require results to be reproduced in the future. A release version should include complete documentation of the model's formulation, the code that matches the formulation, all the input and parameter files, example output, and documentation of the code and its testing. A setup program that automates installation helps avoid such mistakes as replacing some but not all program and data files, and wraps the whole release in a tidy package. Producing and archiving release versions is worthwhile even if the model is used by only one or two people: it is a simple way to guarantee that results can be reproduced in the future, even if the need is unanticipated.

8.8   SOME FAVORITE SOFTWARE MYTHS

After working with both ecological modelers and software professionals (many specializing in agent-based simulation software) for many years, we cannot resist listing some common misperceptions by ecologists about software and the software development process. These "myths" mainly result from the unfortunately low exposure of most ecologists to modern software engineering; the classical models that ecologists have traditionally used do not require much software expertise. Here are some of the misconceptions we have found common (and sometimes discovered ourselves, the hard way) and why they can be counterproductive.

   1. I know how to program in FORTRAN (or C, Java, . . .) so I can implement my model myself, from scratch. This statement is analogous to saying "I know spelling and grammar, so I can write a novel." One problem with this myth is that software for an IBM requires a competent design, and software design is quite a different skill from programming. The second problem is that the modeler rejects a priori the many benefits of software platforms that are now available. Experienced developers never write code from scratch when existing code is available—learning to use existing libraries or platforms is faster, cheaper, and safer.

2. I shouldn't use a programmer to develop my software because it is too much work to explain the model . . . and besides, many important modeling decisions are made during programming. The most unproductive and dangerous habit of modelers that develop their own software is writing the model directly into code, without first formulating and documenting it in written form and in simple test codes. A model that is accurately described only in its software lacks important elements of science: the model cannot be reviewed and is certainly not reproducible (unless the code is exceptionally simple and clear, and is published). With models of any complexity, the modeler too easily loses track of what the code is supposed to do, making it worthless until the modeler goes back and deciphers and documents his own code. A model needs to be fully documented on paper anyway, and this documentation is what the programmer can work from. Working out modeling details on paper and in test codes is more efficient than working them out in the final software: it lets the modeler focus on biology instead of programming, avoids spending time testing and debugging code that is then thrown away as the formulation is refined, and encourages the modeler to document *why* formulation decisions are made.

Some modelers clearly hesitate to work with a software professional because they fear losing control over the software and model. This fear would be legitimate if the modeler expected to simply hand the formulation to a programmer and get final code back. However, under the development cycle we describe in section 8.6.2, modelers remain deeply involved in, and responsible for, code development.

3. Low-level programming languages are better because they run faster. If execution speed was the sole software design consideration, a good software engineer would indeed implement an IBM in a low-level, non-object-oriented programming language like FORTRAN or C. However, the rate at which we can produce good science using an IBM is almost never limited just by the software's execution speed. Far more important is the time required to develop and test the software, and having the tools to test and experiment with the IBM easily and thoroughly. Higher-level platforms are designed to reduce software development times while providing the tools to test and use the IBM efficiently.

4. I don't think there are any important bugs in the software. From extensive experience with IBMs of various complexity, we can assure modelers that their model is almost certainly *not* bug-free until a comprehensive search for bugs is completed and documented. If the modeler cannot provide comprehensive evidence that there are no important bugs, the code is not ready for use.

5. GUIs are just doo-dads that slow the model's execution. GUIs are indeed sometimes used as gimmicks to entertain users while a model runs. However, most IBMs (especially spatially explicit ones) produce important results that can only be understood using visual output. For many IBMs, GUIs are essential for understanding, testing, and communicating results. Whether to use GUIs should no longer even be an issue because modern platforms make GUIs available with almost no extra effort.

6. My IBM is unique so I must design and write the code from scratch. A surprising number of IBMs have been implemented with no attempt to find out what software tools or platforms were available to help. The attitude that there is nothing to learn from others is rarely productive and has never been true for IBM software. When Huston, DeAngelis, and Post (1988) published their landmark article that inspired interest in IBMs, books on discrete-event simulation theory and software had already been around for many years (e.g., Fishman 1973; Zeigler 1976). Now, there is indeed much to learn from the experience of others; for example, Swarm has had an active user community since at least 1997. The amount of literature, theory, and software now available for implementing IBMs is far too great to ignore, and no IBM is so unique that it cannot benefit from being implemented in an agent-based modeling platform. These tools do not take away the creative process of implementing a model; in contrast, they let modelers focus on the unique aspects of their model and spend less time on mundane tasks.

8.9   SUMMARY AND CONCLUSIONS

Individual-based modeling requires a level of sophistication about software that few ecologists have been prepared for by their academic training or experience. This chapter may seem intimidating because we recommend many practices that ecologists experienced only with simpler models might think excessive. However, the long history of simulation modeling makes it clear that projects are very unlikely to meet their objectives if they start without a solid understanding of, and planning for, the inevitable software challenges. Several large, early programs in IBE were far less productive than expected because software (developed from scratch) consumed more resources than expected without providing the reliability and experimentation capabilities needed to do science. Individual-based approaches are routinely criticized as irreproducible because software is inadequately tested and documented. At the same time, though, many IBMs have been developed quite smoothly and successfully, often by scientists that started with little knowledge of software but were smart enough to seek out the right tools and help. Clearly, software is not an insurmountable problem for a well-managed project. Our goal is to help individual ecologists and ecology as a whole prepare for, manage, and reduce the software challenges posed by IBMs.

Most of what we say in this chapter can be summarized in four points.

- Software development for IBMs is not just a matter of "implementing" the model so it executes on a computer; we must also develop a laboratory for observing and experimenting on the implemented model.
- Few ecologists start with the software engineering skills needed for most IBMs; these skills are different from just knowing how to program. Potentially successful ways to proceed include taking the time to learn the software skills, collaborating with software professionals, and keeping the IBM simple enough to implement in a high-level platform that requires little software expertise.
- There are many resources for making software development more likely to succeed; these include specialized platforms, simulation theory and literature, and user communities. Skilled developers avoid designing or programming anything they do not have to.
- Testing software continually and thoroughly may seem onerous, but for IBMs the consequences of not testing are far worse. The same is true of documenting a model and its software thoroughly.

What can ecological modelers and research program managers do to ensure that software development moves an IBM project forward instead of eating up all its resources? It should be clear that software development needs to be planned carefully from the start of the project. Key planning issues are identifying the observer and experimentation capabilities needed to test the software and meet the study's research objectives; selecting an appropriate platform; deciding who will be responsible for the software's overall design, who will do the programming, and who will independently review the design and code; designing and conducting a hierarchical and comprehensive software testing process; and implementing documentation, version control, and release management procedures. Of these steps, selecting the software platform and the personnel are typically the most important and difficult decisions in getting started, so we summarize our recommendations for them in figure 8.2.

Especially important is starting a project with a realistic projection of the resources needed to do software development adequately, and updating this projection as work proceeds. If it appears that software is consuming more resources than expected, and nothing can be done to improve efficiency, then the scope of the model and research must be reduced. No science can be conducted unless there are resources to produce competent and thoroughly tested software *and then* to analyze and learn from the IBM. Our next chapter is about what to do with an IBM once its software is usable, and the first point we make in that chapter is that the all-important analysis phase also often requires more time and resources than anticipated.

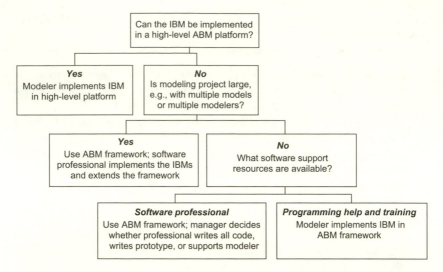

Figure 8.2 Decision tree summarizing recommendations for choosing a software platform and the role of software professionals. The recommendations are from the project manager's perspective, with the goal of providing the essential software capabilities and quality as cost-effectively as possible. This figure is only a general guide, with exceptions likely; for example, a modeler already skilled in a procedural programming language might best use it for a very simple IBM that can be thoroughly tested and analyzed without graphics. Note that the modeler's (i.e., ecologist's) programming skills are not an important decision factor: even if modelers are software experts, their time may not best be spent on model implementation. The terms "high-level ABM platform" and "ABM framework" are defined in section 8.4.

As soon as a project starts, following software quality procedures is of utmost importance. Assume the code will be reviewed and published. Know that mistakes are inevitable and need to be found as soon as possible. Be paranoid about run-time errors. Document software tests assuming reviewers will demand "proof" that your results are valid and not the artifact of code bugs. Remember that the credibility and productivity of your work depends on software quality.

There are also things we need to do, not just for our current research project, but to help ourselves and others in the future by promoting the maturation of individual-based approaches to science. Of primary importance is joining the rapidly growing community of scientists that promote and use agent-based simulation and simulation platforms. The Swarm Development Group's user community (www.swarm.org) has been especially valuable to us and many others, even those who do not use Swarm software.

A second very important thing we need to do is use common software tools. The more we use the same platforms, sharing and contributing software and financial resources, the more rapidly we will reach the point where IBMs require no more specialized software expertise than conducting statistical analyses or analyzing spatial data now do.

Finally, institutions that conduct and teach IBE can promote this appproach by providing more software support. Ecologists who choose to use IBMs will need less training in mathematics but more in simulation technology. Productivity can be enhanced by providing resources such as programmers, training classes, and interdisciplinary collaboration with the software community.

Chapter Nine

Analyzing Individual-based Models

> A model, once it is running reliably on a computer, is like a
> laboratory waiting to be used.
>> —*Anthony Starfield, Karl Smith, and Andrew Bleloch, 1990*

9.1 INTRODUCTION

Analyzing a computer model means studying the model, once it executes, to
understand and improve its performance and then to solve the problems the
model was designed to explain. One consequence of IBMs being less simple
than classical models is that IBMs are not as easy to understand and learn from.
In fact, some ecologists believe simulation models and IBMs are so hard to
understand that they are not useful: if a model is just as complex as nature
itself, why not just study nature instead? Avoiding just this problem was our
primary goal in part 1: readers of chapters 1 to 4 know that a well-designed
IBM is *not* as complex as nature itself. A well-designed IBM captures the
essence of an ecological system with respect to a particular problem and contains
little else.

There are more reasons why IBMs are easier to analyze than natural systems. Everything in an IBM can be completely observed and even manipulated.
With simulation models we can implement any experimental design we can
imagine—including manipulation of the "organisms" themselves—while collecting whatever data we want. Compared with field and laboratory experiments,
simulation experiments are easy and free from ethical and logistical constraints;
they allow us to examine temporal and spatial scales that are not feasible for
real systems (often we simulate thousands of years, repeatedly); and they allow
us to examine conditions (a changed climate perhaps) that are very difficult to
mimic in physical systems.

Our point here is that understanding and learning from IBMs requires special
effort but can be quite efficient and productive. IBMs are like the physical
microcosms used in laboratory ecology. Great effort goes into planning and
building the microcosm—the container, the environmental components like
soil and light, the organisms, and the instrumentation needed to observe the
individual- and system-level processes of interest. Yet the ecologist knows that
building the microcosm is just the start: experiments must be designed and

conducted before anything is learned. When an IBM is built, the modeler is likewise just ready to start doing ecology. This chapter is about what to do at this point.

We start with an overview of the analysis phase of the modeling cycle, identifying four major steps in analyzing an IBM. Then we describe some general strategies for making analysis efficient and a number of specific analysis techniques. In section 9.4 we discuss techniques that are unique to IBMs. Sections 9.5 through 9.9 address techniques that are also used for other kinds of models; we strongly recommend that ecological modelers become familiar with the simulation literature (e.g., Ripley 1987; Kleijnen and van Groenendaal 1992; Law and Kelton 1999; Fishman 2001) for a more complete understanding of these techniques. While there is relatively little literature on analysis of IBMs, analysis of simulation models in general is a highly developed field; almost anything we do can benefit from established methods and software.

Before we start, we need to warn beginners in modeling not to underestimate the amount of work it takes to thoroughly analyze an IBM. Analyzing an IBM may take ten times longer, or more, than developing the model. In the modeling cycle (section 2.3), the tasks preceding analysis are primarily tool building; the analysis task is when we start conducting science, learning about the IBM and the system it represents and drawing conclusions of general interest. Analysis should start as soon as a simple draft model is implemented and proceed as a cycle of testing and revising parts of the IBM, then testing and revising the whole IBM, and finally using the IBM to address ecological problems. Each of these steps can require extensive experimentation and, often, the return to earlier modeling cycle tasks. The analysis task should be the longest but also most exciting and productive part of an IBM project.

9.2   STEPS IN ANALYZING AN IBM

To understand all the reasons why we need to analyze an IBM, think of all the reasons scientists are skeptical about a "black box" model even if the model's formulation is described in great detail. Does the software actually do what the formulation says? Is the formulation "right"? Why should anyone believe the model's predictions—would the model produce similar results (or lead to similar conclusions) if different parameter values or assumptions were used? And *how* did the results arise—what were the individuals doing that caused the system's responses? In this section we discuss what different kinds of analyses are needed to address these kinds of concerns. Analysis, testing, and revision are described in section 2.3 as the second-to-last of six tasks in the modeling cycle; here, we break this task into smaller steps that each have their own objective. The exact kinds of analysis, and the best methods, vary among projects, so we provide general guidelines that modelers should find easy to adapt to their projects. Some kinds of analysis can be skipped for some projects—for example, if an IBM uses individual traits or submodels for environmental processes that are already well tested in similar contexts, then they may not need further analysis. And some analysis techniques can be useful for different analysis objectives: sensitivity or robustness analysis, for example, may help validate an IBM's formulation, find good parameter values, and understand the ecosystem being modeled.

Following are the four analysis objectives that usually need to be addressed in developing an IBM and applying it to a theoretical or applied ecological problem. These objectives can be treated as separate steps in analyzing an IBM.

Software Verification.   Before we can analyze an IBM itself, we must analyze its software to verify that the computer program faithfully implements the model's formulation. This kind of analysis, often called "software verification," is treated extensively in section 8.5 so we do not address it in this chapter. However, modelers must remember that the subsequent analysis steps invariably produce changes to the model's formulation and software, and each change requires documentation and an appropriate level of testing. Software verification is part of the modeling cycle, not a one-time job.

Model Validation and Theory Development.   After software verification, the next analysis task is usually testing and improving the model's design and formulation. Traditionally, this analysis task is called "validation": establishing how valid the model is for the problems it is intended to solve. Note that the goal is establishing *how valid* the model is, not whether the model *is* or *is not* valid: we should establish a number of clearly defined criteria for evaluating an IBM, but rarely is it useful to define a specific standard for accepting or rejecting a model. Instead, we can think of validation as assembling evidence and building a case for why the IBM is valid for its intended application, or as delineating the applications for which the IBM is, or is not, useful.

Because IBMs can be structurally rich, with population behavior arising from traits of the individuals and their simulated environment, validation needs to proceed from the bottom up (section 9.3.2). First, we can test those underlying parts of an IBM that do not include emergent behavior arising from interactions among individuals and their environment. These underlying parts typically include submodels representing the environment and nonbehavioral traits of the individuals. For example, validation can start with showing that an IBM's submodel for food production, and the submodels for how an individual feeds and grows, all produce reasonable results.

Next, developing theory for key individual traits is likely to be a critical part of validation. A cycle for developing and testing theory is presented in chapter 4, so we do not pay special attention to theory development in this chapter.

Only after we have tested lower-level components can we undertake a meaningful validation of the full IBM. At this stage, we might typically parameterize the IBM, undertake systematic analysis of sensitivity and robustness, examine how well alternative versions of the IBM reproduce a variety of observed patterns, and see whether the model can make successful independent predictions (all these kinds of analysis are discussed later).

Parameterization. Validation of many classical ecological models, and some IBMs, is largely a matter of parameter fitting: the models are simple, so their validity depends mainly on finding parameter values that adequately reproduce observed data. However, the validity of many IBMs depends as much or more on the model's structure and mechanisms than on parameter values. Therefore, we treat parameterization as a separate analysis step. This step, also called "calibration," involves finding good values for parameters that cannot be evaluated independently during formulation of submodels (sections 7.5, 9.4.2). Section 9.8 discusses parameterization techniques.

Solving Ecological Problems. The final analysis step is to address the problem the IBM was designed for. This kind of analysis may involve contrasting alternative versions of the IBM to see which best explains observations of a real system, understanding the system dynamics that arise under a variety of conditions, or predicting system responses to environmental management options. These analyses often involve taking an IBM apart and putting it back together in different ways, as illustrated in sections 9.4.4 and 9.4.5. This final analysis step is the most important, but its conclusions will be greeted with skepticism unless the IBM's credibility is built through the previous steps.

9.3 GENERAL STRATEGIES FOR ANALYZING IBMS

This section presents three general analysis strategies that are particularly valuable for IBMs. These strategies not only cope with, but actually take advantage of, the greater complexity of IBMs.

9.3.1 The Main Strategy: Simulation Experiments

The basic, most productive strategy for analyzing simulation models such as IBMs is suggested by the quotation from Starfield et al. (1990) with which we started this chapter: we must think of an IBM as a laboratory system upon which we experiment to gain scientific understanding. But how do scientists develop understanding? We design controlled experiments which give us, step by step, insights into how the system behaves. We usually start with very simple experiments, with the system's complexity reduced so much that we easily can predict the outcome of the experiment. Then we carefully add degrees of freedom to the system, incrementing its complexity while still formulating hypotheses that are easily tested with experiments. Our predictions will sometimes be correct,

316

CHAPTER NINE

but often they will not be, which then makes us design new experiments and continue on, learning as we go.

This is also exactly how IBMs must be analyzed: by carefully designing and performing *controlled simulation experiments*. When a model first runs, it is fun and useful simply to play around with it, exploring how it reacts to changes in parameter values, initial conditions, or assumptions. But it is important to stop playing around before long and ask: what do I want to know about the model, and how can I design experiments so that I learn what I want to know?

The experimental approach to analyzing IBMs corresponds to the inductive inference approach to science first attributed to Francis Bacon and advocated by Platt (1964), which we also referred to in chapter 4. When we have a working IBM the cycle of posing alternative hypotheses and then designing and conducting experiments to test the hypotheses is often just what we need to make rapid progress. When, for example, we are developing IBE theory using the cycle in chapter 4, we use our IBM to find the model of individual behavior that best explains population-level phenomena. We pose alternative theories for the individual behavior as the hypotheses to be tested, implement each hypothesis in the IBM, identify some patterns as the "currency" for evaluating the hypotheses, and then conduct simulations that determine which hypotheses fail to reproduce the patterns. If instead we are using the IBM to understand the cause of some particular system dynamic—cycles in population abundance, for example—we can pose alternative hypotheses for the cause: perhaps environmental fluctuations, density dependence in reproduction, or density-dependent competition among adults. Then we can design simulation experiments that attempt to exclude each hypothesis as an explanation: if we hold environmental conditions constant and the cycles still occur, then environmental fluctuations are excluded as the sole explanation.

However, we need to be aware that system dynamics such as cycles are often caused by complex interactions among processes like environmental fluctuations, reproduction, and competition. This possibility means that if we identify three potential explanations for some dynamic, and then conduct two experiments that exclude two of the explanations, it is risky to just assume that the third explanation must be right. Instead, we also need to test the third explanation because the dynamic could be caused by mechanisms more complex than we assumed. When simple either-or questions (e.g., Are populations regulated by bottom-up or top-down processes?) do not have clear answers (and they rarely do in IBMs and ecosystems), we must design more clever experiments that ask and answer more enlightening questions.

The rest of this chapter assumes that the primary way modelers go about analyzing IBMs is by conducting these kinds of simulation experiments. Once a modeler starts hypothesis-testing experiments with an IBM, the power and fascination of this strategy becomes very evident. Often an IBM can be used in this way to analyze many problems other than the ones it was originally intended for. The analyses of a trout IBM by Railsback et al. (2002), which originated as a graduate class project, began with the objective of testing the IBM's ability to reproduce population-level patterns observed in real trout. However, the authors also analyzed the *causes* of the patterns. For example, density dependence in the size of juvenile trout (real trout were smaller at the end of their first summer when density was higher, a pattern reproduced in the IBM) was hypothesized by the ecologists who observed it in the field and laboratory to be due to feeding competition (Jenkins et al. 1999). However, that hypothesis was excluded in the IBM simulation experiments, during which growth and density were positively related. Railsback et al. then hypothesized three alternative explanations and showed that two of them were also not supported by the simulation results. Considerably more study would be required before these conclusions could be applied to the real trout, but the simulation experiments certainly indicate that the original, most intuitive, hypothesis must be questioned. The simulation experiments also suggest field studies that could test alternative hypotheses. (This kind of analysis rapidly develops belief in the "inverse Occam's razor": simple, obvious explanations for the behavior of a complex system are very often wrong.)

9.3.2 Analyzing from the Bottom Up

There is little sense in analyzing an IBM's system-level behavior before developing confidence that the model's individual-level behavior is acceptable, and there is no reason to expect individual behavior to be acceptable before the environmental processes that drive individual behavior have been tested. It seems self-evident that bottom-up models like IBMs must be validated starting with the bottom levels that system-level behaviors emerge from. Yet one of most common reasons for IBMs failing to develop credibility (besides the failure to analyze the model at all) is attempting to analyze the system level without first testing the validity of individual behavior.

The very bottom of most IBMs is a representation of the individuals' environment, because the behavior of individuals depends in part on environmental conditions. Therefore, model analysis should start with testing and validating how the environment is simulated. Then individual behavior can be analyzed. Especially important is testing behavior that arises from the individuals' key adaptive traits, because the most interesting and important system dynamics arise from these adaptive traits; this problem is addressed using the theory development cycle of chapter 4. Often, it is very effective to start analyzing individual traits by contrasting the behavior of isolated individuals with the behavior of individuals interacting with each other.

An obvious problem with this bottom-up approach is that higher-level processes affect lower levels in many IBMs: not only do system dynamics arise from individual behavior, but individual behavior is affected by system dynamics.

318   CHAPTER NINE

If individuals adapt their feeding behavior to the availability of food, and food
availability depends on consumption by all individuals, then an individual's
feeding behavior will be affected by population density. In some IBMs, even
the bottom-most environmental processes can be affected by how individuals
use resources: food production may be a function of food consumption by individuals, for example. However, the strategy of controlled experimentation can
easily overcome this problem: we can design experiments that isolate behavior
at one level well enough to test it convincingly. These experiments can use
unrealistic scenarios (section 9.4.5) such as freezing reproduction, mortality,
and growth so food consumption is constant.

   The bottom-up analysis strategy is important for efficiency, keeping us from
wasting time analyzing system behaviors when they are adversely affected by
problems at lower levels. However, this strategy is even more important for
an IBM's credibility with reviewers and other "clients" of the model. To keep
skeptics from saying that "a model with that many parameters can be calibrated
to produce whatever results you want," the modeler must show that the IBM was
instead parameterized and tested at all levels, using many kinds of information;
and that system behaviors arose from environmental and individual-level traits
that were analyzed independently before system behaviors were examined.

9.3.3   Analyzing Model Structure Separately

The fact that uncertainty occurs both in model structure and in parameter values is a well-known problem in model analysis. How do we know if a model's
inability to produce some expected result is due to problems with the model's
equations and rules or with its parameter values? Simple classical models cannot
be tested or analyzed meaningfully until their parameter values are fit to data,
so effects of structural uncertainty can be examined only by parameterizing
and analyzing alternative model structures (e.g., Mooij and DeAngelis 2003).
However, this approach can mask some of the effects of model structure. By
fitting each alternative model to the same data, the alternative models are to
some degree *forced* to behave similarly. How do we know whether parameter
fitting hides underlying problems with the model's structure? These problems
are often viewed as intractable; but for IBMs the strategy of pattern-oriented
modeling offers a way to at least partly separate analysis of structure from
analysis of parameter values.

   With IBMs that are rich in structure and process, we can follow a model analysis strategy that lets us analyze an IBM's structure before parameterization,
separating the analysis of structural and parameter uncertainty to some extent.
This strategy is simply the one described in chapters 3 and 4—use observed
patterns to test and contrast alternative model designs, but this structural analysis is done *before* the IBM is calibrated or parameterized in detail. As the IBM
is designed, parameters are given the best values that can be determined without

calibrating the full model. Then the IBM's structural validity can be assessed by testing its ability to reproduce a variety of patterns that capture the structural essence of the system being modeled. These patterns can be qualitative, but the tests can use clear and quantifiable criteria for whether the IBM reproduces the patterns: Do trends go in the right direction? Does some expected response happen or not? If some version of an IBM unexpectedly fails to reproduce some of the test patterns, additional analysis can determine whether the failure is due only to a poor choice of parameter values. Full calibration and parameterization can proceed after this pattern-oriented analysis has identified the most valid model structure.

Highly mechanistic IBMs that have been shown by pattern-oriented analysis to be structurally realistic can often be used to address many problems with little or no detailed parameter fitting. In fact, our experience (including the trout model described in sections 1.2 and 6.4.2 and the beech forest model described in sections 1.2 and 6.8.3) indicates that if an IBM produces reasonable behavior—reproducing general, qualitative patterns observed in the real system—only for a narrow range of parameter values, there is likely something wrong with its structure. (However, we still often use calibration to reproduce detailed patterns; see section 9.8.) The strategy of conducting pattern-oriented analysis of an IBM's structure first can let us understand and reduce structural uncertainty before addressing parameter uncertainty. This ability can be very important for demonstrating that an IBM is more than a black box that can be calibrated to produce any desired results.

9.4 TECHNIQUES FOR ANALYZING IBMS

Now that we have established an overall strategy of analyzing IBMs by conducting simulation experiments, how do we implement this strategy efficiently? Are conventional model analysis techniques the best for IBMs? In this subsection we discuss a variety of analysis techniques, some unique to IBMs and some not.

9.4.1 Currencies for Contrasting Model Versions

The technique of contrasting alternative versions of an IBM is used frequently in the three analysis steps: validation and theory development, parameterization, and solving problems. If we need to know which theory for an adaptive trait of individuals is best, we implement alternative theories in different versions of the IBM and see which performs best. To find the best parameter values, we run the IBM with alternative values and see which produces the best results. To determine which ecological processes are important in the system we are studying, we run different versions that each have different processes turned off. Contrasting model versions requires a currency or standard—we cannot determine which model version is "best" without defining "best at what?" Here, we look at what currencies can be used and conclude that the best currencies for analyzing IBMs are not the ones most commonly used for other population models.

All the types of currencies described in the following sections require that we first define *summary state variables*, which provide a summary description of the state of the system. We must use appropriate summary state variables because the full state of an IBM—all properties of all individuals and of their environment—obviously is not a practical currency for any purpose.

9.4.1.1 Observed Patterns

Anyone who has read chapters 3 and 4 already knows what we recommend as the most important currency for contrasting model versions: the ability to reproduce a variety of patterns, at various levels and in various types of output, that have been observed in the real system represented by the IBM and that capture the system's essential characteristics with respect to the problem the IBM addresses. Their greater "richness"—complexity in model structure and in the kinds of results they produce—is the reason why IBMs are more of a challenge to analyze and understand. We can take advantage of that richness, however, by using the many different kinds of results that IBMs produce as currencies for analysis.

One of a modeler's important jobs is therefore to assemble patterns for use in analyzing the IBM. To do a comprehensive job of collecting patterns, modelers can look at all the different types of results their IBM can produce (individual behaviors; spatial distributions of individuals; relations among population variables, or between population and environmental variables; see also the example IBMs in chapter 6) and look for observations corresponding to these results. Sometimes an IBM produces striking patterns, inspiring us to scan the empirical literature to see if those patterns have been observed in nature, and sometimes striking observed patterns that we discover in the literature will make us look again at the model to see if it produces those patterns. In addition to striking patterns (or in their absence), a variety of weak or general patterns can contain sufficient information to test and calibrate IBMs (Wiegand et al. 2003).

Especially valuable are observed patterns of individual and system response to variables that are inputs to the model (environmental variables, initial conditions, etc.). For example, a disturbance event might have been observed to evoke specific responses. The modeler can easily vary the inputs and attempt to reproduce these patterns. Ideally, modeling projects are conducted in collaboration with field research, so types of patterns particularly powerful for improving the model can be identified and evaluated in the field. Assembling a wide variety of patterns is especially important when some patterns are used to parameterize the IBM, because other patterns can then be used to test the parameterized model.

Using patterns as a currency for analysis does not necessarily mean we need quantitative methods for evaluating the precise fit of simulated or observed pattern. Instead, we can define clear criteria for whether *qualitative* patterns are met (e.g., "the relation between mean weight and population density, measured at the end of each simulated year, has a negative slope"); this approach is often adequate and even preferable, especially in the earlier stages when we are analyzing the IBM's structure and theory. It is often better to contrast alternative model structures and traits via the question, "Which version reproduces more of the qualitative patterns?" than via "Which version reproduces the patterns more quantitatively?" Why? Quantitative assessment of pattern fit is not very meaningful until the IBM has been parameterized carefully, yet we need to contrast model versions before undertaking detailed parameterization. We must also be careful not to overemphasize matching patterns that are themselves uncertain. If two competing versions of an IBM are "tied"—both qualitatively reproduce all the patterns used to contrast them—then it is best, if possible, to break the tie by finding additional patterns that have the analysis "power" to falsify one of the two versions.

9.4.1.2 Census Data

Many ecologists automatically think of goodness-of-fit to observed census data as the primary currency for model analysis: the best model (or best parameter set) is the one that most closely reproduces a time series of observed population abundance. The main reason for using this currency is simply that a census time series is the only output of many classical models. There is also excellent literature on statistical analysis of model fit to observed census data (e.g., E. P. Smith and Rose 1995; Haefner 1996; Hilborn and Mangel 1997; Burnham and Anderson 1998; Kendall et al. 1999; Turchin 2003), including techniques for stochastic models (e.g., Waller et al. 2003).

However, using goodness-of-fit to census data as the *only* currency in model analysis has important limitations. The most obvious is that IBMs produce many kinds of output, so goodness-of-fit against only one or two summary population variables is an incomplete measure of model performance. Another well-known problem is that models with many parameters potentially can be forced, via calibration, to match different data sets. Models that are highly stochastic (as many IBMs are) pose another challenge: we must deal with the variability in model results, which also reminds us that census data are also variable and uncertain. Waller et al. (2003) recommend using Monte Carlo analysis of stochastic models to determine whether the data appear consistent with the model, the inverse of the usual goodness-of-fit question. However, with this approach the degree of fit depends on how stochastic the model is: the more randomness is in the model, the more variable its results will be and the less likely it becomes that data will be found incompatible with the model results.

The most important limitation of goodness-of-fit to census data as an analysis currency is that it provides little understanding of structurally rich models like IBMs: whether the fit is good or poor, we learn nothing about what parts of the model are good and not so good. Especially, we learn nothing about behavior of, or variation among, the model's individuals—two of the defining characteristics of IBMs.

Census data do, however, often contain *patterns* that can be useful for analyzing IBMs. We can look at census data for patterns—such as ranges in abundance, relations between abundance and environmental conditions, and the frequency of events such as population spikes or crashes—and include them in pattern-oriented analyses.

Despite its limitations for contrasting model versions, analyzing an IBM's population-level fit against observed census data is sometimes important, especially for calibration to find good parameter values once the model's structure has been analyzed. Validation against observed census data is an important *last* model analysis step for those IBMs that need to make quantitative population-level predictions, for example in population viability analysis (e.g., Wiegand et al. 1998). Reviewers of such models traditionally think of fit to census data as the primary standard of model validity; whether or not we agree with this standard, we often need to evaluate it. However, it should be clear to the modeler (and probably reviewers) that this kind of validation is meaningful only after the model's structure and lower-level behavior have been tested successfully.

9.4.1.3   *Variability in Results*

When contrasting model versions, important insights can often be obtained from looking at the full range of variation in simulation results—especially, variation among individuals—instead of looking only at mean values. For example, the mean age and size of individuals in an IBM may be reasonable while a few individuals grow to completely unrealistic ages and sizes. In this case, the modeler would need to decide whether the unrealistic individuals are an important problem, considering the IBM's purpose. Variation in many dimensions (over space and time; among individuals) is a fundamental characteristic of IBMs and must be reflected in the currencies used for analysis.

9.4.1.4   *Stability Properties and Diversity*

Two currencies—stability properties and diversity—are so fundamental to ecology that addressing them is often worthwhile (Grimm et al. 1999b; van Nes

2002). The main stability properties studied in ecology are constancy (or its inverse, variability), resilience, persistence, and—to a lesser extent—resistance (or its inverse, sensitivity). These properties never can be assigned to systems, but only to clearly defined summary state variables describing the system, to specific types of disturbances, and to certain spatial and temporal scales (Grimm and Wissel 1997). If an IBM is used to understand "stabilizing" mechanisms (a common issue in ecology), the analysis must first define a specific "ecological situation" to which it applies, a specific stability property, and a specific characteristic of the IBM's population to which the stability property is applied. Then, versions of the IBM can be contrasted to see which contain mechanisms conferring more or less of the specific stability property.

Likewise, deducing mechanisms that enhance or reduce diversity is a common topic of ecological analysis. Most often, the diversity of species is of concern, but the spatial diversity (heterogeneity) of the biotic and abiotic environment (structural diversity; Tews et al. 2004) is also a common topic. As with stability properties, the diversity of a system can be analyzed using an IBM by carefully defining a specific context and definition of diversity, then looking at how that measure of diversity changes among versions of the IBM (e.g., Savage, Sawhill, and Askenazi 2000). The two currencies stability and diversity can also be used simultaneously to study the stability-diversity relationship, one of the most persistent (and diverse) issues in ecology.

For both stability properties and diversity, it is important to be open to both sides: to identify mechanisms that both increase and decrease stability properties or diversity. All too easily, ecologists assume that stability and diversity are inherently good, or that simulated systems with higher stability or diversity must be more realistic; these assumptions are of course nonsense.

9.4.2   Independent Analysis of Submodels

The difficulty and uncertainty in analyzing an IBM can be greatly reduced by treating many parts of the model as independent submodels. This technique is important for counteracting the belief that IBMs are hopelessly uncertain because they have so many parameters. Any part of an IBM's formulation except its general structure and the adaptive traits of its individuals (their rules for making context-specific decisions) can be tested, parameterized, and validated by itself, often using pattern-oriented approaches. One advantage of this technique is that analysis of submodels can often use information that is not useful for testing the full model. Even more importantly, analyzing submodels independently means that analysis of the *full* IBM can focus on the model's general structure and behaviors arising from adaptive traits. This technique is discussed more fully in section 7.6.

9.4.3   Early Analysis of Extremely Simple Patterns

One productive way to get started with analysis of the full IBM is to test its ability to reproduce some extremely simple patterns. This technique can be a way to make the playing-around phase of analysis more productive: identify some very basic things the IBM must be able to do, and then test whether the IBM actually does them. It is essential that these simple patterns include some at the individual level: it is very easy and tempting to start analysis by looking only at whether population results are reasonable, but many problems are most easily identified by looking at the individual level. Some examples of very simple patterns that could be useful at the very start are:

- Do individuals exhibit their adaptive behaviors with any success? Or do they often make decisions that are clearly bad for themselves?
- Do individuals have reasonable values for basic state variables? For example, does growth occur? Are mortality rates reasonable, or do individuals all die very soon or not at all?
- Is there variability among individuals in behavior, or are they all doing the same thing?
- Does the population neither go extinct nor increase without limit? Or does the population rapidly move into an unrealistic state?

A second round of analysis could follow, with incrementally higher standards for acceptable individual and population behavior. This technique is very efficient for finding major problems quickly. The modeler also gains some experience with simulation experiments, and quickly learns whether the IBM's software has the kinds of observability (section 8.3.3) needed to do analysis.

9.4.4   Model Simplification

Because an IBM's complexity is what makes its analysis difficult, why not make analysis easier by reducing the model's complexity? This suggestion seems obvious but it has not been followed as often as it should have. Perhaps once we have worked so hard putting so many things into our model, we feel compelled to analyze all those things at once. This compulsion is likely strongest for "naively realistic" models (section 2.1) that are too complex in the first place. However, setting some of an IBM's complexity aside can help analysis and validation proceed efficiently by letting us understand other parts of the model more easily. This technique can be useful both when we are simply exploring an IBM's behavior and when conducting more rigorous simulation experiments.

Model simplification is most beneficial when used to take a stepwise, controlled approach to analyzing an IBM, examining simplified versions of the model first. Van Nes (2002; van Nes, Lammens, and Scheffer 2002) provides a good example: he started analysis and calibration of a multispecies fish model by first turning off processes such as predation and food competition so they could examine each species by itself. Other tricks include turning off most individual behaviors so the analysis can focus on one behavior at a time, and turning off some system dynamics to focus on individual behaviors—for instance, effects of demographic variability can be bypassed by not letting individuals die or reproduce (e.g., Fahse, Wissel, and Grimm 1998; section 6.6.3).

In many IBMs, there is one key process that produces the behavior of greatest interest. The bottom-up analysis strategy (section 9.3.2) can be implemented by turning that key process off and first analyzing the lower-level processes that drive it. For example, a final objective of the woodhoopoe model described in sections 1.2 and 6.3.1 was understanding how spatial population dynamics depends on scouting forays by members of the multiple social groups making up the population. The modelers first analyzed simulations of a *single* social group to test completely the model implementation and understand fully the group-level dynamics. Only then were multiple groups simulated and the full model analyzed.

9.4.5 Unrealistic Scenarios

The review of IBMs by Grimm (1999) found that there seems to be a psychological barrier to analyzing model scenarios that cannot occur in nature. This barrier is understandable: the purpose of the model is to provide a structurally realistic representation of a system, and it requires a lot of work to achieve and prove structural realism. Then, after all this work, it seems destructive to intentionally simulate extremely unrealistic situations. But analyzing unrealistic scenarios (including unrealistic parameter values) is essential for understanding IBMs. In fact, the ability to simulate things that cannot occur in nature is one of the most powerful advantages of IBMs compared with studying nature directly: if we want to test the hypothesis that process A causes pattern B, we simply reach into our IBM and turn process A off and see whether pattern B still occurs. Deutschman et al. (1997), for example, wanted to know whether spatial processes are important in the SORTIE forest IBM, so they simply eliminated effects of space from the model and reran their simulations (section 11.5.2).

A second example of using unrealistic scenarios is the analysis by Jeltsch, Müller, et al. (1997b; Thulke et al. 1999) of the wavelike pattern of rabies dispersal among red fox in Europe. Initial IBM simulations suggested the hypothesis that this pattern is caused by the small fraction of juvenile fox that disperse long distances before finding a new territory. These juveniles dispersed far into the region uninfected by rabies, then appeared to generate foci of infection that spread and merged into a new peak of the traveling wave of infection (figure 9.1). This hypothesis could easily be tested in the IBM by simply preventing

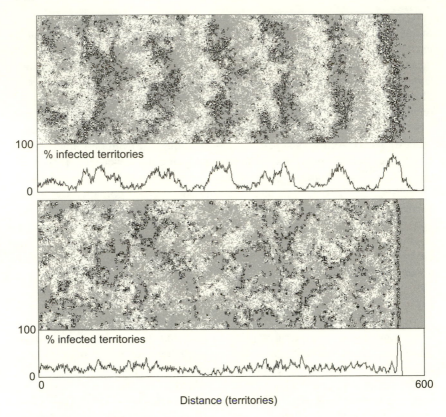

Figure 9.1 Snapshots of the wavelike dispersal pattern of rabies among red fox, as produced by the model of Jeltsch, Müller, et al. (1997). (a) Results of the full model, including long-distance migration of young fox. (b) Results with long-distance migration deactivated. The panels show the grid of 600 × 300 home ranges: home ranges with rabid fox are black, empty ranges are white, and ranges with healthy fox are gray. The graphs display rabies density averaged over the Y dimension. (After Jeltsch, Müller, et al. 1997b.)

young fox from dispersing long distances. Indeed, this unrealistic scenario with no long dispersal did cause the wavelike spreading pattern of rabies to disappear.

A final example of the power of unrealistic scenarios is the analysis by Railsback et al. (2002) of the "−4/3 self-thinning power law." This "law" says that in animal populations the mean weight of a population's age-classes varies with the −4/3 power of the number of individuals in the age classes (Begon, Firbank, and Wall 1986). The extent to which this law applies to real populations, and why, has been a source of debate among ecologists. The self-thinning relationship has been hypothesized to be a population-level consequence of the fact that an individual's metabolic rate (and hence its food demand) varies with its weight to the 3/4 power. To test this hypothesis, Railsback et al. used an IBM for a simple and unrealistic simulation experiment: vary the parameter for how metabolic rate varies with weight from its realistic value of 3/4, and then see if the population's self-thinning relation between weight and abundance responds as predicted by the hypothesis. (The self-thinning relation did not respond completely as predicted.) This very direct analysis of the self-thinning power law was possible only in an IBM because, of course, we cannot control the metabolic characteristics of real organisms.

These examples are from the final analysis phase, in which we use the IBM to understand natural systems. However, unrealistic scenarios can be useful throughout analysis. Modelers should never hesitate to "play god" and manipulate individuals and their environment unrealistically if it helps test hypotheses in simulation experiments.

9.4.6 Multiple Observation Perspectives

In section 5.11 we discuss three different perspectives from which an IBM's results can be viewed: those of an omniscient observer, an individual in the simulation, and a "virtual ecologist" that simulates how a real field ecologist might collect data in the system being simulated. The omniscient perspective is used by far the most: we take whatever observations we want from the whole model and analyze them. However, the other two perspectives can be valuable during analysis. It is often necessary to take the individual's perspective: selecting one individual from the IBM, looking at what data that individual "knows" about itself and its environment, and what decisions it makes and how. This perspective of the individual is often essential for developing the understanding and credibility of individual behavior that must be obtained before system-level analysis can be meaningful. The virtual ecologist perspective allows testing an IBM's ability to reproduce field observations that have known biases (Berger, Wagner, and Wolff 1999; Tyre, Possingham, and Lindenmayer 2001).

9.5 STATISTICAL ANALYSIS

It is tempting to say that statistics should be used for analyzing IBMs just the way that they are used in analyzing field data. Statistical techniques used in field ecology indeed are often very useful in analyzing IBM results. A fundamental difference between field and individual-based ecology, however, affects how we use statistics. In field ecology, data are usually "static": once we have completed the field study and started the analysis, it is difficult or impossible to collect more data, so we must depend on statistical analysis to draw inferences and conclusions. Using IBMs, however, researchers incur almost no cost in conducting more experiments or more replicates. We are not dependent on analyzing a fixed data set to learn what we can from an IBM; instead, we can simply continue our cycle of posing hypotheses and conducting experiments to test them. Once armed with a working IBM, the individual-based ecologist has more powerful techniques than statistical analysis alone.

Still, statistics have many applications in analysis of IBMs, which we discuss here. First, we issue two warnings. We make no attempt to discuss specific methods for statistical analyses, only how the analyses might be used. And our examples in this section focus on methods (especially regression) that assume observations are independent, but many outputs from IBMs are *not* independent over space or time; so time series and spatial methods that consider dependence are sometimes more appropriate.

9.5.1   Summarizing Simulation Results

Perhaps the most basic use of statistics is summarizing data, which is especially necessary for IBMs because IBMs produce results distributed over individuals, time, and space. In fact, for most IBMs it would be impossible to present, to analyze, or (sometimes) even to store results without summarizing them statistically. However, we repeat one caution from section 9.4.1: the full range of results is often very important for analyzing IBMs, so caution should be used in summarizing results. Variances, maximums, and minimum values often should be examined in addition to means or medians. Magnusson (2000), addressing ecological data in general, makes the important point that, as an alternative to statistics, graphs can be designed to provide an informative yet concise summary of results.

9.5.2   Contrast of Treatments

Many field and laboratory experiments are conducted by creating several replicates of each of several treatments (in which one or several independent variables are manipulated), then using hypothesis-testing statistics to analyze for significant differences among treatments. The same approach can be used to analyze an IBM: treatments (or "scenarios," a term used in simulation modeling) are each different versions of the IBM or use different input; and replicates are generated by using different random number sequences for the stochastic events in the IBM. Table 9.1 provides an example.

When we are analyzing how IBM results respond to a continuous independent variable, the sensitivity analysis approach we discuss in section 9.5.3 has advantages over contrasting discrete scenarios. However, we often need to contrast results from scenarios that are inherently discrete. When we are contrasting alternative theories for individual behavior, for example, each theory is

TABLE 9.1

Example of Traditional Study Design Applied to an Analysis of Turbidity Impacts, Using the Trout IBM of Railsback and Harvey (2002)

	Elevated Turbidity	Elevated Turbidity, Reduced Food
Adult trout		
Baseline	*	–
Elevated turbidity		–
Juveniles		
Baseline	–	–
Elevated turbidity		*

Note: Stream turbidity reduces feeding ability but also reduces predation risk; and is also likely to reduce availability of trout food. Three treatments are used: "baseline" conditions for an undisturbed stream; "elevated turbidity," with baseline values increased by twenty turbidity units, typifying a moderately disturbed stream; "elevated turbidity, reduced food," in which turbidity is increased twenty units and food availability decreased by 20 percent. A dash indicates that the treatment identified in the top row of the table produced trout abundance significantly lower than the treatment in the left column; and an asterisk indicates a lack of significant differences. Scenarios were contrasted using analysis of variance followed by Bonferroni *t*-test for differences among means, with $\alpha = 0.05$ and five replicates of each treatment. For abundance of adult trout, only the combination of elevated turbidity and reduced food availability produced a statistically significant decrease in abundance. For juveniles, the turbidity increase by itself significantly reduced abundance, but the additional effect of reduced food was not significant.

a discrete treatment, and we need ways to contrast these treatments. Railsback and Harvey (2002) used analysis of variance followed by *t* tests to contrast population-level results from three versions of an IBM, each version having a different theory for how fish select habitat.

There are several potential pitfalls in using traditional hypothesis-testing statistics to analyze IBM scenarios. The primary problem is that analysis conclusions depend on several arbitrary assumptions, including the choice of how many replicate simulations are used for each scenario (statistical significance can easily be increased by generating more replicates), the α value chosen to define significance, and (sometimes) the degree of difference among scenarios (e.g., two scenarios that contrast food production rates of 0.5 versus 1.0 are less likely to be significantly different than two scenarios that contrast rates of 0.5 versus 5.0). Another kind of arbitrariness is the degree of variability among replicate simulations of the same scenario, or how much "noise" there is in simulation results. We can change the level of variability among replicates by changing assumptions in our model about what processes are stochastic and how (section 5.8); and the more stochasticity we include, the less likely we are to find statistically significant differences among treatments. Another problem is that statistical significance does not necessarily indicate biological significance. (Ecologists savvy in experimental design will no doubt note that each of these problems corresponds to a similar problem with hypothesis-testing statistics in analysis of data from real systems; Suter 1996; Magnusson 2000.)

Fortunately there are ways to supplement (or altogether replace) statistical analysis of discrete treatments in an IBM: using a variety of model results—different state variables describing different aspects and hierarchical levels of the system—to build a case for whether there were important differences among treatments. A first (and sometimes sufficient) step is to present the degree of difference among scenarios in key outputs. We can simply execute a reasonable number of replicates for each treatment, then present the results of all replicates graphically and use our judgment to decide how important the differences are (e.g., figure 9.2). We can also examine many *types* of results to develop a full and convincing picture of the differences (or lack of differences) among scenarios. For example, how big were the differences among scenarios in not just one but all the key population level results? Did individuals behave differently? Were there differences in spatial or temporal patterns? Did different processes dominate the system's dynamics (e.g., did different kinds of mortality dominate in different scenarios)? This comprehensive approach to contrasting treatments is more convincing than depending only on statistics.

Although hypothesis-testing statistics are of limited value for *understanding* differences among IBM scenarios, they can still be useful for *communicating* these differences. Statistical analyses can communicate key results in a familiar and concise way. However, our experience has been that reviewers are touchier about the potential pitfalls discussed here when statistics are used to analyze IBMs, perhaps because some ecologists are uncomfortable with the whole idea of "data" generated by a simulation model. One way to use statistics effectively without risking a battle over methods and interpretation is to use simple statistical analyses (e.g., *t*-tests) to highlight differences (or the lack of differences) that are obvious and explained by other kinds of analysis.

Figure 9.2 Graphical comparison of IBM scenarios. The IBM of Railsback and Harvey (2002) was used to predict the average abundance of (a) juvenile and (b) adult trout under two discrete scenarios. Scenario 1 includes simulation of cannibalism, the risk to juvenile trout of being eaten by adults. Scenario 2 removes this process from the IBM to see if it has important effects. Twenty replicate simulations were conducted for each scenario; each plotted point is the mean abundance for one scenario. (Points are "jiggled" along the x-axis to make them visible.) When analyzed statistically, using the same methods as for table 9.1, the two scenarios were significantly different for both juveniles and adults. However, if only ten replicates are used, the scenarios are not significantly different for juveniles.

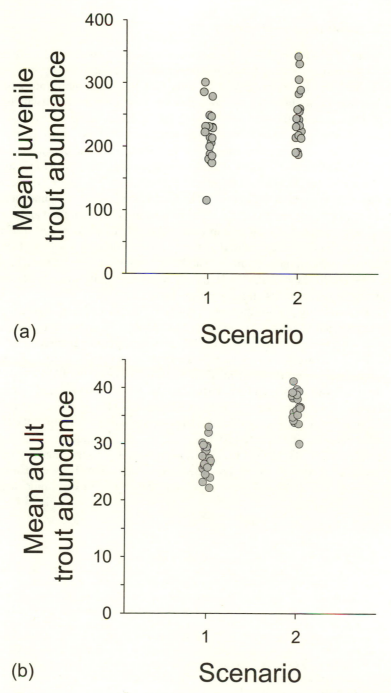

(a)

(b)

9.5.3 Quantifying Relationships

A variety of regression-based methods are commonly used to identify and parameterize relationships between model inputs and outputs, a type of sensitivity analysis. Developing these relationships is often a more informative alternative to the traditional scenario-contrast experimental design discussed in the previous subsection. Instead of testing for significant differences between a small number of scenarios, it is often just as easy to use many scenarios to develop a continuous relationship between independent and dependent variables. Instead of experiments simulating population abundance with food values of 5.0 and 10.0, we can simulate twenty food values from 1.0 to 20.0. This design not only allows us to determine whether food has an effect on abundance; it allows us to determine if there is a significant relation between food and abundance, what shape that relationship has, and (cautiously!) how much of the variation in abundance was caused by food or stochasticity or any other factors varied in the simulation experiment. Another example is the analysis of a brown bear IBM by Wiegand et al. (1999), in which habitat quality was varied over a wide range to examine its effects on bear abundance.

These sensitivity relationships need not be univariate or linear; figure 9.3 is an example analysis of how a simulated trout population responds to variation in both water turbidity and food availability. In fact, a comprehensive analysis of how simulation results vary with a variety of inputs can be used to develop a "metamodel" of an IBM: a statistical model of some of the IBM's system-level behavior (Kleijnen and van Groenendaal 1992). A metamodel can help understand and communicate an IBM by summarizing the model's response to key parameters and can be used (cautiously!) as a simplified version of the IBM for other analyses that require very large numbers of simulations.

Parada et al. (2003; Mullon et al. 2003) demonstrate another, innovative use of regression techniques to analyze IBMs. They started analysis with simple versions of their IBM and used regression techniques to determine which

Figure 9.3 Example bivariate sensitivity analysis of the trout IBM. The average adult (a) and juvenile (b) abundance of adult trout were predicted for forty-eight combinations of stream turbidity and food availability. These graphs show that simulated trout abundance is highly sensitive to turbidity, and especially to the combination of elevated turbidity and decreased food availability. In contrast to the three-treatment statistical analysis of the same issue (table 9.1), the sensitivity analysis shows that turbidity increases of greater than twenty nephelometric turbidity units (NTU) cause sharp decreases in abundance. In fact, juvenile abundance drops to zero at turbidities elevated by twenty-five NTU or more because adults cannot accumulate enough energy to reproduce. This sensitivity analysis shows that one conclusion that could be drawn from table 9.1—that elevated turbidity has insignificant effects on trout abundance unless accompanied by a substantial decrease in food availability—is clearly not robust.

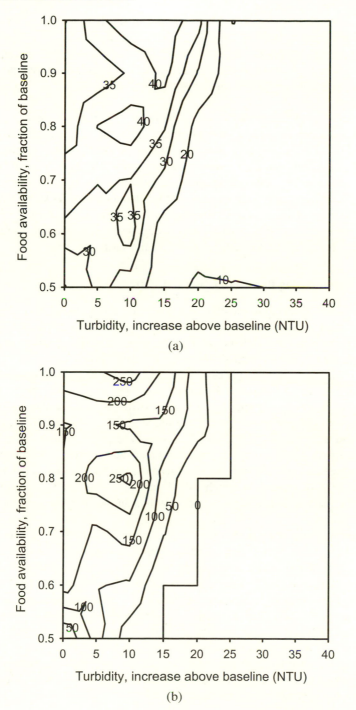

(a)

(b)

parameters explained most of the variation in the IBM's output and which parameters had little effect. The parameters with little effect were then kept constant in the analysis of the next model version, which included additional processes and structures. In this way, the number of parameters that had to be analyzed in detail could be kept nearly constant while the model's complexity was increased.

In addition to relating IBM results to input or parameter values, statistical analysis can be used to look for relations among different outputs of an IBM. For example, we may need to examine the relation between the number of individuals and their growth rate (is growth density-dependent?), how the abundances of competing species are related (is interspecies competition strong?), or how the abundance of adults is related to the number of juveniles of the same generation (is mortality density-dependent?). These kinds of questions can be addressed, cautiously and not conclusively (remember, correlation does not imply causation), using regression analyses.

9.5.4 Comparing Results to Observed Patterns

There is a rich literature on statistical methods for comparing model results to one kind of observed pattern: time series of census data (section 9.4.1). The inverse parameterization technique discussed in section 9.8 is an example of formalized analysis methods that require quantitative comparison between IBM results and observed patterns other than demographic time series. These methods require a computer algorithm to determine how well each version (typically a set of parameter values) of an IBM fits patterns observed in the real system. When the observed patterns are actually field data (e.g., census time series), comparing IBM results to observed patterns can require statistical methods because there is uncertainty in the observed patterns as well as stochasticity in IBM results. To determine whether results for a version of the IBM do or do not match the observed pattern—or to quantify how well the pattern is matched—we need statistical analysis that considers the uncertainties.

Several general approaches to statistical comparison of results to patterns are possible. Uncertainty can be ignored by simply comparing the mean of several replicate IBM runs to the observed pattern (as in the analysis of a bear IBM by Wiegand et al. 1998; Wiegand et al. 2003). Stochasticity in IBM results can be considered by evaluating the range of results produced by replicate simulations; a model version could, for example, be determined to "match" an observed pattern if the data defining the observed pattern fall within this range of results. In addition, statistical methods such as bootstrapping can be used to estimate the uncertainty in the observed data that constitute the pattern. Then IBM results can be determined to match the pattern if the results fall within confidence intervals around the observed data.

9.5.5   Inferring Causality?

A paradox of statistical analysis is that methods very useful for quantifying relationships also tempt us to infer the cause of the relationship when we should not (Huff 1954). With field data, statistical relationships may seem like the only clues we have to explain the cause of observed events. Even in field studies, however, statistical relationships by themselves are usually not the best information for determining causality; Suter (1996), for example, instead advocates a "weight of evidence" approach. This is just the approach suited for analysis of IBMs. When we find relationships between inputs and outputs of an IBM, we should not simply assume that the input caused the output response. Instead, we get busy with our detective work, designing simulation experiments to provide strong evidence for what processes gave rise to the relationship.

9.6   SENSITIVITY AND UNCERTAINTY ANALYSIS

Sensitivity analysis (SA) and uncertainty analysis (UA) are special kinds of simulation experiments designed to analyze a model in standard, rigorous ways. These analyses use many model runs to examine how results vary as inputs— often, parameter values—are varied. Formalized SA and UA are widely used in simulation modeling (examples in ecology include Bartell et al. 1986; Drechsler 1998; Rose et al. 1991), and techniques for it are covered extensively in the simulation literature (e.g., Vose 2000; Gentle 2003; Saltelli et al. 2004). A comprehensive SA and UA, examining all parameters, is often considered a thorough analysis for simple ecological models; these techniques can be valuable for analyzing IBMs as well, but their practicality and sufficiency can be limited.

SA has the objective of evaluating, in a comprehensive and quantitative way, how a particular output of a model responds to variation in selected parameters or inputs. SA can be useful for analyzing and understanding both a model and the system represented by a model. In analyzing a model, SA provides information on what parameters or inputs are most important for calibration, validation, and revision, and also on what parts of the model might be removed because they have little effect. For analyzing the system represented by a model, SA can predict the kinds of change (perhaps via management or natural disturbance) to which the system is more or less sensitive. Drechsler (1998) and Rose (1989) provide guidance for SA of complex ecological models.

Uncertainty analysis (UA) estimates how likely certain model outcomes are. Probability distributions are specified by the modeler for selected model inputs that have particularly important or uncertain values. Then a value is drawn (either randomly or systematically) from the distribution for each input and the resulting simulation executed; this process is repeated until the relative probability of different model results can be estimated.

Both UA and SA require the modeler to make a number of assumptions. For many IBMs it is infeasible to analyze sensitivity or uncertainty for all parameters and input data, so the modeler must decide which to include. For UA, the modeler must also specify probability distributions for the selected inputs, a task that requires a great deal of analysis or judgment by itself. In SA, the modeler must decide whether to examine sensitivity over very limited ranges (which allow linear modeling of sensitivities) or over very wide ranges (which provide a more comprehensive picture of model response).

SA and UA have not yet been widely applied to IBMs; among the examples are Jaworska, Rose, and Brenkert (1997), Huth, Ditzer, and Bossel (1998), Sutton, Rose, and Ney (2000), and Pitt, Box, and Knowlton (2003; see section 6.3.3). Perhaps these techniques have been used rarely because individual-based modeling offers plenty of other challenges, but also perhaps because the techniques provide a limited view of a complex model. Although SA and UA can help us understand an IBM, they are a reversion to "black box" modeling: we look at what goes in and comes out without trying to understand what is going on inside the IBM. In deciding how to use SA and UA, we must make a difficult trade-off: should we use traditional SA and UA techniques to examine the effect of many different inputs on only one or two outputs? (And which outputs should they be?) Or should we use other techniques discussed in this chapter to develop a more comprehensive understanding of how only a few inputs affect the whole IBM?

Our advice is to consider SA and UA as potentially powerful tools that deserve consideration when planning the analysis of an IBM, especially for IBMs that are simple enough that most or all parameters and inputs can be examined. These kinds of analysis are especially important for models used to support management decisions, because they help managers decide how much confidence to have in predictions. Even for more complex IBMs, SA or UA can help identify parts that deserve more or less analysis. However, for several reasons SA and UA will not be sufficient by themselves for analyzing many IBMs. First, a comprehensive analysis of all parameters or inputs is computationally infeasible for complex IBMs. Second, these techniques do not address the primary reason we use IBMs: to understand *how* system dynamics arise from individual traits. Finally, the same kinds of analysis can be included in the more comprehensive framework of robustness analysis, which we discuss next.

9.7   ROBUSTNESS ANALYSIS

The term "robustness analysis" is used in different ways in many fields; here we use it to describe a general strategy of analyzing how robust the outcomes of an IBM are to changes in inputs or assumptions. Conceptually, robustness analysis (RA) is just the inverse of SA, but we discuss RA separately because analyzing and communicating the robustness of a model is more strategic and less of a "black box" approach than traditional SA. The power of terminology must not be underestimated: sensitivity is usually a negative property of a model—we usually hope to prove that sensitivity is low—and analyzing negative properties is not much fun. Robustness, however, is a positive property that we like to communicate to others because it enhances the significance of our model.

Like UA and SA, RA can be used both to understand a model and to understand the system represented by the model. When analyzing an IBM's behavior, the purpose of RA is to identify model results—and, more important, the mechanisms producing the results—that are so robust to changes in parameters and model structure that they are likely to be of general significance. Then, once we have confidence in an IBM, we can use RA to predict how robust certain characteristics of the modeled system are to specific impacts or management practices. Before we discuss ways to do RA, the following examples illustrate what RA is and why it is used.

9.7.1 Examples

1. Tree-grass Coexistence in Savannas. Jeltsch, Milton, et al. (1996, 1997) developed a spatially explicit model of semiarid savannas, with the objective of understanding the spatiotemporal structure of savannas and to devise livestock grazing guidelines. Once the first model version was built, analysis focused on the two main patterns that define savanna ecosystems: the long-term coexistence of trees and grass, and the wide separation of individual trees. These two patterns occur in virtually any semiarid savanna and are thus robust to changes in biological and environmental factors, including properties of the species involved. Therefore, the savanna model needs to reproduce the two patterns in a very robust way.

Jeltsch et al. (1996) showed that the first model version could reproduce the savanna patterns but only for a very limited range of parameter values. This lack of robustness indicated that the model did not capture all the essential mechanisms of savanna dynamics. Exploration with the model predicted that tree-grass coexistence in savannas requires soil microsites where the probability of tree establishment is high. Subsequent field studies identified and quantified mechanisms that generate such microsites. When these processes were added to the model, it reproduced the two savanna patterns very robustly.

2. Spread of Rabies in Fox. The IBM describing the wavelike spread of rabies among red fox (section 9.4.5; Jeltsch, Müller et al. 1997) has a hybrid structure: some elements are very coarse (e.g., rules for what spatial units have healthy, infected, and infectious foxes), whereas others are detailed (the individual-based description of fox dispersal). The model reproduces the observed wavelike pattern amazingly well, but the robustness problem was to show that this result was not highly dependent on parameter values. This problem was important because most parameter values were educated guesses. Model analysis showed that, indeed, the model's ability to produce the wave-like pattern was very robust with respect to parameter values. This robustness was effectively documented and communicated by a matrix of figures, each presenting the sensitivity analysis of one parameter and one specific output variable of the model (Jeltsch, Müller, et al. 1997, fig. 4). The main features are easy to grasp: the curves in all subfigures have broad sections parallel to the x-axis indicating that sensitivity of particular model outputs to changes in particular parameters is low. Clients of the model thus learn within a few seconds that the wavelike dispersal pattern reproduced by the model, and therefore other insights gained with the model, are a robust feature of the structures and processes built into the model.

3. Management of a Fish Population. This example examines robustness characteristics of the ecological system being modeled, not of the IBM itself. Sutton, Rose, and Ney (2000) developed an IBM for the problem of how to manage stocking of hatchery fish to increase fish growth and survival in a lake. They used the IBM to identify good stocking strategies, then analyzed these strategies in the IBM (using Monte Carlo techniques) to determine how robust each strategy was to natural variation in the physical environment and in availability of prey. This analysis predicted that one particular management strategy would be most successful over a wide range of conditions. Similarly, Drechsler (1998) systematically checked the robustness of the ranking of different management options tested in a population viability analysis (PVA) model.

9.7.2   The Questions of Robustness Analysis

These quick examples show that RA is more of a general attitude and strategy than a specific technique. How RA is actually done depends on the details of the IBM and the problem the IBM is designed to solve. In general, RA can be thought of as attempting to answer the following questions while analyzing an IBM.

1. How robust are the model results? This question, which just establishes the goal and attitude of RA, is critical to think about because the most important element of RA is simply *thinking in terms of robustness.* The word "robust" implies forceful manipulations, whereas the word "sensitivity" intuitively refers to minute changes: a robust character is able to withstand extremes, whereas a sensitive character responds strongly even to minor annoyances.

2. What patterns is the model supposed to reproduce? To analyze robustness, we must clearly define what outcome of the IBM we are analyzing the robustness of. The outcome of interest is usually a general pattern—perhaps a type of adaptive individual behavior, or an observed system behavior as in the preceding examples, or even more general and conceptual patterns like stability properties or diversity. Or the pattern may be a time series of observed population data, in which case it is important to remember that the observed data also are uncertain. It might be useful to show, during model validation, that an IBM *can* reproduce a specific, precise set of observed data; but it will probably not be useful to show that the IBM *robustly* reproduces these specific data exactly. If there is some way to estimate uncertainty in the data (how would the data be different with different initial or environmental conditions—or observation techniques?), then we can analyze how robustly the IBM predictions fall within the uncertainty bounds of the observations.

3. How large is the parameter space in which the patterns are reproduced? Often, the range of parameter values over which the IBM reproduces the expected patterns is the most important indicator of robustness. If this range is small, but the patterns ubiquitous, then the model is likely to lack (or poorly represent) an important mechanism. On the other hand, some real patterns may actually be observed under only a small range of conditions. For example, forests show a pattern of trees dying in strips only if there is a clear gradient in the system, as on mountain slopes. Thus, in the model developed by Jeltsch and Wissel (1994; Jeltsch 1992), which reproduces the strip pattern for a restricted range of parameters, the pattern has the same level of robustness as in a real forest.

This kind of analysis can often be conducted using techniques similar to conventional SA: selecting which parameters to analyze and their ranges, then conducting simulations over these ranges. Some cautions are in order. Interactions among parameters are likely in IBMs, so examining robustness to only one parameter at a time is risky. And, of course, in RA we are interested in broad ranges of inputs, not sensitivity over short ranges. The need to look at parameter interactions over broad ranges means that many simulations may be needed for a thorough exploration of the parameter space, making efficient techniques such as Latin hypercube sampling (McKay et al. 1979; Rose 1989; Gentle 2003) likely to be necessary. However, when the patterns of interest are qualitative or difficult to quantify (e.g., spatial or temporal patterns), the results of each simulation require human interpretation. As a practical consequence, the robustness of an IBM's ability to reproduce such patterns can often be analyzed only coarsely.

We should not wait until formal analysis of a "finished" model to look at parameter robustness. An informal evaluation of parameter robustness can be critical at the very start of model analysis, even still in the "playing around" phase. If preliminary exploration shows that critical patterns and behaviors arise only in a very restricted parameter region, we can quickly start looking for what is structurally lacking or wrong.

4. How robust is the ranking of management alternatives? When IBMs are used to support management decisions, the model outcomes of interest are often a *ranking* of decision alternatives. The IBM is used to predict what would happen if each of several management alternatives is followed, and then to identify the "best" alternatives and to exclude unacceptable ones. The robustness question is therefore how robust the ranking of management alternatives is to uncertainty in model inputs (including parameters) and assumptions. A general strategy for this problem has been proposed and explored by Drechsler et al. (2003; see also Drechsler 2000). First, the analyst identifies plausible ranges of input values and assumptions. Second, the analyst creates scenarios that are each a combination of values for the different inputs being examined. The scenarios may be generated randomly or by using more systematic techniques such as Latin hypercube sampling (McKay, Conover, and Beckman 1979; Rose 1989; Gentle 2003). Then, for each input scenario the model is used to rank the management decision alternatives. These rankings are then tabulated over all the input scenarios and analyzed to determine whether they were robust with respect to the input scenarios. Were one or two alternatives almost always best (or acceptable) no matter which combinations of parameter values or model assumptions were used?

The primary advantage of this ranking approach to robustness analysis is that it focuses directly on the ultimate use of the model. Intermediate analyses such as defining the model's sensitivity to each parameter or input are not needed, making the approach much more feasible for IBMs with many uncertain parameters. Instead, the model input scenarios can include (at least initially) only parameter values at the extremes of their plausible ranges.

5. How robust is the model to changes in structure? Often we are also interested in how robust results are to the model's structure: how does an IBM's ability to reproduce expected patterns depend on what entities, variables, and mechanisms are in the model? Obviously, we can address this question by contrasting versions of the IBM with alternative structures. Parameter robustness can also provide important clues to structural robustness: if the patterns of interest are reproduced over a smaller range of parameters than we think they should be, we then look for structural things that need to be added (or changed) to make the model more robust. On the other hand, if the range of successful parameters is very large, we can simplify the model step by step to find a version that is simpler yet still robust.

6. How can the model's robustness be communicated? Establishing an IBM's credibility is a major goal of RA, so results must be communicated effectively. For IBMs, RA is especially important to communicate because it counteracts the widespread belief that IBMs are inherently subject to error propagation and extreme sensitivity to initial conditions. Documenting and communicating analysis of structural robustness is also especially important because this analysis shows that the IBM's level of complexity is not arbitrary but instead carefully limited to what is essential. As with any kind of analysis, visual presentation of results is often most effective. Of course, statistical quantification can also be useful for summarizing RA results, but combining statistics with figures helps clients understand the analyses easily.

9.7.3 Robustness of Highly Variable and Sensitive Systems

A final topic concerning RA is how it can be used when the system being modeled is itself *nonrobust*. Our discussion of RA up to here has assumed that the system behaviors we analyze are robust, but we do not want to imply that more robustness is always better. Ecosystems can have behaviors that are highly sensitive to seemingly small events and may jump from one state into another with little provocation. Some examples are dramatic abundance changes in lower trophic levels when top levels are manipulated (e.g., the lake study of Lammens et al. 2002), forest dynamics driven by fire (Savage, Sawhill, and Askenazi 2000), and the Wadden Sea's tidal flat ecosystem that is highly driven by environmental and biological events such as winter ice, storms, and episodes of extreme macrobenthos recruitment (Grimm, Günther, et al. 1999a). How do we show that we can model such systems usefully?

The key to RA of such systems is to perceive the high variability and sensitivity as a pattern itself and to characterize this pattern with as much detail as possible. For example: What are the mean, variance, and extremes of key state variables in the real system? What does the frequency distribution of fluctuations look like? Are there factors known to strongly moderate the variability and sensitivity of the real system? And we should look hard for observed system-level patterns that are consistent and predictable, even if they are very simple and general. High-level system properties (e.g., measures of diversity or stability of some system characteristics) may be predictable even when many system dynamics are not. For example, Savage, Sawhill, and Askenazi (2000) used an IBM to look at how disturbance levels (lightning strike rates) affect forest dynamics. They analyzed both very sensitive, unpredictable outcomes (e.g., developmental pathways, relative abundance of different species) and more consistent outcomes (e.g., a diversity index, patterns of abundance among guilds of early- and late-successional stage species). When we find robust patterns in highly variable systems, models can be analyzed in the same, pattern-oriented way as more robust systems.

9.8 PARAMETERIZATION

Most models include some parameters with values that are not well known. These parameter values are then determined indirectly via calibration: tuning the unknown parameters until model output fits some observed data. Indirect parameterization is routine and easy for models having few parameters and simple structures: a quick, often ad hoc, exploration of how model results respond to the parameters can identify useful values. Because IBMs often have many parameters, and parameters can have complex and interacting effects on output, this manual calibration approach may be unproductive. In fact, the potential for "error propagation," in which the effects of small changes in parameter values are magnified by the model's interactions, has led some to doubt whether IBMs can be parameterized in any meaningful way. By now, however, so many IBMs have been analyzed that we can conclude that the error propagation concern is rarely justified. In this section we discuss methods that can be used to estimate values for uncertain parameters indirectly by comparing model results with observations.

First, we should be aware that manual calibration may work for complex IBMs. Being structurally rich and having many parameters does not automatically mean that an IBM responds to parameter values in complex ways, or even that many parameters must be calibrated. The more structurally realistic an IBM is, the more parameters can be evaluated directly using many kinds of information (sections 7.6, 9.4.2), so a model with many parameters may still only have a few that must be estimated indirectly. And ecological systems have many negative feedbacks—for example, competition for resources limits population density, energetics limit growth—that can keep a structurally realistic IBM from being particularly sensitive to parameter values. The trout IBM described in sections 1.2 and 6.4.2 provides an example. After—but not before—a successful theory was developed for this IBM's key adaptive trait (how trout select their habitat each day), a coarse calibration of growth and abundance to the limited available data turned out to be surprisingly straightforward (Railsback and Harvey 2002). Four parameters were particularly uncertain yet important: the density of two kinds of food and the risk of two kinds of predation. Each of these parameters was found to have relatively independent effects on the four major results of interest: one food parameter affected only juvenile growth whereas the second food parameter affected juvenile and adult growth; one predation parameter affected juvenile survival whereas the second affected adult survival. Therefore, observed growth and abundance of both juveniles and adults could be matched by manually calibrating these four parameters. The coyote IBM of Pitt, Box, and Knowlton (2003; see section 6.3.3) needed no detailed calibration at all to closely reproduce a number of observed patterns.

However, some IBMs have many uncertain parameters or parameters with complex, interacting effects on results. These models require more systematic techniques for indirect parameterization. In the modeling literature there are many techniques and software tools for the indirect parameter estimation problem. Statistical techniques using Bayesian and maximum likelihood techniques (Hilborn and Mangel 1997) are widely used to fit model parameters; Mooij and DeAngelis (2003) provide an example application of maximum likelihood techniques to parameterization of an IBM. Other widely used parameterization approaches use nonlinear optimization to find parameter values that minimize differences between model results and data. In addition to computational and conceptual challenges (e.g., gradient-based optimization techniques not working for discrete and stochastic models), these techniques are not well suited for looking simultaneously at a variety of model outputs at both individual and system levels, which is important for analyzing IBMs. However, the potential applicability of these conventional calibration methods to IBMs has not yet been explored; there are almost certainly many ways they could be applied to IBMs.

In the remainder of this section we summarize a technique for indirect parameterization that is particularly well suited to IBMs because it can cope with complex interactions and stochasticity and can calibrate parameters against a variety of observed patterns of different types and at both individual and population levels. This approach is simply a rigorous application of the pattern-oriented parameterization process described in section 3.3 (step 4). We illustrate this technique with an analysis by Wiegand and co-workers (Wiegand et al. 1998; Wiegand et al. 2003; Wiegand et al. 2004; see also Hanski 1994, 1999), but the technique is similar to a conventional calibration method called "inverse modeling" or "Monte Carlo filtering" (e.g., Rose et al. 1991 applied a similar technique to a food web model; Saltelli et al. 2004 provide techniques for Monte Carlo filtering).

The general concept of the inverse modeling technique is to execute many simulations while varying the value of uncertain parameters over wide ranges, then to identify the combinations of parameter values that produce acceptable simulation results by excluding the simulations that fail to pass some "filters." The filters are patterns observed in the ecological system of interest. Wiegand et al. (2004) used inverse modeling for an IBM of grassland dynamics, in which the individuals are grass tufts that grow, shrink, die, compete with each other for resources, and reproduce; rainfall is the primary environmental process affecting the individuals. Their analysis had several objectives: (1) finding valid values of parameters considered to be particularly uncertain; (2) examining the IBM's uncertainties, sensitivities, and robustness; (3) identifying any opportunities to simplify the model by removing unimportant processes; and (4) demonstrating a way to make analysis of IBMs more rigorous and reproducible.

In analyzing the grassland model, Wiegand et al. used the following general steps. These steps could be followed in analyzing other IBMs for similar objectives (see also Wiegand et al. 2003; Schadt 2002; section 6.4.1). The analysis starts as soon as a complete draft version of the IBM has been completed and implemented in software, the software tested, and preliminary analyses (of the type discussed in section 9.4.3) conducted to identify and correct any obvious problems with the IBM's structure.

1. Find values for as many parameters as possible using independent analysis (section 9.4.2). Identify the remaining parameters that have highly uncertain values. In the grassland IBM, there were nine such uncertain parameters.
2. Identify ranges of potential values for the uncertain parameters (the minimum and maximum values to be considered). Decide how many values

within the parameter ranges to examine. Wiegand et al. decided to divide each parameter's range into twenty-one equal intervals, with the midpoint of each interval used as a potential parameter value.

3. Create a collection of model parameterizations (each "parameterization" being one set of parameter values for one model run) that represents the full range of values and combinations of values for the uncertain parameters. Wiegand et al. used Latin hypercube sampling to represent the parameter space with 63,194 parameterizations, far fewer than the $\sim 10^{12}$ possible combinations.

4. Define a set of observed patterns that serve as the "filters" that separate acceptable from unacceptable parameterizations. This step requires not only identifying a number of observed patterns that represent different processes (especially processes that depend on the uncertain parameters) at both individual and system levels, but also defining explicit criteria for whether a model run does or does not "match" the pattern. Wiegand et al. used a stability property, persistence, as a primary filter: parameterizations that caused the grass to die completely were excluded from further analysis because the real grassland is highly persistent. They identified five more observed patterns to filter the remaining parameterizations. These patterns generally required "reasonable" values of outputs like growth rate and diversity of tuft sizes, and reproducing specific events (e.g., a postdrought dieback). The criteria for matching these patterns were all yes-no values: a simulation result either did or did not match each pattern.

5. Design an environmental scenario representing the conditions under which the filter patterns were observed. The filter patterns used by Wiegand et al. were all from one nineteen-year field data set, so the environmental scenario used input data from the same period.

6. Run simulations for all the model parameterizations, saving not only the output variables used to evaluate the filter patterns but also output representing other important model predictions for use in the sensitivity and uncertainty analyses (step 8).

7. Identify the parameter combinations that reproduce all the filter patterns. From these combinations, identify—if possible (see the lynx IBM, section 6.4.1)—the range of "good" values for each parameter. A further analysis possible at this point is to analyze the IBM's parameter robustness as we discuss in section 9.7. Signs of low robustness include only a few parameterizations passing all the filters, and good values for any parameter falling within a narrow range.

8. Analyze parameter sensitivity and uncertainty in the simulation output, using conventional techniques. Wiegand et al. (2004) calculated Spearman rank correlation coefficients for the combinations of uncertain parameters and model outputs as an index of sensitivity, and calculated the standard deviation of outputs as a measure of uncertainty. These analyses used

results only from the parameterizations that passed the pattern-oriented filter.

9. Finally, analyze structural robustness. In the grassland model, some of the uncertain parameters had possible values near zero that essentially turn whole processes off. The parameter filtering results were examined to determine how many successful parameterizations turned some process off. Such parameterizations indicate that the process was not essential to reproducing all the patterns and so potentially could be eliminated from the IBM. (The sensitivity and robustness analyses also provide clues about what structural elements of the IBM might be unnecessary; see section 9.7.2.)

This pattern-oriented inverse modeling technique does have some limitations. As with all systematic and statistical analysis techniques, inverse modeling requires some arbitrary assumptions that affect its conclusions. These include choosing the filter patterns and how they are quantitatively defined, the parameters to treat as uncertain, and the range and number of values analyzed for each uncertain parameter. The technique cannot find the "best" parameter value but only ranges of good values. (There may in fact be no single "best" value for some parameters.) And, of course, the computational requirements limit the analysis's precision for particularly large and complex models.

Other limitations of inverse modeling are more specific to IBMs and pattern-oriented analysis. First, this technique could be difficult to use with patterns (e.g., qualitative spatial patterns) that cannot readily be evaluated by a computer. Second, the technique may not always be the best way to use patterns to test a model's structure. Before this technique is applied blindly, the pattern-oriented approaches we discuss in chapters 3 and 4, and also in section 9.3, should be given full consideration because they might help refine the IBM's structure before parameterization is conducted. Finally, the various filter patterns must all occur in the same model run, or else the technique must be modified so that different patterns observed under different conditions (e.g., plant dynamics in rich vs. poor soils) can be used together as filters. Nevertheless, inverse modeling using the protocol of Wiegand et al. has proved powerful and seems particularly useful when extensive data sets contain—in a coded form—empirical information about unknown parameters.

9.9 INDEPENDENT PREDICTIONS

Finally, we address what many consider the "gold standard" in model analysis: using the model to make independent predictions that are then tested against field observations. (This step is also often referred to as "validation.") By "independent," we mean predictions of patterns or variables that were not considered in designing and calibrating the model (e.g., Rykiel 1996). Using a model to make independent predictions, then finding or collecting data on the real system to test those predictions, can be the most convincing demonstration that the model captures the system's essential characteristics.

Testing an IBM with independent predictions is just another kind of pattern-oriented analysis—testing the model's ability to reproduce observed patterns—except that we use patterns that were not used to design or parameterize the IBM. Independent predictions can be tested against existing data, or used to design new field studies specifically to test the predictions, or even tested against patterns gleaned from the literature. The models described in section 1.2 were chosen as examples in part because they illustrate use of independent predictions. The BEFORE beech forest IBM was tested against field observations of the amount and spatial distribution of coarse woody debris (Rademacher and Winter 2003), processes not used to develop and parameterize the model. The trout model was analyzed by its ability to predict a wide range of population-level patterns, at diverse sites, that were found in the literature (Railsback et al. 2002).

Analyzing a model by its ability to make successful independent predictions has its limitations. First, while this technique may be very powerful for validating an IBM, it may also tell us little about the system and problem we designed the model for. For that reason, we often do not bother with independent predictions until after the other analyses are completed, and we are thinking about what else the IBM might be useful for. Second, the precision of the predictions and tests must be compatible with the IBM's purpose and design: predictions that are too precise may not be upheld simply because some processes of the real system were intentionally left out of the IBM, and predictions that are too general may not be convincing. Finally, there is always a strong possibility that a model perfectly good for one system and problem may not be able to make accurate predictions for other systems, conditions, or problems. The solution to these potential problems is the same as for pattern-oriented analysis in general: use a wide range of predictions and observations to make a convincing overall case instead of focusing too much on comparing one particular type of model output with one set of observations.

9.10   SUMMARY AND CONCLUSIONS

New modelers often see building a model as their main task, but analyzing a model is every bit as essential as building it. We might learn things in the process of building an IBM, but the odds of learning things of general interest—and convincingly communicating them to others—are low unless we undertake systematic analysis of the model.

What do we mean by analyzing an IBM? Analyzing a model means *doing research* (often, closely following the inductive scientific method of posing and testing hypotheses) on a model to learn about its behavior and to learn about the system the model represents. Objectives of this research typically include verifying that the software does what we want it to (covered in chapter 8), finding good model structures and theory for individual traits, finding good parameter values, and finally solving the problems we designed the model for in the first place and learning something about ecosystems.

Model analysis requires time and resources, often much more than required to build the model. Therefore, if time and resources are limited (and they always are), then a model's scope and complexity must be sufficiently limited to make sure analysis can be completed. But when we start model analysis as early as possible and integrate it with model development (i.e., make the modeling cycle really a cycle), we push the modeling process forward and make it more efficient. Analysis helps us rapidly find the problems that must be fixed before progress can be made and helps us start learning from the model as soon as we can.

Several overall strategies make analysis of IBMs more efficient, and some of these are quite different from strategies used for other kinds of ecology and even for analyzing other kinds of simulation model. The most important strategy is to base analysis on carefully designed simulation experiments. Perhaps the greatest advantage of IBE is that experiments are cheap, easy, and fast. Instead of feeling constrained to analyzing whatever data first comes out of our model, we should continually be playing an all-powerful, omniscient, and often destructive god in the virtual universe we created. Powerful analysis requires that we be *creative*—that we *create* organisms, environments, and interactions so that we learn, step by step, how the full model works. To understand our IBM, we should freely rewind and rerun time, move mountains, manipulate the physiology and behavior of one individual or whole populations, and observe anything anywhere. And instead of depending on observed correlations to infer why something occurred, we should build convincing cases from many different kinds of evidence.

Other analysis strategies cope with the greater complexity and structural richness of IBMs. These strategies use the pattern-oriented methods we discuss in chapter 3 to take advantage of the many kinds of information available. With IBMs we can use qualitative pattern-oriented analysis to quickly get a feel for whether the model's structure captures the essence of the system and problem we are modeling. A final strategy seems obvious but is easily neglected: analyze the bottom levels—individuals and their environment—before trying to understand and test system-level results. Until we develop confidence that the individual traits we put in an IBM produce the adaptive behaviors we expect them to, there is no benefit to studying the system behaviors that emerge.

An exciting example of experimental analysis of an IBM is the exploration by Deutschman et al. (1997) of the SORTIE forest model. This example is especially valuable because the experiments were captured as "movies" that can be viewed in the on-line version of the paper (currently at www.sciencemag. org/feature/data/deutschman/). To understand the model itself and real forest dynamics, Deutschman et al. fearlessly altered major elements of the model's structure, traits of individual trees, and forest management practices, and then they looked at how these alterations affected specific patterns of forest diversity. From the example experiments in this paper (discussed in section 11.5.2), it is easy to imagine whole sequences of rigorous, hypothesis-based simulation analyses delving deeply into a variety of model dynamics and forest management issues.

Model analysis is what turns modeling into a cycle, telling us if, and how, a model needs to be changed before we can use it to solve problems. When analysis tells us our model is ready to use, we then do more analysis to learn about the system and problem we modeled. Then we are ready to move on to the final phase required to do science: communicating what we learned to our fellow scientists. Model analysis and communication are closely linked. Convincing others to accept the results of our IBM-based research requires communicating the analyses we did to demonstrate the model's validity and the robustness of our results. This final communication phase of the modeling cycle is what we address next.

But before proceeding, we offer a final caveat about analyzing IBMs. If we do analysis well we will gain some understanding of how the model works and why, and under which conditions the model reproduces certain patterns observed in real systems. All this analysis takes a great deal of effort and creative thinking, and can indeed be a major scientific achievement. However, we must never forget that we are experimenting with a model, not the real world. We understand the model, and we have some evidence that the model captures essential features of its real counterpart, so indeed we learn—indirectly—something about the real world. But the real system may still work differently from the model system because it includes mechanisms and structures we neglected in our model. It is natural for all scientists to fall in love with their models over time and treat them as reality (Crick 1988). But the insights we gain about the real world with an IBM are always indirect and should always be open to falsification by new observations. We must see our modeling effort as part of a larger cycle of science, with field studies to test whatever we think we learned via modeling. The strongest case we can make, in demonstrating a model's validity and its value to science, is showing that the model helps devise new, independent predictions, which are then tested and upheld by new observations and experiments in the real world.

Chapter Ten

Communicating Individual-based Models and Research

> Some [journal articles] contain model descriptions so incomplete
> or vague that independent checking of the results is impossible.
> Occasionally this obfuscation seems deliberate.
> —*Diane Beres, Colin Clark, Gordon Swartzman,*
> *and Anthony Starfield, 2001*

10.1 INTRODUCTION

As we progress through the cycle of building and analyzing an IBM, we finally
feel like we have learned important things that need to be communicated to
the "clients" of our research: the scientific community, our sponsors, the agencies that manage the ecosystems we study. Along the way we discover many
differences between IBE and traditional ecology—we often address different
problems, employ a different kind of theory and a different conceptual framework, and use many kinds of information in our IBMs instead of relying only
on classical models or field studies. Now it is time to look at one last difference:
IBE often poses unique challenges to scientific communication.

Communication—incomplete model description, especially—has always
been a concern with IBMs. For many IBMs the only complete description
has been the computer program (Lorek and Sonnenschein 1999; Ford 2000),
but computer programs are a very poor means of communication: they are
too long, often camouflage important concepts in complex detail, and are in
languages foreign to most readers. Articles were published with only a verbal
or graphical metadescription of the IBM (Ford 2000). But clients know that
such metadescriptions are incomplete, so they are left wondering whether the
reported results could really be reproduced. Ideally, an IBM is so interesting
that it "hooks" readers and makes them want to apply it themselves to new
questions. How can we convince our clients that an IBM is trustworthy and
even suitable for them to use themselves? Or will IBMs always be considered
too complex to be reproducible or reusable?

Communication is thus an important challenge that needs to be considered throughout an IBE project. This chapter presents some of the issues that

scientists have encountered in publishing early work with IBMs (and with agent-based models in other fields). Being aware of these issues should help scientists navigate the communication process by preparing for the obstacles likely to be encountered. This chapter may also help reviewers and editors of IBE work understand ways in which IBE is different and ways to accommodate those differences. While its main focus is on journal publication, this chapter is also intended to apply to such other forms of scientific communication as proposals, presentations, reports, and websites. First we set the stage by identifying four types of IBE work that are likely to be published. Then we address what historically has been the biggest communication challenge: describing an IBM with adequate completeness within the constraints imposed by most outlets. Next we discuss several reviewer concerns that are particularly common with IBM-based publications. Finally, we address the very important links between software implementation and communication: ways to communicate executable models and other software aspects of our IBE work.

10.2 TYPES OF IBE WORK TO COMMUNICATE

Over the life of an IBE project, four general kinds of work are likely to be worth communicating to other scientists via presentations and publications.

Model Descriptions.   Publications that simply describe a model's formulation (chapter 7) can be useful for two purposes. First, they make innovative modeling approaches available for others to evaluate and use. Second, publishing a model description by itself facilitates later publications focused on model *applications* (section 10.3). If a model description is sufficiently innovative or interesting, it may be publishable in a modeling journal such as *Ecological Modelling* or *Natural Resource Modeling*. An alternative is to publish the model description in a "gray literature" report that meets the requirements for being cited in journal articles. Journals typically allow reports to be cited only if it is reasonable to expect that future readers will be able to find the report; many research institutions have report series that meet this requirement. Such reports can avoid the length limits and some of the hassles of journal publication. Another alternative, increasingly available, is documenting the IBM in a digital appendix to a model application or analysis paper (discussed later) published in a journal that accepts extensive digital appendixes. Among journals currently accepting digital appendixes are those of the Ecological Society of America and most on-line journals such as *Ecology and Society*. Finally, making model descriptions available on a website can be convenient for both author and reader, an approach that may be suitable to supplement technical reports or conference presentations. However, websites are generally not citable in journal publications.

Management Applications.   This kind of work describes how a model was applied to an ecological management problem. Typically, a model application publication describes the parts of the model that are especially important to the application, the management problem, how the model was applied, and results of the applied analysis. Model applications are often presented in technical reports, conference presentations, or ecological management journals (e.g., *Biological Conservation, Conservation Biology, Ecology and Society, Ecological Applications, Journal of Applied Ecology*).

Modeling and Research Method Discussions.   Publications that analyze methodological issues in designing and using IBMs can be very valuable, especially because IBE is so new and undeveloped. Appropriate forums for this kind of communication include ecological and modeling conferences and journals. Making computer techniques and software available to others can be among a project's most important contributions to ecology (section 10.6).

Model-based Analyses.   This category includes ecological research that uses IBMs as a tool. Examples are development of IBE theory and explaining how population dynamics emerge from individual traits and interactions. Research of this type should be communicated through the same outlets as other basic research: ecological conferences and journals.

10.3   COMPLETE AND EFFICIENT MODEL DESCRIPTION

Probably the most common and difficult problem in communicating IBE is providing an adequate description of the IBM without exceeding a journal's page limit, the time available for a conference presentation, or the patience of readers and listeners. Many reviewers and readers understand that a complete model description is necessary because the results of an IBM can depend on any of its assumptions. A common criticism of IBMs has been that they are not truly reproducible unless described in complete detail, which they rarely are in publications. But it is rarely feasible to describe fully an IBM within a journal article or conference presentation, especially when the focus of the article or presentation is on an *application* of the model. It is quite common for reviewers to say both that (1) they lack confidence in an article's conclusions because some key parts of the IBM were not described in sufficient detail, and (2) the description that was provided was too lengthy and boring.

Often the only way to deal with the model description dilemma is to publish a detailed description of the IBM separately, then cite this description in subsequent work. When available, the alternative of publishing the model description as a digital appendix to an article has the advantages of making it as easy as possible for readers to obtain the description. Even when a complete description of an IBM is published separately, a journal article still usually must summarize the parts of the IBM most relevant to the current work—typical readers want to know that they *could* read a full description of a model, but few want to actually read the description (just as most readers of equation-based models are happy to know that they could, in principle, solve the equations themselves but do not). Every communication describing research based on an IBM will need to include at least a succinct description of model elements most relevant to the research.

One reason why describing an IBM can be painful to readers is a specific problem we can do something about: there is no standard format for describing IBMs. A standard format makes reading and understanding easy because readers are guided by their *expectations*. In their valuable article on scientific writing, Gopen and Swan (1990) explain how understanding is facilitated when writers take readers' expectations into account: readers are better able to absorb information if it is provided in a familiar, meaningful structure. When ecologists read a paper describing a traditional model, they expect to see several equations and definitions of the variables, then a table of parameter values. But when ecologists—especially the majority who have never built an IBM—start reading an IBM-based paper they start without such expectations of a familiar structure.

What is a meaningful structure of information for describing an IBM? It would be frustrating to read a detailed description of a model before first knowing its purpose; readers would have no idea why some things are in the model while others are ignored. Likewise, readers need to know what types of entity (e.g., individuals, habitat units) are in the model, and what state variables characterize these entities before they can understand details of the model's processes. A structure that promotes understanding starts with fundamental information and builds into detail. Therefore, we can help readers understand our IBMs efficiently by always using a familiar, meaningful structure: a standard protocol that provides the information readers need in an order that allows them to build understanding easily.

We propose the following seven elements as a standard protocol for describing IBMs. To promote its use we call it the PSPC+3 protocol (after the initials of the four most important elements—purpose, structure, processes, concepts—and "+3" as a reminder that the protocol consists of three further elements). This protocol is very similar to the contents we recommend for an IBM's written formulation (section 7.2), so a formulation document following those recommendations conforms to the PSPC+3 protocol. (It is also similar to the model description elements of the "Protocol for Model Disclosures" proposed by Beres et al. 2001.) However, the protocol is designed primarily with journal publication in mind.

1. *Purpose.* What problems is the IBM designed to address? This question may be particularly important if the IBM was originally designed for problems other than the topic of the current publication.

2. *Structure*. What kinds of entity (habitat units, individuals, etc.) are represented in the IBM? What state variables represent these entities? What environmental variables (e.g., weather, habitat, disturbance, management) drive the modeled system? What temporal and spatial scales—extent and resolution—are used?

3. *Processes*. What environmental and individual processes (e.g., food production, feeding, growth, mortality, reproduction, disturbance events, management, etc.) are included? Which state variables do they affect? At this stage, a verbal, conceptual description of the processes and their effect is sufficient; the main purpose of this element is to give a concise overview. Details of the submodels describing the processes are presented in element 7.

4. *Concepts*. These are the relevant items from the checklist of section 5.13. What essential system dynamics emerge from individual behaviors? What traits model those behaviors? How do individuals interact with, and what information do they have about, each other and their environment? How are any collectives represented? How is time modeled, and how are concurrent events scheduled? How are data collected from the IBM for testing, understanding, and analyzing it?

5. *Initialization*. How are the model's entities created at the start of a simulation?

6. *Input*. What input data are used to represent environmental conditions over space and time, or as other inputs? How were data obtained?

7. *Submodels*. How is each process in the IBM modeled? What specific assumptions, equations, rules, and parameter values were chosen, and why? How were submodels tested and calibrated?

Elements 1–4 provide the basic information readers need to quickly grasp the IBM's structure and processes and its nature as a system of adaptive individuals. Elements 5–7 provide the detailed information needed for complete understanding and reproducibility.

The PSPC+3 protocol can be followed in any kind of communication or publication, but the focus and level of detail of each element must of course be customized. For a document intended as a complete description of an IBM, then all elements deserve equal and full attention. For journal publication of research that used the IBM, then the protocol must be focused only on the most relevant parts of the model's design; complete details and even the less relevant major assumptions will need to be relegated to an appendix or separate document.

Following the PSPC+3 protocol—explicitly addressing each of its seven elements—routinely will give readers the familiar context they need to understand IBMs (and other bottom-up simulation models) with more efficiency and less pain. Using the protocol is one more step we can take to alleviate the concern that IBMs are too complicated to be useful for science.

10.4   COMMON REVIEW COMMENTS

Reviewers of model-based publications express similar concerns many times: is the model described completely and accurately, was the software tested thoroughly enough, how sensitive are results to assumptions and parameter values, how well do results match observations? Experienced modelers anticipate these comments, and in fact much of this book deals with common concerns like these. However, several additional concerns are commonly expressed by reviewers of publications based on IBMs, in the recent experience of our colleagues and ourselves. At least until IBE becomes more established, authors can anticipate the following three types of review comment.

10.4.1   Comparison of IBM Results with
                Classical Model Results

A remarkably common comment by reviewers and editors has been that results from an IBM should be compared with results from a classical model—a differential or difference equation model, or sometimes a matrix model. Reasons for this request include:

- To "validate" the IBM—if both models produce similar results, then readers may have more confidence in the IBM.
- To illustrate important differences between the two modeling approaches, perhaps interpreted as weaknesses of the classical model if it produces very different results.
- To determine whether an IBM was really needed for the study, or whether a "simpler" model could have sufficed (even though the classical models often require assumptions considerably messier than those of the IBM).

This kind of comparison can certainly be interesting for questions that can be addressed with both kinds of models. Comparing an IBM to a more conventional model may also make it more likely an article gets accepted in an ecological journal, by providing a link to more familiar approaches.

We show in section 6.6, however, that many attempts to compare IBMs and classical models have ultimately been unsatisfactory because the two kinds of model are so different. For many studies, the comparison would be meaningless because IBMs are used to address questions that simply cannot be addressed by classical models; a direct comparison would require simplifying the IBM to the point it could no longer address the study's questions. In other cases, it may be necessary to argue that, while a comparison of results to those from a classical model is feasible, the comparison is an inappropriate distraction because the research has completely different objectives. And it is often not clear what a comparison between IBM and classical model might "prove"—if results are similar, the comparison certainly does not prove that both models are "right"; and if results differ, the comparison does not tell us which (if either) of the two approaches is better.

Finally, it is important to remember that IBMs can be tested against the behavior of real organisms and ecosystems in many ways. We hope that authors of IBE publications can successfully argue that they found it more compelling to compare results of their IBM to reality than to other models. As reviewers and editors become more familiar with individual-based approaches, we expect their focus will change to comparing IBM results with reality instead of comparing modeling approaches with each other.

10.4.2   Generality and Robustness of Results

A common opinion of simulation models in general, and IBMs in particular, is that model results are so dependent on initial conditions, parameter values, and site-specific input that they cannot be robust or of general interest. This opinion may be encouraged by the mistaken beliefs that complex systems are inherently very sensitive to initial conditions (a characteristic of *dynamical* or chaotic systems, but not necessarily of complex systems or IBMs) and that errors inevitably propagate and multiply. However, these concerns often have some merit, and concerns about the robustness and generality of simulation results are always legitimate. It is generally wise to anticipate these concerns.

The primary way to deal with this issue is to design studies so that they address specific hypotheses (or patterns) in a way that produces information about the robustness and generality of conclusions. These study design issues are the focus of chapters 3 and 4. Journal editors have told us that submissions following a conventional hypothesis-testing format—building a case that a particular IBE theory or result explains some particular observations better than alternatives— are much more likely to be accepted and published, especially in ecological journals instead of modeling journals.

The second way to address concerns about generality of results is to use the robustness analysis methods presented in section 9.7. Publications that take simulation uncertainties, sensitivities, and robustness into consideration are more likely to be accepted as making valid, general contributions. Robustness or sensitivity analyses are also widely considered essential for acceptance of ecological management recommendations based on models (Bart 1995; Beres et al. 2001).

10.4.3   Readability

Making IBM-based publications concise and readable seems to be a challenge often noted by reviewers. Such publications tend to be innovative in many ways so authors can be tempted to describe all the exciting new things about their work. A more serious problem, though, is that IBE typically involves more research steps than other kinds of ecology. For example, if a team of ecologists has worked its way once through our IBE theory development cycle (chapter 4), a thorough communication of its work would need to describe (1) the alternative theories posed as models for an individual trait that contributes to population-level phenomena, and how these theories were developed or identified; (2) the patterns of observed behavior used to test the theories, and the field data or literature that document the patterns; (3) the IBM used to implement and test the theories; (4) how simulation experiments were designed to test the IBM's ability to reproduce each pattern; (5) results of the simulation experiments for each pattern and each theory; and (6) the overall conclusions drawn from the simulation experiments. A single article describing this work is likely to be long and complicated.

We can offer only commonsense suggestions for keeping IBE-based publications concise and readable. (1) Keep the article tightly focused on its specific objectives. (2) Do not attempt to fully describe the IBM in the same article that describes experiments using it; instead, only summarize the IBM's most relevant parts and publish a full description elsewhere. (3) The standard Introduction-Methods-Results-Conclusions organization has the advantage of familiarity, but it often is clumsy for IBE. An alternative organization for the previous paragraph's example publication might include an introduction, a short section describing the general analysis approach, a summary description of the IBM, multiple sections that each contain the methods and results for one of the pattern-oriented simulation experiments, and finally a summary and conclusions section. (4) Try even harder than usual to make the text clear, well organized, and well written.

10.5   VISUAL COMMUNICATION OF
EXECUTABLE MODELS

A complete and unambiguous written description of an IBM is essential for making the research based on the IBM reproducible, but it does not make the research *easily* reproduced. The effort required to develop software and input for an IBM usually discourages anyone from actually reproducing IBE experiments from scratch. The same is true of any kind of science requiring an elaborate apparatus. However, simulation research has a unique advantage: once we have assembled and tested our apparatus we can easily copy it and send it to others to use. We can provide clients with our IBM, along with its documentation, and input files, so they can run the model themselves. If they are very skeptical, they simply might want to see if the model really produces the results we reported; or they might want to perform their own experiments, perhaps to test the robustness of model results for themselves, to better understand results by looking at additional outputs, or to address new questions. Instead of forcing clients either to accept our results on faith or to rebuild the IBM from scratch, we can let them test the IBM as much as they want and to reproduce the simulation *experiments*, not the simulator.

Equipping an IBM's software with graphical user interfaces (GUIs; section 8.6.5) makes it especially easy for others to run, understand, and experiment with the IBM. GUIs facilitate what Grimm (2002) called "visual debugging," which includes many elements of communication. The benefits of making an easily used, easily observed version of an IBM available to clients can be dramatic. In our experience, graphic controls and displays can magically transform an IBM from a black box into something that even modeling skeptics such as field ecologists can understand, evaluate, and develop interest and belief in. Even our more model-savvy colleagues take more interest because the IBM seems more understandable and testable, and therefore more useful for new experiments.

The following tools for visually communicating the computer-executable version of an IBM are recommended (Grimm 2002).

Graphical Representations of State Variables. The goal is to make state variables of all hierarchical levels of the model—usually individuals and population—easily observable. However, the gap between these two levels often is so large that their mutual relationship is hard to illustrate. The trick is finding intermediate state variables: individual-level state variables aggregated to an intermediate level so that emergent system-level properties are more easily observed and understood. Distributions of key individual variables (size, age, etc.) provide such intermediate-level information, as do summary statistics (e.g., mean, maximum, and minimum size) broken out by categories of individual: age, sex, etc. During model development and analysis, modelers can try displaying different intermediate state variables to see which provide the best understanding of the system.

Input Screens for Changing Model and Control Parameters. Graphical screens can let observers see and change parameter values, not only for equation coefficients but also for parameters that control execution. To keep clients from feeling that parts of the IBM are hidden, all parameters should be accessible, even those never changed during model analysis. It should be possible to change any parameter at any time.

Input Screens to Select Model Versions. Model analyses typically use several different versions of an IBM—for example, implementing alternative traits for a particular individual behavior. These different versions need to be easily available so others can reproduce the experiments.

Input Screens to Manipulate Low-level State Variables and Processes. The ability to select specific individuals (or other entities such as habitat cells) and observe or manipulate their state variables during a simulation allows users to perform powerful controlled experiments. Similarly powerful are tools allowing users to execute specific low-level model processes, such as telling one individual to execute one of its behaviors.

Control of Random Processes. Conducting controlled experiments on an IBM sometimes requires the ability to control the series of pseudorandom numbers used to model stochastic processes; usually this is done by controlling how random number generators are used (section 8.7.3). Sometimes it is possible and helpful to have a version of an IBM in which stochastic processes are replaced by deterministic equivalents.

Trace Mode. Individual model actions are controlled and executed step by step by the user. Ideally, the user can manually execute a full time step or even a single schedule action.

File Output of Raw Data. Graphical observation cannot provide complete understanding of an IBM by itself. Clients will want to analyze the model using other variables and analyses than the graphical outputs. Therefore, it must be possible to write selected variables to output files.

To an ecologist, providing all these visual communication tools may seem like an enormous software task. However, software platforms for agent-based simulation (section 8.4) were designed to support just this kind of communication and can provide most of these capabilities with little additional programming effort.

Even when we choose not to provide an executable version of an IBM to clients, we can obtain some benefits of visual communication by making "movies" of key model runs available (e.g., the digital appendices of Deutschman et al. 1997; and Railsback and Harvey 2002). Screen graphics from a model run can be captured as animation (e.g., .GIF) or movie (.AVI) files—some software platforms provide this ability. Screen capture programs such as SnagIt can also be used. While movies do not allow clients to run and test an IBM, the ability to "see" a simulation they provide can increase interest and belief in a model.

10.6 COMMUNICATING SOFTWARE

The greater importance of software to IBE often extends even to communication. Here we briefly mention several kinds of software communication that researchers may need, or choose, to conduct; these are discussed more fully in chapter 8.

First, it may be necessary or desirable to publish an IBM's software—usually, the source code that prescribes (in a programming language that both people and computers can interpret) exactly how a model is implemented; but sometimes also the executable code that allows others to run the model (section 10.5). Usually the only practical way to make an IBM's software available in conjunction with a publication is via the web, using either the authors' own site or a journal's site for digital appendixes.

Second, it is often beneficial to provide documentation of software testing to reviewers and readers. Especially as ecologists become more savvy about the software aspects of IBE, we anticipate greater need to provide evidence that a model's software has been tested thoroughly. Such evidence can include very large files containing many thousands of test cases. As with model code, the most practical way to make these files available is via a website.

Finally, practitioners of IBE can contribute to the science by publishing new software methods and codes that others can use. Many projects will produce new software techniques and reusable code (especially, classes and libraries of object-oriented code) that could benefit other modelers. These benefits are especially likely for models implemented in one of the software platforms specifically designed to support individual-based and agent-based simulation (section 8.4.3). Software conferences are important opportunities both to share and to acquire useful software: some agent-based simulation platforms have regular users' conferences and there are general conferences on software for agent- and individual-based modeling.

10.7 SUMMARY AND CONCLUSIONS

For a new scientific approach like IBE, it is especially important and fun to publish methods, theory, and applications. However, there are special challenges to communicating IBE, some also due to the newness of IBE: editors and reviewers are often interested in comparisons to classical approaches even if such comparisons are not appropriate; and (an issue not discussed here) it can be difficult to find qualified reviewers. Other fundamental characteristics of IBE will always pose publication challenges. The tendency of models, software, and the entire research process to be more complex makes it difficult to communicate IBE thoroughly while still being concise and readable. Research using IBMs often combines aspects of both modeling and field studies: after an IBM is developed, we conduct experiments on it, collecting and analyzing observations of model individuals. This combination makes it difficult to use the traditional communication formats that readers are most comfortable with. And IBE is fundamentally interdisciplinary: in the same publication, we may need to address issues of a particular species' autecology and population ecology, complex adaptive systems, and software engineering.

IBE projects (and, especially, dissertation projects) are likely to benefit from a scientific communication strategy that is implemented as progress is made. A communication strategy could include the following tasks.

1. Maintain a Project Website. Many people now turn to the web for information before they even visit the library. A website has many advantages, even for individual research projects: no other communication technique makes it so easy to reach so many people with so much information. Websites are especially valuable for communicating informal, preliminary, or supplemental information: publications that have been submitted but not yet accepted, presentations, study site descriptions and photos, animations and other graphics that illustrate simulations, and complete descriptions of the IBM, its software, and the tests conducted to verify software quality.

Routinely updating its website helps make your project appear active, productive, and important. You can draw attention to your site by establishing links from the popular websites that contain links to individual-based modeling sites. (A bit of web searching will find these sites; current examples are maintained by Craig Reynolds and the Swarm Development Group.)

2. Anticipate Reviewer Concerns. Think about the issues discussed in section 10.4 as you design your research and publications.

3. Publish a Model Description. Publishing a document containing only a complete description of your IBM helps keep subsequent, more important, publications concise and readable. Section 7.1 lists important reasons to maintain a thorough written description of an IBM, and one additional reason is so this document can easily be turned into a model description publication.

4. Make Software and Documentation Available. Communicating your efforts on the software front is an important way to enhance your project's credibility. Among the materials that you may want to make available are the IBM's source code, its executable code (preferably equipped with GUIs), input files, software documentation, and documentation of your tests of the software. Typically, the only practical way to make such materials available is by posting them on your own website or by publishing them as digital appendixes to a journal article. An alternative to posting such materials where they can be freely downloaded is to post an announcement that you will provide the materials upon request.

5. Publish Modeling Analyses. This step is usually the ultimate communication goal of a research project: publishing the theoretical or applied analyses that the project and IBM were designed for. The previous elements of the publication strategy should make success at this step more likely. Even within this step, you can follow a strategic approach by first publishing analyses that test and validate the model and the theory it is based upon, and then publishing applications of the model to basic or applied problems.

6. Communicate Software Products. If your project produces useful software or computational techniques, you can promote ecological science (and yourself) by making these products available to others. Active publication of software products is especially important for interdisciplinary projects that include computer professionals: your project will be more likely to attract talented computer scientists and software engineers if it provides them with publication opportunities.

Naturally, we expect communication issues to change as IBE becomes better established and as technology continues to change. How? First, it seems reasonable to expect that as IBE becomes more commonplace there will be less need for us to justify the approach and less expectation that results from IBMs need to be compared with results from classical models to be interesting or valid.

Perhaps the greatest change we anticipate should result (we hope, very much!) from ecologists cooperating to establish and improve common formats for describing IBMs. One common format is the PSPC+3 model description protocol we propose in this chapter: its widespread use will make it easier to write—and read—descriptions that successfully communicate essential characteristics of an IBM.

In the longer term, even greater advantages will result from use of common software platforms for IBMs. Currently, other kinds of ecological model can be succinctly described *and executed* in specialized software: matrix models (in RAMAS software), mass flow models (in EcoPath), and simple IBMs (in NetLogo or EcoBeaker). For IBMs, platforms like Swarm (section 8.4.3) already provide "shorthand" that makes it easier to describe models and translate the description into working software. As we continue to improve and adopt these platforms, we will approach the goal laid out in section 8.1: a common, concise language for describing IBMs in a way that we ecologists can easily understand and that can be translated directly into the IBM's computer code.

Finally, we anticipate that the trend toward digital communication will continue to alleviate some challenges. Digital journals and digital appendixes to journal articles will make it easier to support our IBE research by providing readers with complete model descriptions, software and evidence of software testing, and graphical displays of simulation runs. Another form of digital communication not yet exploited well by ecologists is on-line forums and "homes" for technologies such as IBMs. We hope that soon there will be a small number of email lists and websites where ecologists using IBMs can go for information on theory and techniques, and where they can share their results and methods with each other.

PART 4

Conclusions and Outlook

Chapter Eleven

Using Analytical Models in Individual-based Ecology

A computer program is essentially a chunk of automated
mathematics, and consequently many of the goals and methods of
computational modeling are shared with traditional mathematical
modeling. The fundamental goal of both is to elucidate mechanisms
and to make predictions.
—*Richard K. Belew, Melanie Mitchell, and David H. Ackley, 1996*

11.1   INTRODUCTION

This book is about individual-based modeling and how it can be used, within
the framework of individual-based ecology, to address ecological problems. Of
course, other modeling approaches are widely used in ecology. In particular,
analytical models that use mathematical formulations are the backbone of classical ecology. Individual-based and analytical approaches are both important
tools for ecology, each designed for specific purposes and each having specific
strengths and weaknesses. Too often, modelers see alternative approaches as
competing or mutually exclusive and get caught up in unproductive debates
over which approach is right or wrong. Instead, in this chapter we explore
the question of how individual-based and analytical modeling approaches can
work together. For the ecologist interested in relations between individuals
and populations, are there advantages to thinking about and using analytical
models?

   Individual-based modeling uses computer simulation, which in principle
imposes no limits on the complexity of ecological situations that can be
described. However, we have seen in this book how even well-designed IBMs
require considerable effort to implement, verify, analyze, understand, and communicate. Analytical models, on the other hand, are based on mathematics,
mainly on ordinary and partial differential equations. The requirement of analytical tractability imposes a strong limitation on the complexity of ecological
situations that can be analyzed. In particular, many problems concerning the
effects of individual variability and adaptive behavior cannot be addressed effectively with analytical models. Nevertheless, analytical models have specific

strengths that make them important for individual-based ecologists to keep in mind and even use.

Therefore, in this chapter we summarize the most important benefits of analytical modeling approaches, how these benefits can be used in IBE, and how the approaches are related to each other. We do not provide detail on the rationale, techniques, and results of analytical approaches because this information is presented in numerous monographs and textbooks, for example, May (1973, 1981b), Hallam and Levin (1986), Yodzis (1989), Wissel (1989), DeAngelis (1992), Levin (1994), Gurney and Nisbet (1998), Roughgarden (1998), Caswell (2001), and Murray (2002). Rather, we show how in some cases analytical approximations can be distilled from IBMs and how analytical approaches can help, directly or indirectly, analyze and understand IBMs. These topics are closely related to the subject of chapter 9, how to analyze IBMs; but the relation between analytical models and IBE is important enough that we treat it separately. First, however, it is important to understand in more detail the different types of existing models and their specific purposes.

11.2 CLASSIFICATIONS OF ECOLOGICAL MODELS

Ecological models have been classified according to many different criteria. Some criteria indicate whether a certain factor is included in the model or not: whether a model is deterministic versus stochastic depends on whether randomness is included; whether a model is spatially explicit depends on whether space is included. A more fundmental criterion for distinguishing models is the purpose they are designed for. In general, there are three purposes of models: description, understanding, and prediction (Hall and DeAngelis 1985). Descriptive (including statistical) models describe data in an aggregated way and therefore allow relationships between variables to be predicted. Descriptive models do not refer to understanding at all, but nevertheless can provide important clues for explaining strong relationships. Models intended for understanding are referred to as "conceptual" (e.g., Wissel 1992a), "heuristic," or "explanatory" (Hall and DeAngelis 1985). These models are usually designed neither to describe specific systems nor to deliver specific, testable predictions. Finally, models intended for prediction often try to mimic nature in more detail, leading to so-called systems models (see the discussion of "naive realism" in chapter 2). Obviously, these categories are not mutually exclusive: models are often used for more than one of these purposes.

The distinction between statistical and other models is well defined because statistics is a well-defined methodology. Classifying models by whether their purpose is understanding or prediction is less clear-cut. An important and widely used classification scheme was proposed by Holling (1966) and May (1973), who contrast *strategic* and *tactical* models. Strategic models try to ignore details while capturing the essential dynamics of a system (Murdoch et al. 1992). Strategic models are believed to provide general insights because many different systems (e.g., different populations) are believed to have the same "essence." In contrast, tactical models focus on the detailed dynamics of particular systems with the purpose of making specific predictions, often to address management problems.

The distinction between strategic and tactical models was introduced to justify simple analytical models, which deliberately ignore detailed empirical knowledge (see also Levins 1966; Levin 1981; May 1981b; Caswell 1988; Wissel 1992a). This justification is important for dispelling the myth that useful models must be "realistic" in the sense of including all known details. The distinction between strategic and tactical modeling was certainly useful at the time it was made: in the 1970s, tactical simulation models were not designed for general understanding, and strategic analytical models were not designed to deliver testable predictions.

Today, however, the borders between strategic and tactical are more porous. In the 1970s, both simulation models and strategic models were based on equations. Today, most simulation models in ecology are bottom-up: individual-based or grid-based. Many bottom-up models are intended, as we show in this book, to provide simultaneously both understanding and prediction. Therefore, classifying models as top-down versus bottom-up has become more useful than trying to distinguish between strategic and tactical models. Top-down models focus on the system level and are based on highly aggregated state variables such as ecosystem function, population density, or number of species. The main design criterion of top-down models is that they can be formulated with equations.

Roughgarden et al. (1996) proposed a very useful classification of models into three types. "Minimal models for ideas" are intended to explore a concept without reference to a particular species or place. These models are not intended to make testable predictions, or even to be applied to specific real systems. Most early models of theoretical ecology are of this kind; examples include Lotka-Volterra models, the logistic equation, Levins's metapopulation model (Levins 1970), and community matrix models (May 1973). "Minimal models for a system" are intended to explain phenomena of certain classes of systems or species, while ignoring many characteristics of the real system in the hope they are not essential. These models are also not designed for specific, detailed predictions. Today, most mathematical models in ecology are of this type. Finally, a "synthetic model for a system" is a synthesis of detailed descriptions of a system's components. Early synthetic models consisted of large sets of differential equations. Modern synthetic models are bottom-up, representing many small spatial units or individuals and their behavior. In contrast to minimal models, synthetic models do not have system dynamics imposed by system-level equations; instead, system dynamics emerge from the interaction of the components.

This chapter addresses use of minimal models. We nevertheless refer to these as "analytical models" (AMs), the more common terminology. It should be noted, however, that "analytical" refers to the formulation of a model, not necessarily to its implementation. Many, if not most, contemporary analytical models are solved at least partly by using computerized numerical techniques.

11.3   BENEFITS OF ANALYTICAL MODELS

The strengths and weaknesses of IBMs and AMs are to a large degree inversely related. IBMs are designed for analyzing complex systems, but the more complex IBMs are, the harder they are to formulate, implement, analyze, understand, and communicate. AMs, on the other hand, have very limited ability to deal with complex systems of autonomous individuals. However, they are strong exactly in the ways that IBMs are inherently cumbersome.

Formulation.   The defining characteristic of AMs is being formulated in the language of mathematics. This language is general, universal, unambiguous, and concise. Computer programs, the ultimate language of bottom-up simulation models, are just the opposite: many different programming languages, compilers, and computers exist, all of which are outdated after a few years; and verbal descriptions of computer code are often ambiguous and not at all concise. (There are, as we discuss in chapters 5, 7, and 8, many "languages" for formulating simulation models but none are well established.) Furthermore, the apparent limitation that analytical models must be very simple in fact can be a decided advantage: the analytical modeler is forced to simplify and to ignore virtually every characteristic of the real system except one or two factors considered most essential. Individual-based modelers should also simplify as much as they can, but they must force themselves to do so (chapter 2).

Implementation.   By "implementation" we refer to the methods needed to "run" a model and produce results. Very simple AMs are solved analytically, which implies all the advantages of the language of mathematics. Most often, however, not the full model but an even more simplified scenario (e.g., an equilibrium) is solved. For example, community models based on generalized Lotka-Volterra models of competition or predator-prey interactions are usually not solved numerically to obtain the population dynamics of all component populations (but see Huisman and Weissing 1999; McCann 2000). Instead, a linear stability analysis of the equilibrium solution is performed (May 1973). Thus, the need to solve AMs is another very strong incentive to simplify. Today, analytical solutions of simplified scenarios are often augmented by numerical solutions of some other scenarios that use the full model. Like IBMs, the exact results of such numerical solutions depend on the computer algorithms used, but (unlike for IBMs, yet) standard algorithms and software are widely used.

Comprehensive Understanding. AMs are designed to capture the essence of a system or a class of systems (e.g., May 1973), but the same is true for IBMs (chapter 2). The difference is that AMs also have the characteristic of being easy to analyze and understand. According to Bossel (1992), typical minimal models have two to five state variables, whereas synthetic models (forest models, for example) typically have ten to thirty state variables. Having few state variables means an AM has few parameters. Therefore it is possible to explore AMs fully by looking at wide ranges and combinations of parameter values. Moreover, closed formulas can often be derived; this technique clearly shows how state variables depend on one or more parameters. As a result of these advantages, AMs are particularly useful for identifying and understanding basic processes governing system dynamics, for example, negative feedback loops leading to equilibria or extinction thresholds in metapopulations.

Communication. Being entirely mathematical, the formulations of many AMs are easy to communicate concisely and unambiguously. For more complex AMs, however, solutions may require formidable mathematics or numerical methods that are not easily understood by many ecologists.

Generality. AMs that are "minimal models for ideas" are general in the sense that they address concepts (e.g., density dependence or metapopulation effects), not specific systems. "Minimal models for systems" are also general in the sense that they describe classes of systems, not a particular system or place.

All these benefits have made analytical modeling the dominant approach in theoretical ecology, although there have always been critical voices (Pielou 1981; Simberloff 1981, 1983; Hall 1988, 1991; Krebs 1988; Grimm 1994; Weiner 1995; den Boer and Reddingius 1996). How can these benefits be combined with strengths of IBMs such as the abilities to represent effects of adaptive behavior and reproduce complex patterns? In the following two sections we first describe how analytical modelers try to incorporate elements of IBMs in AMs, and then how individual-based modelers can, at least indirectly, try to adopt the benefits of AMs.

11.4 ANALYTICAL APPROXIMATION OF IBMS

One way that the benefits of AMs can be applied to IBE is by developing analytical approximations of IBMs. Why build an IBM and then attempt to approximate its population-level behaviors with an analytical model? First, of course, this kind of research attempts to expand the ecologists' toolbox by expanding the range of ecological problems that can be addressed with AMs to include those for which individuals must be considered. But, perhaps just as important, this research helps bridge the perceived gap between individual-based and classical approaches to ecological modeling. Analytical modelers develop an interest in problems for which individuals are acknowledged as important, and individual-based ecologists develop a better understanding of potentially valuable mathematical techniques.

Much of the work on analytical approximation of IBMs has addressed IBMs which include discrete individuals and explicit spatial distributions. The research program underlying these approximations is characterized by the title of an article by Durrett and Levin (1994): "On the importance of being discrete (and spatial)." The approximations provide insights into which elements of the approximated IBMs contain an ecological "signal" and which are "noise." Further examples of analytical approximations of simple IBMs (or of other simulation models that address problems in which individuals are important) include Levin and Durrett (1996), Bolker and Pacala (1997, 1999), Wilson (1998), Grünbaum (1998), Law and Dieckmann (2000), Picard and Franc (2001), and Law, Murrell, and Dieckmann (2003). The volume edited by Dieckmann, Law, and Metz (2000) gives a comprehensive overview of examples and techniques, in particular the so-called pair approximations (Sato and Iwasa 2000) and moment methods (Bolker, Pacala, and Levin 2000; Dieckmann and Law 2000). These methods try to capture the essence of the distribution of individuals over time and space in second-order spatial moments and to formulate dynamical equations not only of the first moment (the common "mean-field approximation") but also of the second moment. Dieckmann et al. (2000) also provide numerous examples of problems for which the mean field assumption of traditional analytical models breaks down (e.g., Wissel 2000).

Often, this kind of research has followed a general protocol exemplified by Bolker and Pacala (1997). First, a general problem is defined; Bolker and Pacala addressed spatial pattern formation in single-species plant populations that experience density-dependent mortality, in uniform habitat. Second, a simulation model is developed to more completely define the problem and provide a baseline against which analytical approximations are compared. Bolker and Pacala (1997) used a simulation model that was spatially explicit and stochastic but not fully individual-based. Third, an AM that addresses the problem is developed. In fact, more than one AM may be developed using different degrees of simplifying approximation. Bolker and Pacala developed and compared AMs using only a mean-field approximation of spatial structure, a second-moment approximation, and a simplification of the second-moment approximation; Levin and Durrett (1996) similarly compared mean-field and second-moment approximations. Finally, experiments are conducted (using the kinds of techiques discussed in chapter 9) to compare the AMs' behavior with that of the simulation model. Bolker and Pacala (1997), for example, compared their models by examining predictions of how equilibrium plant density, and the spatial covariance of density, varied with the parameter controlling the scale of density dependence. Conclusions can then be drawn about the ability of the AMs to approximate the simulation model and, therefore, to address the original ecological problem.

A second example analytical approximation of an IBM is by Flierl et al. (1999), who addressed the phenomenon of aggregation in marine organisms. We describe this example in some detail because it closely resembles the IBMs of group living (fish schools, social animals) reviewed in section 6.2.

Flierl et al. used an IBM as the baseline simulation model. In the IBM, each individual i is characterized by its position x_i and velocity v_i which change due to acceleration a_i:

$$\delta x_i = v_i \delta t + \delta X_i,$$

$$\delta v_i = a_i \delta t + \delta V_i.$$

The term δV_i represents stochastic variability in velocity. Acceleration is modeled as driving velocity toward some preferred velocity V:

$$a_i = \alpha(V - v_i).$$

Different assumptions can be made about the preferred velocity. For example, the animals may try to move to higher concentrations of their neighbors, described by a preferred velocity V_i:

$$V_i = \sum W_1(x_j - x_i)$$

where W_1 is a weighting function:

$$W_1(z) = \begin{cases} Vz(1 - z \cdot z) & |z| < 1 \\ 0 & |z| > 1 \end{cases}.$$

For this W_1 the resulting groups are very compact. An alternative weighting function can be used to describe repulsion among individuals at short distances:

$$W_2(z) = \begin{cases} 3.3Vz(1 - z \cdot z)(z \cdot z - \frac{1}{4}) & |z| < 1 \\ 0 & |z| > 1 \end{cases}.$$

Flierl et al. (1999) simulated these and other simple IBMs and showed how aggregations form and move.

The analytical approximation of this IBM starts with a simple bookkeeping equation describing the change of the organisms' density ρ due to the convergence or divergence of the flux of organisms, J:

$$\frac{\partial}{\partial t}\rho = -\nabla \cdot J. \tag{11.1}$$

Obviously, an appropriate model of J cannot be expressed as a function only of the density but will also depend on the detailed shape of the probability distributions for the individual's positions. Thus, in the case of social animals with aggregation behavior, the spatial occurrence of individuals cannot be explained only by the density ρ.

Flierl et al. (1999) employed the following strategy to overcome this problem: they assumed a specific relationship between density and higher joint probabilities of neighbors, for example, that the distribution of neighbors is random (Grünbaum 1994). Alternatively, more realistic relationships could be measured from the IBM's simulation. The assumption of randomly distributed neighbors certainly is unrealistic in many cases, but may be sufficient if neighborhoods are large relative to typical spacing between individuals and if there is enough randomness in behavioral responses of the individuals.

The analytical approximations by Flierl et al. (1999) are quite complex, derived using a variety of assumptions and techniques from statistical physics. The resulting expression for the flux J is inserted in equation 11.1 and then this equation solved numerically to obtain the organisms' density ρ over time. The results match the output of the IBM quite well, although there are also some discrepancies regarding the timing of merging and splitting of groups. Flierl et al. also derive analytical approximations for another population-level quantity, the group size distribution. This approximation is feasible when group size distributions become stable quickly enough that the average rate at which groups merge and split can be defined as a function of group size. (The IBM of Fahse, Wissel, and Grimm 1998 addressing lark flocks makes an interesting comparison; see section 6.6.3.)

The Flierl et al. example and the other approximation studies mentioned in this section show that analytical approximations can capture essential features of at least simple IBMs. Powerful and easy-to-use approximation methods could be very useful but these methods, like IBMs themselves, are in their infancy. Currently, most approximation methods are too demanding mathematically to be applied by nonmathematicians, and our expectations must be limited by the fact that even in physics there are many problems (e.g., fluid turbulence, behavior of complex and digital circuits) for which analytical techniques are considered impractical. It will be interesting to see how analytical techniques develop and how useful they become for ecological problems in which interactions among individuals and their environment are important.

11.5 USING ANALYTICAL MODELS TO UNDERSTAND AND ANALYZE IBMS

In this section we discuss ways that some of the advantages of AMs, such as more complete understanding, ease of communication, and generality, can be obtained in IBE without actually implementing AMs. The individual-based ecologist can adopt several useful things from the analytical modeler: system-level concepts, a framework for simplifying models, and mathematical constructs for analyzing models.

11.5.1 Adopting System-level Concepts

Even a well-designed, parsimonious IBM may still contain so much informa-tion that it is difficult to understand how system-level phenomena emerge. In contrast, AMs are designed specifically to think about and explain system-level behaviors. Classical theoretical ecology includes a number of concepts concerning population behaviors and their causes, and these concepts can help us understand population behaviors of our IBMs.

An example is provided by our old friend, the beech forest model BEFORE (section 6.8.3; Rademacher et al. 2004). A typical mosaic pattern of forest structure emerges in BEFORE sooner or later, no matter what initial conditions are assumed (Neuert 1999). Moreover, this forest structure pattern is rather robust to changes in demographic parameters. What causes this robustness, a kind of equilibrium? The system-level concepts of AMs and classical ecology provide a strong clue: we know from minimal models that negative feedback loops often lead to equilibrium, either locally or globally. Such feedback loops exist in forests at the local scale: high recruitment into the upper-canopy layer increases mortality in younger, lower layers by shading them (Neuert 1999). This negative feedback between recruitment and survival is partly responsible for the quasi equilibrium of the forest structure pattern. Knowing about negative feedback loops (or positive feedback loops in other cases) from analytical modeling helps quickly figure out this important element of the forest's robustness.

Other examples of system-level concepts that should be borne in mind while analyzing IBMs include the existence of extinction thresholds in metapopulations (Bascompte and Solé 1998), the importance of environmental noise for the survival of small populations (Wissel, Stephan, and Zaschke 1994), the relationship between disturbance intensity and species diversity (Connell 1978), and the relationship between the complexity of interactions within a community and the community's stability properties (May 1973). An awareness of such concepts is a productive starting point for understanding even complex models, and talking about these concepts often helps communicate IBMs and their results by putting them in a context familiar to all ecologists.

11.5.2 Using Analytical Models as a Framework for Simplifying IBMs

One very productive technique for analyzing IBMs is to simplify them in a stepwise way (section 9.4.4). Analytical modeling approaches can be used as a framework to guide this analysis technique. Instead of attempting to directly compare an IBM with an AM of the same problem (section 6.6), the idea is to simplify an IBM, step by step, toward what an AM of the same problem might look like, while learning what is gained and lost at each step. The purpose is similar to that of the analytical modeler who tries to distill analytical approximations from an IBM (section 11.4): to understand which elements of an IBM are essential for a certain phenomenon, to distinguish between the ecological "signal" and "noise."

Implementing such simplifications is straightforward and can follow steps such as:

- Eliminating spatial heterogeneity by making the habitat homogeneous.
- Reducing temporal variability by holding inputs constant, for example, by using constant weather conditions.
- Reducing the importance of local interaction by expanding the distance over which individuals interact.
- Eliminating individual variability by making all individuals identical.
- Deactivating detailed processes by deleting them from the IBM's schedule or setting parameters to zero.
- Replacing stochastic processes with deterministic ones.

Simulation experiments (chapter 9) can then be used to determine what capabilities of the model are lost, and what ability to understand the model is gained, as each simplification makes the model more closely resemble a simple AM.

One of the most instructive analyses of a complex IBM via simplification used this approach. This example is the analysis by Deutschman et al. (1997) of SORTIE, an individual-based, spatially explicit forest model. SORTIE describes growth, mortality, and recruitment of nine tree species in a detailed way. For example, growth depends on local light availability, which is determined by a detailed submodel of shading by neighboring trees and movement of the sun. The purpose of SORTIE is to understand forest composition and structure under different disturbance and harvesting regimes and altered climate conditions.

First, Deutschman et al. ran baseline simulations of the full model, contrasting scenarios with and without disturbances (circular clearcuts). The main result was that the undisturbed forest is dominated by shade-tolerant beech, whereas yellow birch is much more abundant in the disturbed forest. Then, as part of a comprehensive model analysis, Deutschman et al. employed two simplifications that made SORTIE much more like a classical AM. First, the effect of spatial relationships was removed by replacing local light availability of each tree by the average light availability of the entire forest and by making the distribution of seedlings global and random. Second, the number of processes and parameters was sharply reduced. In SORTIE, each of the nine species is characterized by ten parameters that govern six key traits. A principal components analysis

(PCA) showed that two factors accounted for 69 percent of the total variance in species position in parameter space. These two factors are combinations of model parameters with strong, biologically meaningful interpretations: shade tolerance and growth strategy. Simulations with individual traits simplified to these two factors were then compared to the baseline simulations.

In the "mean-field" forest without spatial information, biomass was considerably lower and competitive exclusion accelerated. In the disturbed scenario, yellow birch was completely missing. Thus, spatial relationships were found *essential* for explaining the baseline results. Results using the simplified, PCA-defined species traits were quite similar to those of the baseline model in the undisturbed case, but failed to reproduce the dominance of yellow birch in the disturbed case. Deutschman et al. (1997) concluded that the PCA by itsclf indicated that the IBM could likely be simplified, but the simplified representation of species traits failed to reproduce essential features of the full model. The 30 percent of variance in the parameter space unexplained by the two PCA factors turned out to have a "strong dynamical signature." Emergent forest dynamics are thus sensitive to the details of each species' traits—at least in SORTIE.

By simplifying *toward*, but not all the way to, the simplicity of AMs, we remove detail from an IBM and check whether this detail was essential for system-level phenomena. We do not obtain new analytical approximations of the IBM, but we develop an understanding of what is gained and lost as we traverse a gradient between IBM and simple AM.

11.5.3   Using Analytical Constructs to Analyze IBMs

Simple AMs can provide ideas for evaluating more complex simulation models in a unifying way because AMs use general, well-known, and well-understood underlying mathematical constructs. One way we can analyze an IBM is by seeing whether its population-level dynamics resemble the dynamics produced by such familiar constructs as the Lotka-Volterra equations and random walk models. Again, the idea is not to approximate the IBM's dynamics in an AM; instead, here the question is whether an IBM's dynamics include mathematical properties familiar from analytical modeling. If so, we can take advantage of what we know from analytical modeling to better understand the IBM.

One example of this approach has already been described in section 6.6.3. Fahse, Wissel, and Grimm (1998) developed an IBM to study searching and flocking strategies of larks. After building this bottom-up model entirely from biological considerations, Fahse et al. then analyzed whether its population-level dynamics followed the properties of any mathematical constructs known from analytical modeling. This analysis found that the IBM did in fact have dynamics inherently resembling those of the logistic equation, a discovery that greatly increased the IBM's understanding and usefulness.

A second example of using analytical mathematical constructs to understand bottom-up simulation models addresses the stochastic models of small populations frequently used in population viability analysis. In these models, persistence and viability are fundamental quantities to be analyzed. Surprisingly, however, there is no consensus on how to quantify persistence and viability, especially in simulation models. The arithmetic mean time to extinction in, say, 1,000 simulations might be biased by the specific choice of initial conditions; the distribution of extinction times is skewed so that the median time to extinction is sometimes a better measure than the mean; viability—whether the probability of a population's extinction over a certain time horizon is acceptably low—is more meaningful than extinction times, but the choice of time horizon and acceptable extinction risk are arbitrary.

However, a familiar analytical construct, the first-order Markov process, was found to reveal an underlying structure of extinction processes (Wissel, Stephan, and Zaschke 1994; Grimm and Wissel 2004). In such a Markov process, the current change of a population only depends on the population's current state, not on earlier stages. Population dynamics thus are described as lacking "memory." We discuss later why this assumption is less restrictive than it seems. Consider a population of n individuals in which the probability $P_n(t)$ of having n individuals at time t is determined by the so-called master equation (Markov process of birth and death type):

$$\frac{dP_n(t)}{dt} = b_{n-1}P_{n-1}(t) + d_{n+1}P_{n+1}(t) - b_n P_n(t) - d_n P_n(t). \qquad (11.2)$$

This equation contains the probabilities $d_n dt$ and $b_n dt$ that the population size n decreases (via death) or increases (via birth), respectively, by one individual in the infinitesimal time interval dt. For the birth and death rates b_n and d_n, several submodels can be chosen, but this choice is irrelevant for the following considerations. Equation 11.2 can be rewritten with the help of a tridiagonal transition matrix A, with the matrix elements $A_{n,m}$ defined by b_n and d_n:

$$\frac{dP_n(t)}{dt} = \sum_m A_{n,m} P_m(t)$$

with m extending from 1 to some ceiling population size. It can be shown that after a short transient time, the probability of extinction, $P_0(t)$, can be approximated by:

$$P_0(t) = 1 - c_1 \exp(-\omega_1 t) \qquad (11.3)$$

where ω_1 is to the first eigenvalue of the transition matrix A, and c_1 is the inner product of the corresponding left-hand eigenvector with the initial condition, $P_n(t=0)$. Equation 11.3 provides a general structure for how probability of extinction depends on time, using two constants, c_1 and ω_1, that have a clear ecological meaning explained later. This structure provides the basis for a general protocol that (1) finds values for the two constants from simulations and (2) checks whether equation 11.3 holds for specific simulation models.

The protocol is very simple. The simulation model is repeatedly run to extinction. Then, the probability of extinction by time t, $P_0(t)$, is determined by the ratio of runs in which the population is extinct at time t to the total number of runs. Then, according to equation 11.3, which can be rewritten as:

$$-\ln(1 - P_0(t)) = -\ln(c_1) + \frac{t}{T_m},$$

a plot of $-\ln(1 - P_0(t))$ versus time t should yield a straight line when the time T_m is defined as:

$$T_m = \frac{1}{\omega_1}.$$

The constant c_1 can be determined from the intercept with the y-axis, which is $-\ln(c_1)$, and the characteristic time T_m is given by the inverse of the slope of the plot. Grimm and Wissel (2004) show that T_m is a mean time to extinction which does not depend on initial conditions, just as the intrinsic rate of population increase in a matrix model does not depend on the initial state of the population. Therefore, T_m may be referred to as "intrinsic mean time to extinction." On the other hand, c_1 reflects initial conditions. It turns out that c_1 can be interpreted as the probability that a population reaches the so-called established phase, which is characterized by quasi-stationary fluctuations of the population's state variables (including population size) and by the short-term probability of extinction being constant and equal to $1/T_m$.

The $\ln(1 - P_0)$ versus time plot has been used for all kinds of stochastic simulation models of populations (figure 11.1). In all cases, the plot was linear, reflecting the general structure of equation 11.3. It is perhaps surprising how well the plot works even for complex models which seem to violate the Markov assumption that systems have no "memory," for example, models with age structure or succession. However, possible memory effects in population dynamics can be described by a Markov process if the model includes state variables that carry this memory. Typical examples include age, weight, size, or other covariates of fitness. If these additional state variables carry the memory of earlier states of the population, the basic assumption of Markov processes that the current change only depends on the current state still can be used. Grimm and Wissel (2004) discuss theoretical and applied implications of the $\ln(1 - P_0)$ plot.

The lesson from these examples is that population-level dynamics of bottom-up models may have underlying mathematical structure which is hard to perceive directly. Simple models reveal such structure more clearly and can provide ideas

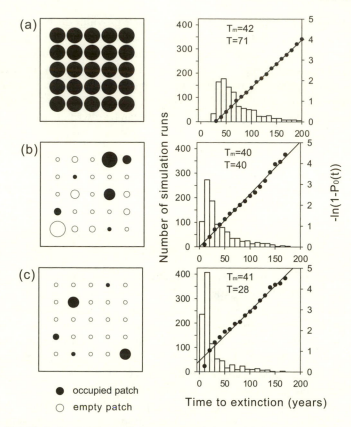

Figure 11.1 The $\ln(1 - P_0)$ plot applied to a metapopulation model (Stelter et al. 1997) for different initial conditions. The left panels show the initial state of the population and of the landscape (patches; patch capacity ranges from five to thirty individuals and is indicated by the size of the patch symbols). (a) All patches occupied and with maximum capacity. (b) A state of intermediate quality. (c) An extremely poor initial state. The right panels show the corresponding frequency distribution of times to extinction from 1,000 runs of the model, the arithmetic mean time to extinction, T (in years), the $\ln(1 - P_0)$ plot, and the intrinsic mean time to extinction, T_m, which are largely independent of the initial conditions. Note that the model includes environmental variation and catastrophes, and that patch capacity is not static but may become maximal after catastrophes and then decrease linearly due to succession. (Modified after Grimm and Wissel 2004.)

for what kind of structure to look for and how to use the underlying structure if it is found. But note that we here are talking about underlying mathematical structures, not specific models. The general structure underlying equation 11.3 is independent of the specific submodels chosen for the birth and death rates. We can view minimal models as demonstrations of general structures and mechanisms and search for these general features in IBMs. Sometimes we might find the same structures, but we should not be surprised when we do not.

11.6   SUMMARY AND DISCUSSION

Analytical modeling is unquestionably a fundamental tool of ecology. This approach has produced many important insights on the most fundamental problems of science, and perhaps its greatest benefit is the universality of mathematics as a language for scientific problems and their solutions. Therefore, whenever ecologists tackle new problems they should think about whether analytical modeling is a feasible and practical approach. When the problems involve the dynamics that arise from individuals interacting with each other and their environment, the answer to this question will often be no—which is of course why IBMs exist. But if the answer is yes, or even perhaps, then analytical models should be given a try.

As individual-based simulation has arrived as another tool of ecology, conflict between users of these two approaches has unfortunately (but, perhaps, naturally) arisen. At best, many ecological modelers and theorists see IBMs and AMs as fundamentally different tools for fundamentally different problems. This view is largely true, but not completely. We hope that this chapter shows that instead of conflict there can be a productive overlap between the two approaches. Many IBMs are population models, so we should expect that the long tradition of analytical population modeling can contribute to the development and analysis of IBMs. In particular, individual-based ecologists should try to borrow some of the key advantages of analytical modeling: simplification and established, well-known population-level concepts and mathematical constructs. Doing so can help us understand IBMs more quickly and thoroughly and help us communicate models and results to other ecologists.

Any scientist should be thankful for a more diverse set of tools and for more than one perspective on complex problems. Individual-based and analytical approaches each have their own momentum that can be unproductive if unchecked: AMs tend to drift into realms where mathematical interest supplants ecological relevance, whereas IBMs tend to drift into excess detail that makes understanding difficult. Constantly referring to each other can help both approaches maintain a productive direction.

Chapter Twelve

Conclusions and Outlook for Individual-based Ecology

> A new tool does not merely increase the number of ways to attack
> old problems, but also changes the nature of these existing problems,
> and, in an extreme case, may reveal whole new classes of problems
> to systematic enquiry.
> —*Ezequiel Di Paolo, Jason Noble, and Seth Bullock, 2000*

12.1 INTRODUCTION

In the preface and chapter 1 we explained why we wrote this book: to establish
an effective and coherent framework for using individual-based modeling, a
new approach to ecology that we refer to as individual-based ecology (IBE).
The strategic elements of IBE include fundamentals of good modeling (chapter
2), "pattern-oriented" modeling (chapter 3), an approach to theory (chapter 4),
and a conceptual framework for modeling systems from an individual-based
perspective (chapter 5). In chapter 6 we illustrated, with more than thirty IBMs,
how we can conduct IBE and what we have already learned from the approach.
Then, in chapters 7–10, we descended into the "engine room" to discuss four
important technical aspects of IBMs: formulation (chapter 7), software (chapter
8), model analysis (chapter 9), and model communication (chapter 10). In chapter 11, we discussed the relationship of more traditional analytical modeling
approaches with IBE.

   Now, in this final chapter it is time to ascend from the engine room and return
to the strategic level. We boldly assume that we were more or less able to convey
the main elements of effective and coherent individual-based modeling and
teach readers how to run the "engine": how to formulate, implement, analyze,
and communicate IBMs. But where will the new research vessel of IBE take us?
What course should we take, and what will we learn on the voyage? How will
our explorations differ from those undertaken with more traditional approaches
to ecology? And, as we steam ahead, what will we contribute not only to ecology
but to science in general?

   To plot the route we envision, this chapter first discusses the kind of problems
we can solve with IBE and how IBE differs from traditional approaches to

ecology. Then we discuss the role of IBE in the larger context of complexity science. Finally, just for fun we take a quick look into the individual-based ecology laboratory of the future to see who works there and what they are doing. A primary purpose of all this is to emphasize a point we made starting in chapter 1: IBMs are a tool for fundamentally different approaches to solving ecological problems—both theoretical and applied. IBMs are not an end in themselves but part of a process through which we can develop mechanistic understanding of ecological patterns and solve urgent environmental problems.

12.2   WHY DO WE NEED IBE?

Individual-based ecology is ecology from the perspective of individual organisms and their behavior, but IBE is still ecology and addresses the same problems as ecology in general. A widely accepted definition of ecology (Krebs 1972; but see Peters 1991) is that ecology tries to detect and explain patterns in the distribution and abundance of organisms. Within this general framework there is a plethora of subdisciplines, all narrowing ecology down by adding qualifiers: population, community, landscape, plant, forest, insect, arctic, microbial, food web, metapopulation, wetland, and many others.

   In our view, IBE does not just add to this list but is as generic as "ecology" itself: each of these subdisciplines of ecology can—and should, sometimes— be approached in an individual-based way. And IBE, like ecology in general, attempts to explain patterns of distribution and abundance. However, the way IBE goes about this fundamental problem is different in many ways. The example models and studies in chapter 6 show that IBE can be used to address many kinds of questions and problems that other "ecologies" cannot:

- How do system-level patterns of distribution and abundance emerge from the interaction of individuals with each other and with their environment?
- How does population structure, not just abundance, affect ecology? In IBE, "abundance" is treated as only a summary state variable of a system, not sufficient by itself to explain its own value and dynamics. Because individuals are different, populations can be structured by many individual characteristics: sex, age, size, life stage, social rank, location, and the like. Completely different structures may underlay the same abundance, so considering only abundance ignores information that is decisive for most problems. To explain patterns in distribution and abundance we must also explain the structure of populations. In many example IBMs (e.g., of woodhoopoes, marmots, canids, social spiders, and even the hypothetical species models of Uchmański and Donalson and Nisbet), populations were strongly affected by how they were structured by size, age, space, or social rank.

- What mechanisms determine "distribution"? Often, IBE tells us that these mechanisms include the adaptive dispersal and habitat selection traits of individuals, the structure of the habitat or landscape, and interactions among individuals (e.g., the marmot, woodhoopoe, lynx, and trout IBMs). Thus, "distribution" is an outcome of individual behavior, environment, and population status.

- What is the significance of local interactions to dynamics of plant abundance and distribution? Because plants are sessile, their local spatial configuration determines local competition; and mortality due to local competition is the key to explaining such general patterns as the linearity of the self-thinning trajectory.

- What are the individual-level processes that explain abundance and distribution patterns at the community and ecosystem level? Addressing such diversity problems requires community and ecosystem IBMs to focus on the "driving" species instead of including every species.

- How can we develop and test management models that are mechanistic and structurally realistic enough to let applied ecologists analyze a wide array of stressors and management scenarios? IBMs that are relatively simple yet mechanistic at the individual level can explain and predict complex population responses (especially to habitat alteration) that have proved elusive with population-level models.

- How do the environment and other species (e.g., predators) affect the behavior, life history, and population dynamics of a species? IBMs include the full life cycle of individuals and can include complex environmental dynamics. Especially by using techniques such as artificial evolution of adaptive traits, we can even look at how ecosystems are affected by individual variability in behavioral strategies.

Thus, we need IBE to solve traditional questions of ecology in new ways, but we also need it to address new questions that cannot be asked in the framework of traditional ecology (hence, the motto of Di Paolo et al. at the beginning of this chapter).

12.3   HOW IS IBE DIFFERENT FROM TRADITIONAL ECOLOGY?

Most ecologists are now trained in what we refer to as "traditional ecology." We refer both to classical theoretical ecology, which is based on analytical models (chapter 11), and to empirical ecology strongly influenced by classical theoretical ecology—for example, population studies where individuals are censused but not considered in more detail. Especially in part 3 we identify technical ways in which IBE is different from traditional ecology: in IBE our models use many kinds of information, we use computer simulation instead of calculus, we test and analyze models differently, and we communicate our models and research in different ways. But there are more fundamental, strategic ways that IBE is different from traditional ecology. Here we list eight general ways in which the *conduct* of IBE, as we envision it, is different from the way traditional ecology is usually conducted.

12.3.1   How We Address Complexity

Ecological systems are traditionally viewed as complex systems because they consist of a large number of unique entities subject to many kinds of processes and interactions. Taking this complexity directly into account in a model, without applying any filters, certainly would render understanding impossible. Traditional theoretical ecology uses mathematics as the complexity filter, simplifying the description of ecological systems until we can use the classical tools of theoretical science: analytical models. This approach means modeling highly aggregated state variables such as abundance or production, and assuming that these variables depend only on themselves and other aggregate variables, not lower-level processes (but see the approximation techniques discussed in section 11.4). For example, "density dependence" means that the rate of change in abundance depends on abundance itself—which of course is only a metaphor because changes in abundance emerge from what individuals do and what their environment does to them (Grimm and Uchmański 2002).

With IBE, we deal with complexity in a completely different way. The main difference is that we explicitly model *across levels* of organization. In IBE, as in real ecosystems, individuals and their behavior are the essential drivers of any ecological phenomenon. The description of unique individuals, with their full life cycle, is not sacrificed for the sake of a conceptual framework (differential equations) that was designed for simple dynamic systems. At the same time, we do not focus only on individuals, trying to understand their behavior in detail. Instead, we look for models of *individual* behavior that explain *system* dynamics and complexities. This across-level approach has already been very successful for modeling complex physical systems: scientists and engineers use it every day to predict the behavior of spacecraft, electronic circuits, and building structures.

IBE does not ignore complexity from the outset but combines simple theories (or models) of individual behavior in computer models that serve as our laboratories. Then, we try to understand the dynamics emerging from individual behavior using the hypothesis-testing experimental approach to science (Platt 1964). But what filter do we use to know which models of which behaviors we need to explain ecosystem complexities? The answer is in the pattern-oriented approach: patterns indicate structure and organization instead of amorphous complexity. Multiple patterns at different hierarchical levels can shed much light into the dark: we filter complexity by keeping in our models only the processes and behaviors needed to reproduce a set of specific patterns. The resulting IBMs are more complex than classical models, but still orders of magnitude simpler and easier to study than real ecosystems.

12.3.2   We Develop General Understanding by Studying Specific Systems

In traditional ecology, another reason to ignore complexity is generality: classical models attempt to reveal general mechanisms underlying abundance and distribution patterns. Of course, IBE also aims—as does any science—to gain general insights, but must follow a different strategy because complexity is no longer ignored. When we address individuals explicitly, we must be explicit about assumptions that are typically hidden or unstated in classical models (section 6.6). Most IBMs are designed to represent real species and systems, either as a way to obtain reasonable assumptions about processes ignored in classical models or simply to solve real-world problems. The focus of IBMs on real systems can be seen as meaning they are not general, if one assumes that specificity is the opposite of generality. But if we think of generality as "applying to systems in general," then to develop a truly general model (or understanding) we must first show, convincingly, that it is useful and predictive for at least *some* specific, real, systems—which has not been a focus of traditional theoretical ecology.

In IBE we follow the advice of Schopenhauer to seek the general in the particular. There are two ways to seek generality by applying IBE to specific systems. First, the theory development cycle of IBE (chapter 4) is used to identify theories for adaptive individual traits that are useful in IBMs. The constraints shaping the evolution of these traits are likely to be similar for similar behaviors of similar organisms, and well-designed theories of traits can be general to many environmental conditions. Therefore, instead of hundreds of theories for some behavior such as habitat selection, it is more likely that we end up with a toolbox of general theories that can be used for a wide array of species and systems. The toolbox itself can become an object to be analyzed: are there patterns in the theories for a certain adaptive trait, or for multiple traits? Can we, for example, identify broad categories of adaptive decisions that can be modeled using similar approaches? There are many decisions, for example, involving trade-offs between mortality risk and growth that potentially can be modeled with fitness-seeking theory.

The second way we can seek generality in IBE is by looking for common characteristics of the different systems we study. When we apply IBMs to a system we can learn much more about the system's internal structure and processes than we can by using classical models. We can also develop much more confidence in what we learn by testing the IBMs in a variety of ways (chapter 9). As we study more systems, it becomes possible to use comparative approaches to address general questions about the significance of certain system "ingredients" such as environmental variability and disturbance, phenology, structural diversity, and energy and nutrient inputs.

12.3.3 Behavior and Population Ecology Are Tightly Linked: Evolution Underlies All

Ecology is "made" by individual behavior, which in turn is shaped by ecology through evolution: for real organisms, behavior and ecology are thus inseparably bound to each other. Traditional population ecology largely ignores individual behavior, and traditional behavioral ecology often ignores ecological dynamics. In IBE we focus on both levels and, especially, on theory linking the levels. And often in IBE, evolution is the key concept underlying this theory: we can often explain ecological dynamics by assuming individual behaviors act, directly or indirectly, to improve fitness.

12.3.4 Modeling and Empirical Research Are Closely Linked

Field ecologists are very good at observing what the individuals of their systems do and how they adapt to changing conditions. Empirical knowledge about individual behavior and autecology is a rich resource for developing predictive models and theories, yet it is almost completely ignored in traditional modeling, which focuses more on numbers (e.g., abundance) than on the processes and structures underlying abundance. In IBE, we put empirical knowledge to work in two important ways. First, empirical knowledge can be used directly in individual traits: we can base our models of what individuals do on detailed laboratory studies, large-scale observations, and even informal, infrequent observations of rare events. Second, we use a variety of observed patterns to design and test our models and their theories (section 12.3.7). These patterns may be observed via field research or from controlled laboratory experiments.

12.3.5 Environmental Processes Are Integral to Models and Thus Explicit

Traditional ecology often minimizes the importance of environmental variability and heterogeneity. Simple classical models treat environmental effects as "noise," deviations from standard conditions that are acknowledged but otherwise ignored. In other population-level models, environmental effects show up only as parameter values that must be calibrated. Obviously, the inability to consider and explain effects of environmental variability and heterogeneity is a serious limitation. First, these effects are very common; very many populations and communities are affected by spatial and temporal variation in food availability or mortality risks; natural or human disturbances; and, now, climate change. Second, environmental effects are often exactly the problem we build ecological models to address: what happens when habitat is lost, exotic species invade, climate changes? In IBMs we can easily represent the local interactions between individual and environment that, along with interactions among individuals, drive populations. The effects of environmental change emerge from an IBM just as naturally as do the effects of population state.

12.3.6   "Theory" Is Not Separate from Real-world Problem Solving

In traditional ecology, the pursuit of theory is isolated from real-world problem solving: theoretical models are so abstract that they can rarely be applied directly to management problems. (Suter 1981, for example, discusses how ecosystem theory has not been useful for managing ecosystems.) The theory toolbox we envision for IBE (chapter 4), in contrast, is designed to let ecologists assemble IBMs for both management and research applications. A theory for how individual plants compete for light, for example, can be used to explain structural diversity of natural forests and to manage timber harvests and to predict the effects of exotic plant invasions. Inversely, developing IBMs for management applications also produces theory: for an IBM to be credible for management we must test its traits for individual behavior, and these traits, once tested, become IBE theory.

12.3.7   Testing and Analysis Are Integral to Modeling

Testing and analysis of models are not important concerns of traditional theoretical ecology. Testability is "sacrificed" (Levins 1966) to generality: models designed to be so general that they apply to all populations end up being so lacking in structure that they make few testable predictions. And many classical models are so simple that they require little analysis to understand. Most IBMs, however, address specific systems and produce a variety of results that are both complex and testable. Consequently, extensive analysis with simulation experiments (chapter 9) is an essential and profitable undertaking. From the analysis we can learn much about the IBM; and by testing its predictions against a variety of observed patterns, we can develop a level of confidence in what the model tells us about the real system.

12.3.8   Research Is Inherently Interdisciplinary

Traditional ecology is typically a collaboration of mathematics and biology; in fact, theoretical ecology is largely a mathematical enterprise. IBE requires a greater diversity of expertise because it considers more aspects of the system being modeled. Mullon et al. (2003) describe a typical large IBE project, which addressed recruitment variability in cape anchovy off the west coast of South Africa. Recruitment is highly dependent on ocean currents, so a hydrodynamic model was coupled with individual-based models of the anchovy. The models produced huge amounts of data to be analyzed. Thus, this project needed experts in fish biology, oceanography, simulation modeling, and statistics. In general, IBE projects addressing real systems will require expertise in field biology to understand the system level; physiology, behavior, and natural history to model the individuals; physical science or engineering, and geographic data, to model the environment; software engineering to design and build the IBM's software; and model analysis to finally test and learn from the IBM. For the ecologists, working in IBE is likely to mean we need less expertise in mathematics and more in the system's biology and more in simulation modeling and software.

12.4 WHAT CAN ECOLOGY CONTRIBUTE TO THE SCIENCE OF COMPLEX SYSTEMS?

This book borrows much from the new science of Complex Adaptive Systems. The CAS movement of the past several decades started with the recognition that models based on classical mathematics cannot capture essential dynamics of many systems of adaptive individuals, and it now searches for models and theory to explain such dynamics. Ecologists were among the pioneers of CAS; the earliest IBMs (section 1.4) were among the first attempts to find productive ways to model and study complex adaptive systems. Now, scientists in fields that specialize in complex systems (e.g., physics, mathematics, and computational science) come to biology—especially, evolutionary ecology—in search of interesting complexity problems. (A young scientist at the Santa Fe Institute, which specializes in complexity, once wrote a history of the evolution of bad theories of evolution developed by physicists.) What can ecology, and especially IBE, contribute now to CAS? Our problems are certainly still among the most complex, but can we still be pioneers in figuring out how to deal with complex problems?

So far, much of the theoretical work in CAS has had a "what-if" nature: what system dynamics emerge if we turn simple adaptive individuals loose to evolve in simple digital worlds? What happens to emergent dynamics if we change a trait of the individuals? The contribution of IBE to complexity science is tackling the inverse of the "what-if" approach, which could be referred to the "if-what" approach: "if" we look at specific patterns of real complex systems, "what" are the individual traits that explain them? The "what-if" studies have been extremely important for getting complexity science off the ground; classic

CAS models such as Boids (section 6.2.1) and Axelrod's (1984, 1997) models of social interaction convinced science that complex and important system behaviors can emerge from simple individual traits. But, as we argue in section 12.3, to develop predictive theory of how complex systems work scientists must get busy applying the approaches we describe in parts 2 and 3 to real systems: they must also ask the "if-what" question. Ecology has already made more progress than most other complex sciences in linking individual traits and complex system behaviors. We ecologists are certainly in a position to regain leadership in not only understanding how complex systems work but also in learning how to study complex systems.

12.5   A VISIT TO THE INDIVIDUAL-BASED
ECOLOGY LABORATORY

We have now said what we want to say in this book. Like all authors of science books, we hope you read this book from cover to cover, recommend it to all your colleagues, and decide to enter IBE and follow all our recommendations. But, to be more serious, it is important to think about the consequences of more and more ecologists making more and more use of individual-based methods. Or, as we asked in the introduction to this chapter: where will the research vessel IBE take us? Of course, we cannot precisely answer this question. As Niels Bohr said, predictions are difficult, especially if they are about the future. Science is exploration, so whatever we will find will be new and unexpected. Nevertheless, we would like to end this book with a *vision* of what the practice of IBE might become, which we do by peeking into an imaginary ecology lab.

So now let us push open the door and enter the ecology laboratory of Professor S., at a university of the not-too-distant future. The first thing we see in the main room is the all-important conference table in the center. Along the walls are the graduate students' desks, piled high with field equipment and monitors connected to the computer cluster humming in the corner. At one of the desks, a graduate student studies for his class in simulation software, the last of three core classes (along with introductory classes in modeling and programming) that are offered jointly by the ecology and computer science departments.

The bookshelf on one wall includes the old annual editions of the *Handbook of Individual-based Ecology*, which collected all available models of individual behavior and environmental processes, along with information on how each model has been tested in what contexts. On the same shelf are the old, well-worn users' manuals for EcoSwarm, the software package that implements all the models in the handbook. Now, because the national science funding agency finally decided to support these essential tools of IBE, the handbook and

EcoSwarm Manual are on-line so scientists can find—and add to—a complete and up-to-date collection of models and software. In fact, Dr. S.'s laboratory continues to contribute many of the models in the handbook and the corresponding EcoSwarm code. Originally she kept a software engineering graduate student on staff, but now that her ecology students are better trained in software and the new software platforms are so much easier to use, she did not replace the last software engineer after he graduated.

The next bookshelf holds a complete collection of the short-lived *Journal of Individual-based Ecology*; *JIBE* grew rapidly in influence for several years but as the quality of individual-based research improved and its methods became established, scientists like Dr. S. published more and more in the mainstream ecology journals. On the bottom shelf a tattered and outdated copy of Grimm and Railsback (2005) lies under a pile of field instrument catalogs.

In her adjacent office, Dr. S. works preparing the next lecture for her IBE class. She will discuss how plants sense and respond to potentially competing neighbor plants, and how this adaptive behavior affects community diversity. The students of her lectures are interested and enthusiastic as they see how the theory she teaches directly relates to what they see in the field. The students are especially excited by the seemingly limitless range of interesting and unsolved, yet imminently solvable, problems in IBE. One of her students now is using published data and an existing IBM to test a new theory for how large herbivores adapt their fecundity to temporal variation in habitat conditions; the theory was modified from the approach used by an earlier student to model how the herbivores select habitat at large scales. Another student is in the field evaluating fitness elements for a bird species that seems threatened by habitat loss: how do mortality risks, feeding and growth, and nesting success vary with what habitat characteristics? Even the simple IBMs that students develop as class projects often provide a fascinating new perspective on an old problem.

But Dr. S. realizes it is time for her meeting on a new project, a study to evaluate alternative ways to control an exotic plant's invasion of a nearby national park. Around the lab's conference table is the team she assembled to help design an IBM and the analyses it will be used for. A naturalist from the park represents the client and provides field data and anecdotal observations on the invasion problem. A professor from the botany department is familiar with the park's native community and plans, as part of the project, to conduct laboratory studies to see if a suspected allelopathic trait partly explains the invader's success. Wind dispersal of seeds is clearly important, so Dr. S. has invited a fluid dynamicist from the mechanical engineering department to sit in and help decide if and how the team should model local wind patterns. Her graduate student on the project is prepared to discuss what remote sensing data are already available or need to be acquired. Finally, Dr. S. has invited an assistant professor in mathematics to consult because he has worked on mathematical models of invasion. Although his "dispersal kernels" are still obscure to Dr. S., she knows from earlier projects that his different perspective on modeling is productive and helps improve the design and analysis of her IBMs.

As the meeting proceeds, Dr. S. bites her tongue as the park naturalist and the other scientists identify detail after detail that they think must be in the IBM to make it realistic—she knows that most of these details must be weeded out as they reduce the model to the most essential structures and processes so they can analyze it and explain it to park management.

Finally, after the meeting wraps up, Dr. S. decides to clear her mind with a bicycle ride. As she wheels across campus on the busy bikepaths, the cyclists unconsciously collect in groups for safety as they cross the streets, each cyclist staying just far enough from its neighbors to avoid collision. As she reaches the edge of town and the traffic danger wanes, the cyclists disperse and Dr. S. eventually rides alone through the fields. She mulls over the new project, thinking about all the information and data the team has collected on the individual plants and the natural community. "Now," she thinks, "in all that information, what are the key patterns that will tell us what needs to be in the model?"

Glossary

These are terms defined throughout the book. Most of these terms are used with various meanings throughout ecology, so we clarify how we use them in individual-based ecology. Other terms are adopted from software engineering and the field of Complex Adaptive Systems. (Italicized words are defined elsewhere in the glossary.)

Action — An element in an IBM's *schedule*. An action is defined by a list of model objects, the *methods* of these objects executed by the action (e.g., *traits* of individuals, updating the environment; producing output), and the order in which the objects are processed.

Adaptive behavior — Individual *behavior* that results from *adaptive traits* instead of being directly specified. Adaptive behavior therefore is an outcome of both the trait and the conditions occurring at the time the trait is executed.

Adaptive trait — A *trait* that includes some kind of active choice among alternative behaviors, with the decision depending on environmental or internal conditions.

Agent-based model (ABM) — The term for individual-based models in fields other than ecology. ABM is a more widespread and generic term than IBM.

Behavior, individual behavior, system behavior — What a model *individual* or *system* actually does during a simulation. A behavior is an outcome of an IBM, whereas a *trait* is a set of model rules that the individuals use to select their behaviors.

Classical models — The modeling approaches most commonly portrayed in ecology texts and courses. Classical models typically operate at the population level and use analytical equations or matrices.

Collective — In an IBM, an aggregation of *individuals* that exhibits some behavior of its own (e.g., a social group, pack, flock). Individuals belong to a collective, and the *state* of the collective affects the individuals. Collectives are an intermediate level of organization between individuals and populations.

Complex Adaptive Systems (CAS), complex adaptive system — When capitalized, CAS refers to the scientific study of systems made up of interacting, adaptive agents. Such systems are referred to as complex adaptive systems (not capitalized).

Discrete-event simulation — A general category of modeling that includes IBMs. Discrete-event simulators are simulation models that represent *system behavior* over time as a series of discrete events happening to the system's components, as opposed to *classical models* that use system-level rates. Much of the extensive literature and software for discrete-event simulation is useful for IBMs.

Emergent behavior — *System behavior* (or, sometimes, *individual behavior* that is affected by interactions among individuals) that is not directly specified by individual *traits*. Instead, emergent behavior arises from individuals' *adaptive traits* and their interactions with each other and their environment. System behaviors are characterized more as emergent (instead of *imposed*) if they (1) are not simply the sum of the properties of the individuals, (2) are of a different type than the properties of the individuals, and (3) cannot be predicted by examining only the individual traits.

Expected fitness — An *individual*'s current estimate of its future *fitness*, used as a way of evaluating decision alternatives in some *adaptive traits*: individuals make decisions in a way that increases expected fitness. Expected fitness at any one time may have little relation to the individual's eventual actual fitness.

Fitness — The actual success of an *individual* in passing its genes on to succeeding generations; in an IBM, fitness is an outcome. An individual's fitness can be evaluated as, for example, the number of its offspring that survive and reproduce; or the fraction of the simulation's final population that descended from the individual.

Fitness element — In *direct fitness-seeking traits*, a target that must be met for fitness to be high. Example fitness elements are future survival, growth to reproductive size, and attaining the social rank needed to reproduce.

Fitness measure, completeness, directness — A specific model of *expected fitness* that model *individuals* use as the basis of *fitness-seeking adaptive traits*. Decisions are based on the value of the fitness measure for each alternative. A fitness measure models how one or several *fitness elements* depend on the decision. A fitness measure's completeness increases with the number of fitness elements it considers: a fitness measure that models how expected fitness varies with survival probability, growth to reproductive size, and mate selection is more complete than a fitness measure that considers only growth. Directness refers to how explicitly the fitness measure reflects fitness consequences of a decision alternative: assuming that decisions are made to maximize growth is a less-direct fitness measure than one that considers how growth affects the expected number and viability of offspring.

Fitness-seeking, direct fitness-seeking, indirect fitness-seeking — A type of *adaptive trait* that assumes *individuals* make decisions to improve their *expected fitness*. Traits using direct fitness-seeking assume individuals evaluate decision alternatives using a *fitness measure*. Indirect fitness-seeking traits are designed to reproduce *behaviors* that are observed in real organisms and assumed to contribute indirectly to fitness.

Formulation — A complete and detailed written description of a model (typically called a "specification" by software engineers).

Imposed behavior — *System behavior* that is strongly determined by, and predictable from, individual *traits*; in contrast to *emergent behavior*.

Individual — In an IBM, the organizational unit at which behavior is modeled; the modeled *system* is a collection of individuals. "Individual" is a modeling term, not a biological one. In most IBMs, an individual represents one organism; but sometimes a model's individuals represent a *collective*, a *superindividual*, or all the organisms within a spatial grid cell.

Individual-based ecology (IBE) — The study of ecological systems from the perspective that *system* properties arise from unique, independent, *individuals* and the interactions of the individuals with each other and with their environment.

Individual-based model (IBM) — A model of a *system* of *individuals* and their environment, in which system *behavior* arises from *traits* of the individuals and characteristics of the environment. IBMs do not include system-level models that consider individual variation, nor do they include models of a single individual.

Interaction, direct interaction, mediated interaction, interaction fields — Mechanisms by which model *individuals* communicate with each other or otherwise affect each other. Direct interactions involve an encounter among individuals in which information is exchanged or one individual directly affects the other (e.g., by killing it). Mediated interactions are mechanisms by which individuals affect each other indirectly by consuming or producing a common resource. Food competition is easily modeled as a mediated interaction: by consuming food, an individual affects others by reducing their food availability. Interaction fields represent how an individual is affected by the total or average effect of multiple neighboring individuals.

Method — In *object-oriented software*, a block of code that executes one particular *trait* or process. Methods are similar to subroutines in non-object-oriented software.

Observation, observer tools — The process of collecting data and information from an IBM; typical observations include graphical display of patterns over space and time and file output of summary statistics. Observer tools are software tools such as graphical user interfaces that make certain kinds of observation possible.

Platform — The programming language or software environment used to convert a model into executable code and run it. Platforms range from procedural or *object-oriented* programming languages to high-level environments in which specific kinds of model can be built and executed with little programming.

Prediction, tacit prediction, overt prediction — The way an IBM represents how individuals foresee the future outcomes of their decisions. Tacit prediction includes simple, implicit assumptions about decision outcomes. Overt prediction explicitly forecasts the consequences of each decision alternative.

Sensing — The way an IBM represents how *individuals* obtain information about their environment and neighboring individuals. What kinds of information do individuals sense (what variables do they "know")? How much information can individuals sense: over what distances, or from many neighbors, can they obtain information? How accurate is the sensed information?

Schedule, dynamic schedule, fixed schedule — In an IBM's *formulation*, a description of the order in which events are assumed to occur: the schedule defines the *actions* and the rules for executing them. In an IBM's software, the schedule is the code that defines actions and controls when they are executed. Fixed schedules define a single order in which events always occur, a cycle repeated each time step. Dynamic schedules allow the number and order of actions to be determined by the model as it executes.

State, state variable — A measure of the status of some part of a model (*individuals*, habitat units, the *system*) that typically can be described using a single number. A state variable is a model variable describing a particular state of some model component. Example individual states are weight, sex, and location; example system states are population biomass, number of species, and mortality rate (number of individuals dying per time step).

Stochasticity — The use of pseudorandom numbers to represent a process or *trait*. Often, traits or processes are represented as a stochastic model: whether an event occurs is stochastic, but the event's probability is modeled deterministically. For example, an individual's death is stochastic but its probability of dying depends on its age and size. Stochasticity is often used to represent processes that are assumed to be variable but are not sufficiently important, or not understood well enough, to be represented deterministically; but processes in IBMs are often variable without being stochastic.

Submodel—A part of an IBM's *formulation* that represents one *trait* or process. An IBM can be divided into submodels so each process can be modeled, calibrated, and tested separately.

Superindividual—An *individual* that represents multiple organisms, assuming they all have identical states and behavior. Superindividuals do not represent natural aggregations of individuals but are instead a modeling technique for simulating very large numbers of individuals.

System—All the *individuals* in an IBM, which may represent a population or community of organisms. The system has properties of different types than individuals have, for example, abundance, mortality and reproduction rates, persistence, diversity, and spatial patterns.

Theory, IBE theory—In IBE, a "theory" is a *trait* that has been tested and shown useful for explaining *system behavior* under a known range of ecological contexts. "IBE theory" includes both the collection of theories that have been developed and the process for developing theories in *IBE*.

Trait—A model of a particular *behavior* of an *individual*. A trait is typically a set of rules for what individuals do at particular times or in response to specific situations in an IBM; an individual is represented as a collection of *state variables* and traits. Example traits include models of short-term activities (foraging, response to predators), phenotypic expression, or changes in life stage.

References

Adami, C. 2002. Ab initio modeling of ecosystems with artificial life. *Natural Resource Modeling*, 15, 133–146.

Adler, F. R. 1996. A model of self-thinning through local competition. *Proceedings of the National Academy of Sciences of the USA*, 93, 9980–9984.

An, G. 2001. Agent-based computer simulation and SIRS: Building a bridge between basic science and clinical trials. *Shock*, 16, 266–273.

Anderson, J. J. 2002. An agent-based event driven foraging model. *Natural Resource Modeling*, 15, 55–82.

Andersson, M. 1994. *Sexual selection*. Princeton, NJ: Princeton University Press.

Antonsson, T., and S. Gudjonsson. 2002. Variability in timing and characteristics of Atlantic salmon smolt in Icelandic rivers. *Transactions of the American Fisheries Society*, 131, 643–655.

Aoki, I. 1982. A simulation study on the schooling mechanisms in fish. *Bulletin of the Japanese Society of Scientific Fisheries*, 48, 1081–1088.

Arnold, W., and J. Dittami. 1997. Reproductive suppression in male alpine marmots. *Animal Behaviour*, 53, 53–66.

Arthur, W. B. 1994. *Increasing returns and path dependence in the economy (economics, cognition, and society)*. Ann Arbor: University of Michigan Press.

Arthur, W. B., S. Durlauf, and D. A. Lane. (eds.). 1997. *The economy as an evolving complex system II*. Reading, MA: Addison-Wesley.

Auyang, S. Y. 1998. *Foundations of complex system theories in economics, evolutionary biology, and statistical physics*. New York: Cambridge University Press.

Axelrod, R. 1984. *The evolution of cooperation*. New York: Basic Books.

——. 1997. *The complexity of cooperation: Agent-based models of competition and collaboration*. Princeton, NJ: Princeton University Press.

Axelrod, R., R. L. Riolo, and M. D. Cohen. 2001. Beyond geography: Cooperation with persistent links in the absence of clustered neighborhoods. *Personality and Social Psychology Review*, 6, 341–346.

Banks, J. 2000. *Discrete-event system simulation*. Upper Saddle River, NJ: Prentice-Hall.

Bart, J. 1995. Acceptance criteria for using individual-based models to make management decisions. *Ecological Applications*, 5, 411–420.

Bartell, S. M., J. M. Breck, R. H. Gardner, and A. L. Brenkert. 1986. Individual parameter perturbation and error analysis of fish bioenergetics models. *Canadian Journal of Fisheries and Aquatic Sciences*, 43, 160–168.

Bascompte, J., and R. V. Solé. 1998. Models of habitat fragmentation. In J. Bascompte and R. V. Solé (eds.), *Modelling spatiotemporal dynamics in ecology*, 127–149. New York: Springer-Verlag.

Bauer, S., U. Berger, H. Hildenbrandt, and V. Grimm. 2002. Cyclic dynamics in simulated plant populations. *Proceedings of the Royal Society of London B*, 269, 2443–2450.

Bauer, S., T. Wyszomirski, U. Berger, H. Hildenbrandt, and V. Grimm. 2004. Asymmetric competition as a natural outcome of neighbour interactions among plants: Results from the field-of-neighbourhood modelling approach. *Plant Ecology*, 170, 135–145.

Beckmann, N., H. P. Kriegel, R. Schneider, and B. Seeger. 1990. The R*-tree: An efficient and robust access method for points and rectangles. In H. Garcia-Molina and H. V. Jagadish (eds.), *Proceedings of the 1990 ACM SIGMOD International Conference on Management of Data, SIGMOD Record* 19(2), 322–331. New York: ACM Press.

Begon, M., L. Firbank, and R. Wall. 1986. Is there a self-thinning rule for animal populations? *Oikos*, 46, 122–124.

Begon, M., J. L. Harper, and C. R. Townsend. 1990. *Ecology: Individuals, populations and communities*. Oxford: Blackwell.

Beissinger, S. R., and M. I. Westphal. 1998. On the use of demographic models of population viability in endangered species management. *Journal of Wildlife Management*, 62, 821–841.

Belew, R. K., M. Mitchell, and D. H. Ackley. 1996. Computation and the natural sciences. In R. K. Belew, and M. Mitchell (eds.), *Adaptive individuals in evolving populations*, 431–440. SFI Studies in the Sciences of Complexity, vol. 26. Reading, MA: Addison-Wesley.

Bender, C., H. Hildenbrandt, K. Schmidt-Loske, V. Grimm, C. Wissel, and K. Henle. 1996. Consolidation of vineyards, mitigations, and survival of the common wall lizard (*Podarcis muralis*) in isolated habitat fragments. In J. Settele, C. Margules, P. Poschlod, and K. Henle (eds.), *Species survival in fragmented landscapes*, 248–261. Dordrecht: Kluwer.

Beres, D. L., C. W. Clark, G. L. Swartzman, and A. M. Starfield. 2001. Truth in modeling. *Natural Resource Modeling*, 14, 457–463.

Berger, U., and H. Hildenbrandt. 2000. A new approach to spatially explicit modelling of forest dynamics: Spacing, ageing and neighourhood competition of mangrove trees. *Ecological Modelling*, 132, 287–302.

———. 2003. The strength of competition among individual trees and the biomass-density trajectories of the cohort. *Plant Ecology*, 167, 89–96.

Berger, U., G. Wagner, and W. F. Wolff. 1999. Virtual biologists observe virtual grasshoppers: An assessment of different mobility parameters for the analysis of movement patterns. *Ecological Modelling*, 115, 119–128.

Bernstein, C., A. Kacelnik, and J. R. Krebs. 1988. Individual decisions and the distribution of predators in a patchy environment. *Journal of Animal Ecology*, 57, 1007–1026.

Bissonette, J. A. 1997. Scale-sensitive ecological properties: Historical context, current meaning. In J. A. Bissonette (ed.), *Wildlife and landscape ecology: Effects of pattern and scale*, 3–31. New York: Springer.

Bjørnstad, O. N., J. M. Fromentin, N. C. Stenseth, and J. Gjøsæter. 1999. Cycles and trends in cod populations. *Proceedings of the National Academy of Sciences of the USA*, 96, 5066–5071.

Blackmore, S. 1999. *The meme machine*. Oxford: Oxford University Press.

Blasius, B., A. Huppert, and L. Stone. 1999. Complex dynamics and phase synchronization in spatially extended ecological systems. *Nature*, 399, 354–359.

Bolker, B. M., D. H. Deutschman, G. Hartvigsen, and D. L. Smith. 1997. Individual-based-modelling: What is the difference? *Trends in Ecology and Evolution*, 12, 111.

Bolker, B. M., and S. W. Pacala. 1997. Using moment equations to understand stochastically driven spatial pattern formation in ecological systems. *Theoretical Population Biology*, 52, 179–197.

Bolker, B. M., and S. W. Pacala. 1999. Spatial moment equations for plant competition: Understanding spatial strategies and the advantages of short dispersal. *American Naturalist*, 153, 575–602.

Bolker, B. M., S. W. Pacala, and S. A. Levin. 2000. Moment methods for ecological processes in continuous space. In U. Dieckmann, R. Law, and J.A.J. Metz (eds.), *The geometry of ecological interactions: Simplifying spatial complexity*, 388–411. Cambridge: Cambridge University Press.

Booth, G. 1997. Gecko: A continuous 2-D world for ecological modeling. *Artificial Life Journal*, 3, 147–163.

Bossel, H. 1992. Real-structure process description as the basis of understanding ecosystems and their development. *Ecological Modelling*, 63, 261–276.

———. 1996. TREEDYN3 forest simulation model. *Ecological Modelling*, 90, 187–227.

Botkin, D. B. 1977. Life and death in a forest: The computer as an aid to understanding. In C.A.S. Hall and J. W. Day Jr. (eds.), *Ecosystem modeling in theory and practice: An introduction with case histories*, 3–4. New York: John Wiley and Sons.

———. 1993. *Forest dynamics: An ecological model*. Oxford: Oxford University Press.

Botkin, D. B., J. F. Janak, and J. R. Wallis. 1972. Some ecological consequences of a computer model of forest growth. *Journal of Ecology*, 60, 849–872.

Brang, P., B. Courbaud, A. Fischer, I. Kissling-Naf, D. Pettenella, W. Schönenberger, W. Spörk, and V. Grimm. 2002. Developing indicators for sustainable management of mountain forests using a modelling approach. *Forest Policy and Economics*, 4, 113–123.

Breckling, B., and K. Mathes. 1991. Systemmodelle in der Ökologie: Individuen-orientierte und kompartiment-bezogene Simulation, Anwendung und Kritik. *Verhandlungen der Gesellschaft für Ökologie*, 19, 635–646.

Breckling, B., and H. Reuter. 1996. The use of individual based models to study the interaction of different levels of organization in ecological systems. *Senckenbergiana maritima*, 27, 195–205.

Breitenmoser, U., C. Breitenmoser-Würsten, H. Okarma, T. Kaphegyi, U. Kaphegyi-Wallmann, and U. M. Müller. 2000. *Action plan for the conservation of the Eurasian lynx (Lynx lynx) in Europe*. Nature and Environment, no. 112. Strasbourg: Council of Europe Publishing.

Briggs, C. J., S. M. Sait, M. Begon, D. J. Thompson, and H.C.J. Godfray. 2000. What causes generation cycles in populations of stored-product moths? *Journal of Animal Ecology*, 69, 352–366.

Bruun, C. 2001. Prospect for an economic framework for Swarm. In F. Luna and A. Perrone (eds.), *Agent-based methods in economics and finance: Simulations in Swarm*, 3–35. Dordrecht: Kluwer.

Bull, C. D., N. B. Metcalfe, and M. Mangel. 1996. Seasonal matching of foraging to anticipated energy requirements in anorexic juvenile salmon. *Proceedings of the Royal Society of London B*, 263, 13–18.

Burgman, M. A., S. Ferson, and H. R. Akcakaya. 1993. *Risk assessment in conservation biology*. London: Chapman and Hall.

Burgman, M. A., and H. P. Possingham. 2000. Population viability analysis for conservation: The good, the bad and the undescribed. In A. G. Young and G. M. Clarke (eds.), *Genetics, demography and viability of fragmented populations*, 97–112. Cambridge: Cambridge University Press.

Burnham, K. P., and D. R. Anderson. 1998. *Model selection and inference: A practical information-theoretic approach*. New York: Springer.

Camazine, S., J.-L. Deneubourg, N. R. Franks, J. Sneyd, G. Theraulaz, and E. Bonabeau. 2001. *Self-organization in biological systems*. Princeton, NJ: Princeton University Press.

Carter, J., and J. T. Finn. 1999. MOAB: A spatially explicit, individual-based expert system for creating animal foraging models. *Ecological Modelling*, 119, 29–41.

Casagrandi, R., and M. Gatto. 1999. A mesoscale approach to extinction risk in fragmented habitats. *Nature*, 400, 560–562.

Casti, J. L. 1998. *Would-be worlds*. New York: John Wiley and Sons.

Caswell, H. 1988. Theory and models in ecology: A different perspective. *Ecological Modelling*, 43, 33–44.

———. 2001. *Matrix population models: Construction, analysis and interpretation*. Sunderland, MA: Sinauer.

Chitty, D. 1996. *Do lemmings committ suicide? A beautiful hypothesis and ugly facts*. New York: Oxford University Press.

Claessen, D., A. M. de Roos, and L. Persson. 2000. Dwarfs and giants: Cannibalism and competition in size-structured populations. *American Naturalist*, 155, 219–237.

Clark, C. W., and M. Mangel. 2000. *Dynamic state variable models in ecology*. New York: Oxford University Press.

Clark, M. E., and K. A. Rose. 1997. Individual-based model of stream-resident rainbow trout and brook char: Model description, corroboration, and effects of sympatry and spawning season duration. *Ecological Modelling*, 94, 157–175.

Cohen, M. D., R. L. Riolo, and R. Axelrod. 2001. The role of social structure in the maintenance of cooperative regimes. *Rationality and Society*, 13, 5–32.

Connell, J. H. 1978. Diversity in tropical rain forests and coral reefs. *Science*, 199, 1302–1310.

Cornforth, D., D. G. Green, D. Newth, and M. Kirley. 2002. Do artificial ants march in step? Ordered asynchronous processes and modularity in biological systems. In R. K. Standish, M. A. Bedau, and H. A. Abbass (eds.), *Artificial Life VIII*, 28–32. Cambridge, MA: MIT Press (available at: http://parallel.hpc.unsw.edu.au/complex/alife8/proceedings.html).

Corten, A. 1999. A proposed mechanism for the Bohuslän herring periods. *ICES Journal of Marine Sciences*, 56, 207–220.

Cosgrove, D. J., S. Gilroy, T. Kao, H. Ma, and J. C. Schultz. 2000. Plant signalling 2000. Cross talk among geneticists, physiologists, and ecologists. *Plant Physiology*, 124, 499–505.

Cowell, R. G., A. P. David, S. L. Lauritzen, and D. J. Spiegelhalter. 1999. *Probabilistic networks and expert systems*. New York: Springer.

Crawley, M. J. 1990. The population dynamics of plants. *Philosophical Transactions of the Royal Society of London B*, 330, 125–140.

Crick, F. 1988. *What mad pursuit*. New York: Basic Books.

Crone, E. E., and D. R. Taylor. 1996. Complex dynamics in experimental populations of an annual plant, *Cardamine pennsylvanica*. *Ecology*, 77, 289–299.

Czárán, T. 1984. A simulation model for generating patterns of sessile populations. *Abstracta Botanica*, 8, 1–13.

———. 1998. *Spatiotemporal models of population and community dynamics*. New York: Chapman and Hall.

Czárán, T., and S. Bartha. 1989. The effect of spatial pattern on community dynamics: A comparison of simulated and field data. *Vegetatio*, 83, 229–239.

———. 1992. Spatiotemporal dynamic models of plant populations and communities. *Trends in Ecology and Evolution*, 7, 38–42.

Dean, W.R.J. 1995. Where birds are rare or fill the air: The protection of the endemic and nomadic avifaunas of the Karoo. Ph.D. diss., University of Cape Town.

DeAngelis, D. L. 1992. *Dynamics of nutrient cycling and food webs*. London: Chapman and Hall.

DeAngelis, D. L., L. W. Barnthouse, W. Van Winkle, and R. G. Otto. 1990. A critical appraisal of population approaches in assessing fish community health. *Journal of Great Lakes Research*, 16, 576–590.

DeAngelis, D. L., D. K. Cox, and C. C. Coutant. 1980. Cannibalism and size dispersal in young-of-the-year largemouth bass: Experiment and model. *Ecological Modelling*, 8, 133–148.

DeAngelis, D. L., and L. J. Gross (eds.). 1992. *Individual-based models and approaches in ecology: Populations, communities and ecosystems*. New York: Chapman and Hall.

DeAngelis, D. L., and W. M. Mooij. 2003. In praise of mechanistically-rich models. In C. D. Canham, J. J. Cole, and W. K. Lauenroth (eds.), *Models in ecosystem science*, 63–82. Princeton, NJ: Princeton University Press.

DeAngelis, D. L., W. M. Mooij, and A. Basset. 2003. The importance of spatial scale in the modeling of aquatic ecosystems. In L. Seuront and P. G. Strutton (eds.), *Handbook of scaling methods in aquatic ecology: Measurement, analysis, simulation*, 383–400. Boca Raton, FL: CRC Press.

DeAngelis, D. L., K. A. Rose, and M. A. Huston. 1994. Individual-oriented approaches to modeling ecological populations and communities. In S. A. Levin (ed.), *Frontiers in mathematical biology*, 390–410. New York: Springer.

den Boer, P. J., and J. Reddingius. 1996. *Regulation and stabilization paradigms in population ecology*. London: Chapman and Hall.

de Roos, A. M., E. McCauley, and W. G. Wilson. 1991. Mobility versus density-limited predator-prey dynamics on different spatial scales. *Proceedings of the Royal Society of London B*, 246, 117–122.

Deutschman, D. H., S. A. Levin, C. Devine, and L. Buttel. 1997. Scaling from trees to forests: analysis of a complex simulation model. *Science*, 277, 1688 (available at: http://www.sciencemag.org/feature/data/deutschman/index.htm).

Dieckmann, U., and R. Law. 2000. Relaxation projections and method of moments. In U. Dieckmann, R. Law, and J.A.J. Metz (eds.), *The geometry of ecological interactions: Simplifying spatial complexity*, 412–455. Cambridge: Cambridge University Press.

Dieckmann, U., R. Law, and J.A.J. Metz (eds.). 2000. *The geometry of ecological interactions: Simplifying spatial complexity*. Cambridge: Cambridge University Press.

Diekmann, O., and J.A.J. Metz (eds.). 1986. *The dynamics of physiologically structured populations*. Lecture Notes on Biomathematics, vol. 68. New York: Springer Verlag.

Di Paolo, E. 2000. Ecological symmetry breaking can favour the evolution of altruism in an action-response game. *Journal of Theoretical Biology*, 203, 135–152.

Di Paolo, E. A., J. Noble, and S. Bullock. 2000. Simulation models as opaque thought experiments. In M. A. Bedau, J. S. McCaskill, N. H. Packard, and S. Rasmussen (eds.), *Artificial Life VII: The Seventh International Conference on the Simulation and Synthesis of Living Systems, Reed College, Portland, Oregon, USA, 1–6 August*, 497–506 Cambridge, MA: MIT Press/Bradford Books.

Doak, D. F., and W. Morris. 1999. Detecting population-level consequences of ongoing environmental change without long-term monitoring. *Ecology*, 80, 1537–1551.

Donalson, D. D., and R. M. Nisbet. 1999. Population dynamics and spatial scale: Effects of system size on population persistence. *Ecology*, 80, 2492–2507.

Dorndorf, N. 1999. Zur Populationsdynamik des Alpenmurmeltiers: Modellierung, Gefährdungsanalyse und Bedeutung des Sozialverhaltens für die Überlebensfähigkeit. Ph.D. diss., University of Marburg, Germany.

Drechsler, M. 1998. Sensitivity analysis of complex models. *Biological Conservation*, 86, 401–412.

———. 2000. A model-based decision aid for species protection under uncertainty. *Biological Conservation*, 94, 23–30.

Drechsler, M., K. Frank, I. Hanski, B. O'Hara, and C. Wissel. 2003. Ranking metapopulation extinction risk: From patterns in data to conservation management decisions. *Ecological Applications*, 13, 990–998.

Dunning, J., D. J. Stewart, B. J. Danielson, B. R. Noon, T. L. Root, R. H. Lamberson, and E. E. Stevens. 1995. Spatially explicit population models: Current forms and future uses. *Ecological Applications*, 5, 3–11.

du Plessis, M. A. 1992. Obligate cavity-roosting as a constraint on dispersal of green (red-billed) woodhoopoes: Consequences for philopatry and the likelihood of inbreeding. *Oecologia*, 90, 205–211.

Durrett, R., and S. Levin. 1994. The importance of being discrete (and spatial). *Theoretical Population Biology*, 46, 363–394.

Ek, A. R., and R. A. Monserud. 1974. Trials with program FOREST: Growth and reproduction simulation for mixed species even- or uneven-aged forest stands. In S. Fries (ed.), *Growth models for tree and stand simulation*, 56–73. Royal College of Forestry, Sweden, Research Notes, no. 30. Stockholm.

Elliot, J. A., A. E. Irish, C. S. Reynolds, and P. Tett. 2000. Modelling freshwater phytoplankton communities: An exercise in validation. *Ecological Modelling*, 128, 19–26.

Ellner, S. P., E. McCauley, B. E. Kendall, C. J. Briggs, P. R. Hosseini, S. N. Wood, A. Janssen, M. W. Sabelis, P. Turchin, R. M. Nisbet, and W. M. Murdoch. 2001. Habitat structure and population persistence in an experimental community. *Nature*, 412, 538–543.

Enquist, B. J., J. H. Brown, and G. B. West. 1998. Allometric scaling of plant energetics and population density. *Nature*, 395, 163–165.

Enquist, B. J., and K. J. Niklas. 2001. Invariant scaling relations across tree-dominated communities. *Nature*, 410, 655–660.

Ermentrout, G. B., and L. Edelstein-Keshet. 1993. Cellular automata approaches to biological modeling. *Journal of Theoretical Biology*, 160, 97–133.

Fagerström, T. 1987. On theory, data and mathematics in ecology. *Oikos*, 50, 258–261.

Fahse, L., C. Wissel, and V. Grimm. 1998. Reconciling classical and individual-based approaches of theoretical population ecology: A protocol for extracting population parameters from individual-based models. *American Naturalist*, 152, 838–852.

Firbank, L. G., and A. R. Watkinson. 1985. A model of interference within plant monocultures. *Journal of Theoretical Biology*, 116, 291–311.

Fishman, G. S. 1973. *Concepts and methods in discrete event simulation*. New York: Wiley.

———. 2001. *Discrete event simulation: Modeling, programming and analysis*. Berlin: Springer Verlag.

Flierl, G., D. Grünbaum, S. Levin, and D. Olson. 1999. From individuals to aggregations: The interplay between behavior and physics. *Journal of Theoretical Biology*, 196, 397–454.

Ford, E. D. 2000. *Scientific method for ecological research*. New York: Cambridge University Press.

Ford, E. D., and P. J. Diggle. 1981. Competition for light in plant monocultures modelled as a spatial stochastic process. *Annals of Botany*, 48, 481–500.

Ford, E. D., and K. A. Sorrensen. 1992. Theory and models of inter-plant competition as a spatial process. In D. L. DeAngelis and L. J. Gross (eds.), *Individual-based approaches in ecology*, 363–407. New York: Chapman and Hall.

Franklin, A. B., D. R. Anderson, R. J. Gutiérrez, and K. P. Burnham. 2000. Climate, habitat quality, and fitness in Northern Spotted Owl populations in northwestern California. *Ecological Monographs*, 70, 539–590.

Frey-Roos, F. 1998. Geschlechtsspezifisches Abwanderungsmuster beim Alpenmurmeltier (*Marmota marmota*). Ph.D. diss., University of Marburg, Germany.

Fromentin, J.-M., R. A. Myers, O. N. Bjørnstad, N. C. Stenseth, J. Gjøsæter, and H. Christie. 2001. Effects of density-dependent and stochastic processes on the regulation of cod populations. *Ecology*, 82, 567–579.

Garshelis, D. L. 2000. Delusions in habitat evaluation: Measuring use, selection, and importance. In L. Boitani and T. K. Fuller (eds.), *Research techniques in animal ecology, controversies and consequences*, 111–164. Methods and Cases in Conservation Science. New York: Columbia University Press.

Gentle, J. E. 2003. *Random number generation and Monte Carlo methods* (2nd edition). New York: Springer.

Ghilarov, A. M. 2001. The changing place of theory in 20th century ecology: From universal laws to arrays of methodologies. *Oikos*, 92, 357–362.

Gigerenzer, G. 2002. *Calculated risks: How to know when numbers deceive you*. New York: Simon and Schuster.

Gigerenzer, G., and P. M. Todd. 1999. Fast and frugal heuristics: The adaptive toolbox. In G. Gigerenzer, P. M. Todd, and ABC Research Group (eds.), *Simple heuristics that make us smart*, 3–34. Evolution and Cognition. New York: Oxford University Press.

402
REFERENCES

Gilbert, S., and B. McCarty. 1998. *Object-oriented design in Java™*. Corte Madera, CA: Waite Group Press.

Gilliam, J. F., and D. F. Fraser. 1987. Habitat selection under predation hazard: Test of a model with foraging minnows. *Ecology*, 68, 1856–1862.

Ginot, V., C. Le Page, and S. Souissi. 2002. A multi-agents architecture to enhance end-user individual-based modelling. *Ecological Modelling*, 157, 23–41.

Giske, J., G. Huse, and Ø. Fiksen. 1998. Modelling spatial dynamics of fish. *Reviews in Fish Biology and Fisheries*, 8, 57–91.

Giske, J., M. Mangel, P. Jakobsen, G. Huse, C. Wilcox, and E. Strand. 2003. Explicit trade-off rules in proximate adaptive agents. *Evolutionary Ecology Research*, 5, 835–865.

Gmytrasiewicz, P. J., and E. H. Durfee. 2001. Rational communication in multiagent environments. *Autonomous Agents and Multi-agent Systems*, 4, 233–272.

Gopen, G. D., and J. A. Swan. 1990. The science of scientific writing. *American Scientist*, 78, 550–559.

Goss-Custard, J. D., R. A. Stillman, R.W.G. Caldow, A. D. West, and M. Guillemain. 2003. Carrying capacity in overwintering birds: When are spatial models needed? *Journal of Applied Ecology*, 40, 176–187.

Goss-Custard, J. D., R. A. Stillman, A. D. West, R.W.G. Caldow, and S. McGrorty. 2002. Carrying capacity in overwintering migratory birds. *Biological Conservation*, 105, 27–41.

Goss-Custard, J. D., R. A. Stillman, A. D. West, R.W.G. Caldow, P. Triplet, S.E.A. le V. dit Durell, and S. McGrorty. 2004. When enough is not enough: Shorebirds and shellfish. *Proceedings of the Royal Society of London B*, 271, 233–237.

Goss-Custard, J. D., A. D. West, R. A. Stillman, S.E.A. le V. dit Durell, R.W.G. Caldow, S. McGrorty, and R. Nagarajan. 2001. Density-dependent starvation in a vertebrate without significant depletion. *Journal of Animal Ecology*, 70, 955–965.

Grand, T. C. 1999. Risk-taking behavior and the timing of life history events: Consequences of body size and season. *Oikos*, 85, 467–480.

Grimm, V. 1994. Mathematical models and understanding in ecology. *Ecological Modelling*, 75–76, 641–651.

———. 1999. Ten years of individual-based modelling in ecology: What have we learned, and what could we learn in the future? *Ecological Modelling*, 115, 129–148.

———. 2002. Visual debugging: A way of analyzing, understanding, and communicating bottom-up simulation models in ecology. *Natural Resource Modeling*, 15, 23–38.

Grimm, V., and U. Berger. 2003. Seeing the forest for the trees, and vice versa: Pattern-oriented ecological modelling. In L. Seuront and P. G. Strutton (eds.), *Handbook of scaling methods in aquatic ecology: Measurement, analysis, simulation*, 411–428. Boca Raton, FL: CRC Press.

Grimm, V., N. Dorndorf, F. Frey-Roos, C. Wissel, T. Wyszomirski, and W. Arnold. 2003. Modelling the role of social behavior in the persistence of the alpine marmot *Marmota marmota*. *Oikos*, 102, 124–136.

Grimm, V., K. Frank, F. Jeltsch, R. Brandl, J. Uchmański, and C. Wissel. 1996. Pattern-oriented modelling in population ecology. *Science of the Total Environment*, 183, 151–166.

Grimm, V., C.-P. Günther, S. Dittmann, and H. Hildenbrandt. 1999. Grid-based modelling of macrozoobenthos in the intertidal of the Wadden Sea: Potentials and limitations. In S. Dittmann (ed.), *The Wadden Sea ecosystem—stability properties and mechanisms*, 207–226. Berlin: Springer.

Grimm, V., H. Lorek, J. Finke, F. Koester, M. Malachinski, M. Sonnenschein, A. Moilanen, I. Storch, A. Singer, C. Wissel, and K. Frank. 2004. META-X: A generic software for metapopulation viability analysis. *Biodiversity and Conservation*, 13, 165–188.

Grimm, V., and J. Uchmański. 2002. Individual variability and population regulation: A model of the significance of within-generation density dependence. *Oecologia*, 131, 196–202.

Grimm, V., and C. Wissel. 1997. Babel, or the ecological stability discussions: An inventory and analysis of terminology and a guide for avoiding confusion. *Oecologia*, 109, 323–334.

———. 2004. The intrinsic mean time to extinction: A unifying approach to analyzing persistence and viability of populations. *Oikos*, 105, 501–511.

Grimm, V., T. Wyszomirski, D. Aikman, and J. Uchmański. 1999. Individual-based modelling and ecological theory: Synthesis of a workshop. *Ecological Modelling*, 115, 275–282.

Groeneveld, J., N. J. Enright, B. B. Lamont, and C. Wissel. 2002. A spatial model of coexistence among three *Banksia* species along a topographic gradient in fire-prone shrublands. *Journal of Ecology*, 90, 762–774.

Grünbaum, D. 1994. Translating stochastic density-dependent individual behavior with sensory constraints to an Eulerian model of animal swarming. *Journal of Mathematical Biology*, 33, 139–161.

———. 1998. Using spatially explicit models to characterize foraging performance in heterogeneous landscapes. *American Naturalist*, 151, 97–115.

Gurney, W.S.C., and R. M. Nisbet. 1998. *Ecological dynamics*. New York: Oxford University Press.

Guttman, A. 1984. R-trees: A dynamic index structure for spatial searching. In Diane Smith and Beatrice Yeimark (eds.), *Proceedings of the 1984 ACM SIGMOD International Conference on Management of Data, SIGMOD Record 2*, 47–57. New York: ACM Press.

Haefner, J. W. 1996. *Modeling biological systems: Principles and applications*. New York: Chapman and Hall.

Hall, C.A.S. 1988. An assessment of several of the historically most influential theoretical models used in ecology and the data provided in their support. *Ecological Modelling*, 43, 5–31.

———. 1991. An idiosyncratic assessment of the role of mathematical models in environmental sciences. *Environment International*, 17, 507–517.

Hall, C.A.S., and D. L. DeAngelis. 1985. Models in ecology: Paradigms found or paradigms lost? *Bulletin of the Ecological Society of America*, 66, 339–346.

Hallam, T. G., and S. A. Levin. (eds.). 1986. *Mathematical ecology: An introduction*. New York: Springer Verlag.

Hanski, I. 1994. A practical model of metapopulation dynamics. *Journal of Animal Ecology*, 63, 151–162.

———. 1999. *Metapopulation ecology*. Oxford: Oxford University Press.

Hara, T. 1988. Dynamics of size structure in plant populations. *Trends in Ecology and Evolution*, 3, 129–133.

Harper, J. L. 1977. *The population biology of plants*. London: Academic Press.

Harper, S. J., J. D. Westervelt, and A.-M. Shapiro. 2002. Modeling the movement of cowbirds: Applications toward management at the landscape scale. *Natural Resource Modeling*, 15, 111–131.

Harte, J. 1988. *Consider a spherical cow: A course in environmental problem solving.* Repr., Mill Valley, CA: University Science Books.

Harvey, B. C., and S. F. Railsback. 2004. Elevated turbidity reduces abundance and biomass of stream trout in an individual-based model. U.S. Department of Agriculture, Redwood Sciences Laboratory, Arcata, CA: Draft manuscript.

Hemelrijk, C. K. 1999. An individual-orientated model of the emergence of despotic and egalitarian societies. *Proceedings of the Royal Society of London B*, 266, 361–369.

———. 2000a. Self-reinforcing dominance interactions between virtual males and females. Hypothesis generation for primate studies. *Adaptive Behavior*, 8, 13–26.

———. 2000b. Social phenomena emerging by self-organization in a competitive, virtual world (DomWorld). In K. Jokinen, D. Heylen, and A. Nijholt (eds.), *Learning to behave. Workshop II: Internalising knowledge*, 11–19. Proceedings of the Twente Workshop on Language Technology 18. Enschede, The Netherlands: Department of Computer Science, University of Twente.

———. 2002. Understanding social behaviour with the help of complexity science. *Ethology*, 108, 1–17.

Hengeveld, R., and G. H. Walter. 1999. The two coexisting ecological paradigms. *Acta Biotheoretica*, 47, 141–170.

Hilborn, R., and M. Mangel. 1997. *The ecological detective: Confronting models with data.* Princeton, NJ: Princeton University Press.

Hildenbrandt, H. 2003. The Field of Neighbourhood (FON)—ein phänomenologischer Modellansatz zur Beschreibung von Nachbarschaftsbeziehungen sessiler Organismen. Ph.D. diss., University of Bremen, Germany.

Hildenbrandt, H., C. Bender, V. Grimm, and K. Henle. 1995. Ein individuenbasiertes Modell zur Beurteilung der Überlebenschancen kleiner Populationen der Mauereidechse (*Podarcis muralis*). *Verhandlungen der Gesellschaft für Ökologie*, 24, 207–214.

Hirvonen, H., E. Ranta, H. Rita, and N. Peuhkuri. 1999. Significance of memory properties in prey choice decisions. *Ecological Modelling*, 115, 177–190.

Hogeweg, P. 1988. Cellular automata as paradigm for ecological modelling. *Applied Mathematics and Computation*, 27, 81–100.

Hogeweg, P., and B. Hesper. 1979. Heterarchical, selfstructuring simulation systems: Concepts and applications in biology. In B. P. Zeigler, M. S. Elzas, G. J. Klir, and T. I. Oren. (eds.), *Methodologies in systems modelling and simulation*, 221–231. Amsterdam: North-Holland.

———. 1983. The ontogeny of interaction structure in bumble bee colonies: A MIRROR model. *Behavioral Ecology and Sociobiology*, 12, 271–283.

———. 1990. Individual-oriented modelling in ecology. *Mathematical and Computer Modelling*, 13, 83–90.

Holland, J. H. 1975. *Adaptation in natural and artificial systems.* Ann Arbor: University of Michigan Press.

———. 1995. *Hidden order: How adaptation builds complexity.* Reading, MA: Perseus Books.

———. 1998. *Emergence: From chaos to order.* Reading, MA: Addison-Wesley.

Holling, C. S. 1966. The strategy of building models of complex ecological systems. In K.E.F. Watt (ed.), *Systems analysis in ecology*, 195–214. New York: Academic Press.

Houston, A. I., and J. M. McNamara. 1999. *Models of adaptive behavior: An approach based on state*. Cambridge: Cambridge University Press.

Huberman, B. A., and N. S. Glance. 1993. Evolutionary games and computer simulations. *Proceedings of the National Academy of Sciences*, 90, 7716–7718.

Huff, D. 1954. *How to lie with statistics*. New York: W. W. Norton.

Huisman, J., and F. J. Weissing. 1999. Biodiversity of plankton by species oscillations and chaos. *Nature*, 402, 407–410.

Huse, G., and J. Giske. 1998. Ecology in Mare Pentium: An individual-based spatio-temporal model for fish with adapted behaviour. *Fisheries Research*, 37, 163–178.

Huse, G., J. Giske, and A.G.V. Salvanes. 2002. Individual-based modelling. In P.J.B. Hart and J. Reynolds (eds.), *Handbook of fish biology and fisheries*, 228–248. Oxford: Blackwell.

Huse, G., S. Railsback, and A. Fernö. 2002. Modelling changes in migration pattern of herring: Collective behaviour and numerical domination. *Journal of Fish Biology*, 60, 571–582.

Huse, G., E. Strand, and J. Giske. 1999. Implementing behaviour in individual-based models using neural networks and genetic algorithms. *Evolutionary Ecology*, 13, 469–483.

Huston, M., D. DeAngelis, and W. Post. 1988. New computer models unify ecological theory. *BioScience*, 38, 682–691.

Huth, A. 1992. Ein Simulationsmodell zur Erklärung der kooperativen Bewegung von polarisierten Fischschwärmen. Ph.D. diss., University of Marburg, Germany.

Huth, A., T. Ditzer, and H. Bossel. 1998. *The rain forest growth model FORMIX3*. Göttingen: Verlag Erich Goltze.

Huth, A., and C. Wissel. 1992. The simulation of the movement of fish schools. *Journal of Theoretical Biology*, 156, 365–385.

———. 1993. Analysis of the behavior and the structure of fish schools by means of computer simulations. *Comments of Theoretical Biology*, 3, 169–201.

———. 1994. The simulation of fish schools in comparison with experimental data. *Ecological Modelling*, 75–76, 135–146.

Inada, Y., and K. Kawachi. 2002. Order and flexibility in the motion of fish schools. *Journal of Theoretical Biology*, 214, 371–387.

Iwasa, Y., and J. Roughgarden. 1986. Interspecific competition among metapopulations with space-limited subpopulations. *Theoretical Population Biology*, 30, 194–214.

Jackson, P. 1999. *Introduction to expert systems* (3rd edition). Reading, MA: Pearson-Addison-Wesley.

Jaworska, J. S., K. A. Rose, and A. L. Brenkert. 1997. Individual-based modeling of PCB effects on young-of-the-year largemouth bass in southeastern USA reservoirs. *Ecological Modelling*, 99, 113–135.

Jax, K., C. G. Jones, and S.T.A. Pickett. 1998. The self-identity of ecological units. *Oikos*, 82, 253–264.

Jedrzejewski, W., K. Schmidt, H. Okarma, and R. Kowalczyk. 2002. Movement pattern and home range use by the Eurasian lynx in Bialowieza Primeval Forest (Poland). *Annales Zoologici Fennici*, 39, 29–41.

Jeltsch, F. 1992. Modelle zu natürlichen Waldsterbephänomenen. Ph.D. diss., University of Marburg, Germany.

Jeltsch, F., S. J. Milton, W.R.J. Dean, and N. van Rooyen. 1996. Tree spacing and coexistence in semiarid savannas. *Journal of Ecology*, 84, 583–595.

———. 1997. Analysing shrub encroachment in the southern Kalahari: A grid-based modelling approach. *Journal of Applied Ecology*, 34, 1497–1508.

Jeltsch, F., M. S. Müller, V. Grimm, C. Wissel, and R. Brandl. 1997. Pattern formation triggered by rare events: Lessons from the spread of rabies. *Proceedings of the Royal Society London B*, 264, 495–503.

Jeltsch, F., and C. Wissel. 1994. Modelling dieback phenomena in natural forests. *Ecological Modelling*, 75–76, 111–121.

Jeltsch, F., C. Wissel, S. Eber, and R. Brandl. 1992. Oscillating dispersal patterns of tephritid fly populations. *Ecological Modelling*, 60, 63–75.

Jenkins, T. M., S. Diehl, K. W. Kratz, and S. D. Cooper. 1999. Effects of population density on individual growth of brown trout in streams. *Ecology*, 80, 941–956.

Johst, K., and R. Brandl. 1997. The effect of dispersal on local population dynamics. *Ecological Modelling*, 104, 87–101.

Judson, O. P. 1994. The rise of the individual-based model in ecology. *Trends in Ecology and Evolution*, 9, 9–14.

Kaiser, H. 1974. Populationsdynamik und Eigenschaften einzelner Individuen. *Verhandlungen der Gesellschaft für Ökologie*, 4, 25–38.

———. 1979. The dynamics of populations as result of the properties of individual animals. *Fortschritte der Zoologie*, 25, 109–136.

Kauffman, Stuart. 1995. *At home in the universe: The search for the laws of self-organization and complexity*. New York: Oxford University Press.

Kendall, B. E., C. J. Briggs, W. W. Murdoch, P. Turchin, S. P. Ellner, E. McCauley, R. M. Nisbet, and S. N. Wood. 1999. Why do populations cycle? A synthesis of statistical and mechanistic modeling approaches. *Ecology*, 80, 1789–1805.

Kenkel, N. C. 1990. Spatial competition models for plant populations. *Coenoses*, 5, 149–158.

Kleijnen, J.P.C., and W. van Groenendaal. 1992. *Simulation: A statistical perspective*. Chichester: Wiley.

Knight, J. C., and N. G. Leveson. 1986. An experimental evaluation of the assumption of independence in multi-version programming. *IEEE Transactions on Software Engineering*, SE-12, 96–109.

Köhler, P., and A. Huth. 1998. The effects of tree species grouping in tropical rainforest modelling: Simulations with the individual-based model FORMIND. *Ecological Modelling*, 109, 301–321.

Korpel, S. 1995. *Die Urwälder der Westkarpaten*. New York: Gustav Fischer Verlag.

Kramer-Schadt, S., E. Revilla, T. Wiegand, and U. Breitenmoser. 2004. Fragmented landscapes, road mortality and patch connectivity: Modelling influences on the dispersal of Eurasian lynx. *Journal of Applied Ecology*, 41, 711–723.

Krause, J., and G. D. Ruxton. 2002. *Living in groups*. Oxford: Oxford University Press.

Krebs, C. J. 1972. *Ecology: The experimental analysis of distribution and abundance*. New York: Harper and Row.

———. 1988. The experimental approach to rodent population dynamics. *Oikos*, 52, 143–149.

————. 1996. Population cycles revisited. *Journal of Mammology*, 77, 8–24.

Kreft, J.-U., G. Booth, and J.W.T. Wimpenny. 2000. Applications of individual-based modelling in microbial ecology. In C. R. Bell, M. Brylinsky, and P. Johnson-Green (eds.). *Microbial biosystems: New frontiers (Proceedings of the 8th international symposium on microbial ecology)*, 917–923. Halifax, Nova Scotia: Atlantic Canada Society for Microbial Ecology.

Kreft, J.-U., C. Picioreanu, J.W.T. Wimpenny, and M.C.M. van Loosdrecht. 2001. Individual-based modeling of biofilms. *Microbiology*, 147, 2897–2912.

Kunz, H., and C. K. Hemelrijk. 2003. Artificial fish schools: Collective effects of school size, body size, and form. *Artificial Life*, 9, 237–253.

Lammens, E.H.R.R., E. H. van Nes, and W. M. Mooij. 2002. Differences in the exploitation of bream in three shallow lake systems and their relation to water quality. *Freshwater Biology*, 47, 2435–2442.

Lande, R. 1993. Risks of population extinction from demographic and environmental stochasticity and random catastrophes. *American Naturalist*, 142, 911–927.

Lankester, K., R. C. van Appeldoorn, E. Meelis, and J. Verboom. 1991. Management perspectives for populations of the Eurasian badger *Meles meles* in a fragmented landscape. *Journal of Applied Ecology*, 28, 561–573.

Latto, J. 1992. The differentiation of animal body weights. *Functional Ecology*, 6, 386–395.

Laval, P. 1995. Hierarchical object-oriented design of a concurrent, individual-based, model of a pelagic Tunicate bloom. *Ecological Modelling*, 82, 265–276.

————. 1996. The representation of space in an object-oriented computational pelagic ecosystem. *Ecological Modelling*, 88, 113–124.

Law, R., and U. Dieckmann. 2000. A dynamical system for neighborhoods in plant communities. *Ecology*, 81, 2137–2148.

Law, A. M., and W. D. Kelton. 1999. *Simulation modeling and analysis* (3rd edition). New York: McGraw-Hill.

Law, R., D. J. Murrell, and U. Dieckmann. 2003. Population growth in space and time: Spatial logistic equations. *Ecology*, 84, 252–262.

Laymon, S. A., and J. A. Reid. 1986. Effects of grid-cell size on tests of a spotted owl HSI model. In J. Verner, M. L. Morrison, and C. J. Ralph (eds.), *Wildlife 2000: Modeling habitat relationships of terrestrial vertebrates*, 93–96. Madison: University of Wisconsin Press.

Leibundgut, H. 1993. *Europäische Urwälder: Wegweiser zur naturnahen Waldwirtschaft*. Bern: Haupt.

Leonardsson, K. 1991. Predicting risk-taking behaviour from life-history theory using static optimization technique. *Oikos*, 60, 149–154.

Leps, J. and P. Kindlmann. 1987. Models of the development of spatial pattern of an even-aged plant population over time. *Ecological Modelling*, 39, 45–57.

Letcher, B. H., J. A. Priddy, J. R. Walters, and L. B. Crowder. 1998. An individual-based, spatially-explicit simulation model of the population dynamics of the endangered red-cockaded woodpecker, *Picoides borealis*. *Biological Conservation*, 86, 1–14.

Levin, S. A. 1981. The role of theoretical ecology in the description and understanding of populations in heterogeneous environments. *American Zoologist*, 21, 865–875.

————. 1992. The problem of pattern and scale in ecology. *Ecology*, 73, 1943–1967.

————(ed.). 1994. *Frontiers in mathematical biology*. Lecture Notes in Biomathematics, vol. 1. New York: Springer-Verlag.

————. 1999. *Fragile dominion: Complexity and the commons*. Reading, MA: Helix Books.

Levin, S. A., and R. Durrett. 1996. From individuals to epidemics. *Philosophical Transactions of the Royal Society of London B*, 351, 1615–1621.

Levins, R. 1966. The strategy of model building in population biology. *American Scientist*, 54, 421–431.

————. 1970. Extinction. In M. Gerstenhaber (ed.), *Some mathematical questions in biology*, 75–108. Providence, RI: American Mathematical Society.

Lewellen, R. H., and S. H. Vessey. 1998. The effect of density dependence and weather on population size of a polyvoltine species. *Ecological Monographs*, 68, 571–594.

Li, B.-L., H.-I. Wu, and G. Zou. 2000. Self-thinning rule: A causal interpretation from ecological field theory. *Ecological Modelling*, 132, 167–173.

Lima, S. L., and P. A. Zollner. 1996. Towards a behavioral ecology of ecological landscapes. *Trends in Ecology and Evolution*, 11, 131–135.

Liu, J., and P. S. Ashton. 1995. Individual-based simulation models for forest succession and management. *Forest Ecology and Management*, 73, 157–175.

Liu, J., J. Dunning Jr., and H. R. Pulliam. 1995. Potential effects of a forest management plan on Bachman's sparrows (*Aimophila aestivalis*): Linking a spatially explicit model with GIS. *Conservation Biology*, 9, 62–75.

Loehle, C. 1990. A guide to increased creativity in research—inspiration or perspiration? *BioScience*, 40, 123–129.

Łomnicki, A. 1978. Individual differences between animals and the natural regulation of their numbers. *Journal of Animal Ecology*, 47, 461–475.

————. 1988. *Population ecology of individuals*. Princeton, NJ: Princeton University Press.

————. 1992. Population ecology from the individual perspective. In D. L. DeAngelis and L. J. Gross (eds.), *Individual-based models and approaches in ecology*, 3–17. New York: Chapman and Hall.

Lonsdale, W. M. 1990. The self-thinning rule: Dead or alive? *Ecology*, 71, 1373–1388.

Lorek, H., and M. Sonnenschein. 1999. Modelling and simulation software to support individual-based ecological modelling. *Ecological Modelling*, 115, 199–216.

Lotka, A. J. 1925. Reports of talk. *Journal of the American Statistical Association*, 20, 569–570.

Ludwig, D., D. D. Jones, and C. S. Holling. 1978. Qualitative analysis of insect outbreak systems: The spruce budworm and forest. *Journal of Animal Ecology*, 47, 315–332.

MacArthur, R. H., and E. O. Wilson. 1967. *The theory of island biogeography*. Princeton, NJ: Princeton University Press.

Magnusson, W. E. 2000. Error bars: Are they the king's clothes? *Bulletin of the Ecological Society of America*, 81, 147–150.

Mangel, M., and C. W. Clark. 1986. Toward a unified foraging theory. *Ecology*, 67, 1127–1138.

————. 1988. *Dynamic modeling in behavioral ecology*. Princeton, NJ: Princeton University Press.

Matsinos, Y. G., W. F. Wolff, and D. L. DeAngelis. 2000. Can individual-based models yield a better assessment of population viability? In S. Ferson and

M. A. Burgman (eds.), *Quantitative methods for conservation biology*, 188–198. New York: Springer.

May, R. M. 1973. *Stability and complexity in model ecosystems*. Princeton, NJ: Princeton University Press.

———. 1976. Simple mathematical models with very complicated dynamics. *Nature*, 261, 459–467.

———. 1981a. The role of theory in ecology. *American Zoologist*, 21, 903–910.

——— (ed.). 1981b. *Theoretical ecology: Principles and applications*. Oxford: Blackwell.

Maynard Smith, J. 1989. *Evolutionary genetics*. Oxford: Oxford University Press.

Mazerolle, M. J., and M.-A. Villard. 1999. Patch characteristics and landscape context as predictors of species presence and abundance: A review. *Ecoscience*, 6, 117–124.

McCann, K. 2000. The diversity-stability debate. *Nature*, 405, 228–233.

McKay, M. D., W. J. Conover, and R. J. Beckman. 1979. A comparison of three methods for selecting values of input variables in the analysis of output from a computer code. *Technometrics*, 21, 239–245.

McKelvey, K., B. R. Noon, and R. H. Lamberson. 1993. Conservation planning for species occupying fragmented landscapes: The case of the northern spotted owl. In P. M. Kareiva, J. G. Kingsolver, and R. B. Huey (eds.), *Biotic interactions and global change*, 424–450. Sunderland, MA: Sinauer.

McQuinn, I. H. 1997. Metapopulations and the Atlantic herring. *Reviews in Fish Biology and Fisheries*, 7, 297–329.

Metcalfe, N. B., N.H.C. Fraser, and M. D. Burns. 1999. Food availability and the nocturnal vs. diurnal foraging trade-off in juvenile salmon. *Journal of Animal Ecology*, 68, 371–381.

Minar, N., R. Burkhart, C. Langton, and M. Askenazi. 1996. *The Swarm simulation system: A toolkit for building multi-agent simulations*. Tech. rept., Santa Fe Institute, Santa Fe, NM.

Mitchell, M. 1998. *An introduction to genetic algorithms*. Cambridge, MA: MIT Press.

Mitchell, M., and C. E. Taylor. 1999. Evolutionary computation: An overview. *Annual Review of Ecology and Systematics*, 30, 593–616.

Mollison, D. 1986. Modelling biological invasions: Chance, explanation, prediction. *Philosophical Transactions of the Royal Society of London B*, 314, 675–693.

Mooij, W. M., and D. L. DeAngelis. 1999. Error propagation in spatially explicit population models: A reassessment. *Conservation Biology*, 13, 930–933.

———. 2003. Uncertainty in spatially explicit animal dispersal models. *Ecological Applications*, 13, 794–805.

Mullon, C., P. Cury, and P. Penven. 2002. Evolutionary individual-based model for the recruitment of anchovy (*Engraulis capensis*) in the southern Benguela. *Canadian Journal of Fisheries and Aquatic Science*, 59, 910–922.

Mullon, C., P. Fréon, C. Parada, C. van der Lingen, and J. Huggett. 2003. From particles to individuals: Modelling the early stages of anchovy (*Engraulis capensis/encrasicolus*) in the southern Benguela. *Fisheries and Oceanography*, 12, 396–406.

Murdoch, W. W., E. McCauley, R. M. Nisbet, W.S.C. Gurney, and A. M. de Roos. 1992. Individual-based models: Combining testability and generality. In D. L. DeAngelis and L. J. Gross (eds.), *Individual-based models and approaches in ecology*, 18–35. New York: Chapman and Hall.

Murray, J. D. 2002. *Mathematical biology. I. An introduction* (3rd edition). New York: Springer-Verlag.

Myers, J. H. 1976. Distribution and dispersal in populations capable of resource depletion—a simulation study. *Oecologia*, 23, 255–269.

Neuert, C. 1999. Die Dynamik räumlicher Strukturen in naturnahen Buchenwäldern Mitteleuropas. Ph.D. diss., University of Marburg, Germany.

Neuert, C., M. A. du Plessis, V. Grimm, and C. Wissel. 1995. Welche ökologischen Faktoren bestimmen die Gruppengröße bei *Phoeniculus purpureus* (Gemeiner Baumhopf) in Südafrika? Ein individuenbasiertes Modell. *Verhandlungen der Gesellschaft für Ökologie*, 24, 145–149.

Neuert, C., C. Rademacher, V. Grundmann, C. Wissel, and V. Grimm. 2001. Struktur und Dynamik von Buchenwäldern: Ergebnisse des regelbasierten Modells BEFORE. *Naturschutz und Landschaftsplanung*, 33, 173–183.

Newnham, R. M. 1964. The development of a stand model for Douglas fir. Ph.D. diss., University of British Columbia, Canada.

NeXT. 1993. *Object-oriented programming and the Objective-C language*. Redwood City, CA: NeXT Computer.

Nott, M. P. 1998. Effects of abiotic factors on population dynamics of the Cape Sable seaside sparrow and continental patterns of herpetological species richness: An appropriately scaled landscape approach. Ph.D. diss., University of Tennessee.

Nowak, M. A., S. Bonhoeffer, and R. M. May. 1994. Spatial games and the maintenance of cooperation. *Proceedings of the National Academy of Sciences*, 91, 4877–4881.

Nowak, M. A., and K. Sigmund. 1998. Evolution of indirect reciprocity by image scoring. *Nature*, 393, 573–577.

Odum, E. P. 1971. *Fundamentals of ecology* (3rd edition). Philadelphia: Saunders.

Pacala, S. W. 1986. Neighborhood models of plant population dynamics. IV. Single species and multispecies models of annuals with dormant seeds. *American Naturalist*, 128(6), 859–878.

———. 1987. Neighborhood models of plant population dynamics. III. Models with spatial heterogeneity in the physical environment. *Theoretical Population Biology*, 31, 359–392.

Pacala, S. W., C. D. Canham, and J. A. Silander Jr. 1993. Forest models defined by field measurements: I. The design of a northeastern forest simulator. *Canadian Journal of Forest Research*, 23, 1980–1988.

Pacala, S. W., and J. Silander. 1985. Neighborhood models of plant population dynamics. I. Single species models of annuals. *American Naturalist*, 125, 385–411.

Pachepsky, E., J. W. Crawford, J. L. Bown, and G. Squire. 2001. Towards a general theory of biodiversity. *Nature*, 410, 923–926.

Parada, C., C. D. van der Lingen, C. Mullon, and P. Penven. 2003. Modelling the effect of buoyancy on the transport of anchovy (*Engraulis capensis*) eggs from spawning to nursery grounds in the southern Benguela: An IBM approach. *Fisheries Oceanography*, 12, 170–184.

Parrish, J. K., S. V. Viscido, and D. Grünbaum. 2002. Self-organized fish schools: An examination of emergent properties. *Biological Bulletin*, 202, 296–305.

Peters, R. H. 1991. *A critique for ecology*. Cambridge: Cambridge University Press.

Pfister, C. A., and F. R. Stevens. 2003. Individual variation and environmental stochasticity: Implications for matrix model predictions. *Ecology*, 84, 496–510.

Picard, N., and A. Franc. 2001. Aggregation of an individual-based space dependent model of forest dynamics into distribution-based and space-independent models. *Ecological Modelling*, 145, 69–84.

Pielou, E. C. 1981. The usefulness of ecological models: A stock-taking. *Quarterly Review of Biology*, 56, 17–31.

Pitt, W. C., P. W. Box, and F. F. Knowlton. 2003. An individual-based model of canid populations: Modelling territoriality and social structure. *Ecological Modelling*, 166, 109–121.

Platt, J. R. 1964. Strong inference. *Science*, 146, 347–352.

Porté, A., and H. H. Bartelink. 2002. Modelling mixed forest growth: A review of models for forest management. *Ecological Modelling*, 150, 141–188.

Pretzsch, H., P. Biber, and J. Dursky. 2002. The single tree-based stand simulator SILVA: Construction, application and evaluation. *Forest Ecology and Management*, 162, 3–21.

Rademacher, C., C. Neuert, V. Grundmann, C. Wissel, and V. Grimm. 2001. Was charakterisiert Buchenurwälder? Untersuchungen der Altersstruktur des Kronendachs und der räumlichen Verteilung der Baumriesen in einem Modellwald mit Hilfe des Simulationsmodells BEFORE. *Forstwissenschaftliches Centralblatt*, 120, 288–302.

———. 2004. Reconstructing spatiotemporal dynamics of central European beech forests: The rule-based model BEFORE. *Forest Ecology and Management*, 194, 349–368.

Rademacher, C., and S. Winter. 2003. Totholz im Buchen-Urwald: Generische Vorhersagen des Simulationsmodelles BEFORE-CWD zur Menge, räumlichen Verteilung und Verfügbarkeit. *Forstwissenschaftliches Centralblatt*, 122, 337–357.

Railsback, S. F. 2001a. Concepts from complex adaptive systems as a framework for individual-based modelling. *Ecological Modelling*, 139, 47–62.

———. 2001b. Getting "results": The pattern-oriented approach to analyzing natural systems with individual-based models. *Natural Resource Modeling*, 14, 465–474.

Railsback, S. F., and B. C. Harvey. 2001. *Individual-based model formulation for cutthroat trout, Little Jones Creek, California*. Tech. rept., Pacific Southwest Research Station, Forest Service, U.S. Department of Agriculture, Albany, CA.

———. 2002. Analysis of habitat selection rules using an individual-based model. *Ecology*, 83, 1817–1830.

Railsback, S. F., B. C. Harvey, J. W. Hayes, and K. E. LaGroy. In press. Tests of theory for diel variation in salmonid feeding activity and habitat use. *Ecology*.

Railsback, S. F., B. C. Harvey, R. H. Lamberson, D. E. Lee, N. J. Claasen, and S. Yoshihara. 2002. Population-level analysis and validation of an individual-based cutthroat trout model. *Natural Resource Modeling*, 15, 83–110.

Railsback, S. F., and S. Jackson. 2000. Sacramento River chinook salmon individual-based model, user guide and software documentation. Unpublished report prepared by Lang, Railsback and Associates, Arcata, CA., for Jones and Stokes Associates, Sacramento, CA. December.

Railsback, S. F., R. H. Lamberson, B. C. Harvey, and W. E. Duffy. 1999. Movement rules for individual-based models of stream fish. *Ecological Modelling*, 123, 73–89.

Railsback, S. F., H. B. Stauffer, and B. C. Harvey. 2003. What can habitat preference models tell us? Tests using a virtual trout population. *Ecological Applications*, 13, 1580–1594.

Ratz, A. 1995. Long-term spatial patterns created by fire: A model oriented towards boreal forests. *International Journal of Wildland Fire*, 5, 25–34.

Reiss, M. J. 1989. *The allometry of growth and reproduction*. Cambridge: Cambridge University Press.

Remmert, H. 1991. The mosaic-cycle concept of ecosystems—an overview. In H. Remmert (ed.), *The mosaic-cycle concept of ecosystems*. Ecological Studies 85. New York: Springer.

Renshaw, E. 1991. *Modelling biological populations in space and time*. Cambridge: Cambridge University Press.

Reuter, H., and B. Breckling. 1994. Self-organisation of fish schools: An object-oriented model. *Ecological Modelling*, 75–76, 147–159.

———. 1999. Emerging properties on the individual level: Modelling the reproduction phase of the European robin *Erithacus rubecula*. *Ecological Modelling*, 121, 199–219.

Revilla, E., T. Weigand, F. Palomares, P. Ferreras, and M. Delibes. 2004. Effects of matrix heterogeneity on animal dispersal: From individual behavior to meta-population-level parameters. *American Naturalist* 164.

Reynolds, C. W. 1987. Flocks, herds, and schools: A distributed behavioral model. *Computer Graphics*, 21, 25–36.

Reynolds, J. H., and E. D. Ford. 1999. Multi-criteria assessment of ecological process models. *Ecology*, 80, 538–553.

Ripley, B. D. 1987. *Stochastic simulation*. New York: Wiley.

Ropella, G.E.P., S. F. Railsback, and S. K. Jackson. 2002. Software engineering considerations for individual-based models. *Natural Resource Modeling*, 15, 2–22.

Rose, K. A. 1989. Sensitivity analysis in ecological simulation models. In M. G. Singh (ed.), *Systems and control encyclopedia*, 4230–4234. New York: Pergamon Press.

Rose, K. A., S. W. Christensen, and D. L. DeAngelis. 1993. Individual-based modeling of populations with high mortality: A new method based on following a fixed number of model individuals. *Ecological Modelling*, 68, 273–292.

Rose, K. A., E. P. Smith, R. H. Gardner, A. L. Brenkert, and S. M. Bartell. 1991. Parameter sensitivities, Monte Carlo filtering, and model forecasting under uncertainty. *Journal of Forecasting*, 10, 117–133.

Roughgarden, J. 1998. *Primer of ecological theory*. Upper Saddle River, NJ: Prentice-Hall.

Roughgarden, J., A. Bergman, S. Shafir, and C. Taylor. 1996. Adaptive computation in ecology and evolution: A guide for future research. In R. K. Belew and M. Mitchell (eds.), *Adaptive individuals in evolving populations*. SFI Studies in the Sciences of Complexity, vol. 26. Reading, MA: Addison-Wesley.

Roughgarden, J., and Y. Iwasa. 1986. Dynamics of a metapopulation with space-limited subpopulations. *Theoretical Population Biology*, 29, 235–261.

Ruckelshaus, M., C. Hartway, and P. Kareiva. 1997. Assessing the data requirements of spatially explicit dispersal models. *Conservation Biology*, 11, 1298–1306.

———. 1999. Dispersal and landscape errors in spatially explicit population models: A reply. *Conservation Biology*, 13, 1223–1224.

Ruxton, G. D. 1996. Effects of the spatial and temporal ordering of events on the behaviour of a simple cellular automaton. *Ecological Modelling*, 84, 311–314.

Ruxton, G. D., and L. A. Saravia. 1998. The need for biological realism in the updating of cellular automata models. *Ecological Modelling*, 107, 105–112.

Rykiel, E. J., Jr. 1996. Testing ecological models: The meaning of validation. *Ecological Modelling*, 90, 229–244.

Saltelli, A., F. Tarantola, F. Campolongo, and M. Ratto. 2004. *Sensitivity analysis in practice: A guide to assessing scientific models*. New York: John Wiley and Sons.

Sato, K., and Y. Iwasa. 2000. Pair approximations for lattice-based ecological models. In U. Dieckmann, R. Law, and J.A.J. Metz (eds.), *The geometry of ecological interactions: Simplifying spatial complexity*, 341–358. Cambridge: Cambridge University Press.

Savage, M., B. Sawhill, and M. Askenazi. 2000. Community dynamics: What happens when we rerun the tape? *Journal of Theoretical Biology*, 205, 515–526.

Schadt, S. 2002. Scenarios assessing the viability of lynx populations in Germany. Ph.D. diss., Technical University of Munich, Germany.

Schadt, S., F. Knauer, P. Kaczensky, E. Revilla, T. Wiegand, and L. Trepl. 2002. Rule-based assessment of suitable habitat and patch connectivity for the Eurasian lynx. *Ecological Applications*, 12, 1469–1483.

Schadt, S., E. Revilla, T. Wiegand, F. Knauer, P. Kaczensky, U. Breitenmoser, L. Bufka, J. Cerveny, P. Koubek, T. Huber, C. Stanisa, and L. Trepl. 2002. Assessing the suitability of central European landscapes for the reintroduction of Eurasian lynx. *Journal of Applied Ecology*, 39, 189–203.

Scheffer, M., J. M. Baveco, D. L. DeAngelis, K. A. Rose, and E. H. van Nes. 1995. Super-individuals: A simple solution for modelling large populations on an individual basis. *Ecological Modelling*, 80, 161–170.

Schiegg, K., J. R. Walters, and J. A. Priddy. 2002. The consequences of disrupted dispersal in fragmented red-cockaded woodpecker *Picoides borealis* populations. *Journal of Animal Ecology*, 71, 710–721.

Schmitt, J., A. C. McCormac, and H. M. Smith. 1995. A test of the adaptive plasticity hypothesis using transgenic and mutant plants disabled in phytochrome-mediated elongation responses to neighbors. *American Naturalist*, 146, 937–953.

Schmitz, O. J. 2000. Combining field experiments and individual-based modeling to identify the dynamically relevant organizational scale in a field system. *Oikos*, 89, 471–484.

———. 2001. From interesting details to dynamical relevance: Toward more effective use of empirical insights in theory construction. *Oikos*, 94, 39–50.

Schönfisch, B., and A. de Roos. 1999. Synchronous and asynchronous updating in cellular automata. *BioSystems*, 51, 123–143.

Schultz, J. C., and H. M. Appel. 2004. Cross-kingdom cross-talk: Hormones shared by plants and their insect herbivores. *Ecology*, 85, 70–77.

Seibt, U., and W. Wickler. 1988. Bionomics and social structure of "Family Spiders" of the genus *Stegodyphus*, with special reference to the African speces *S. dumicola* and *S. mimosarum* (Araneida, Eresidae). *Verhandlungen des Naturwissenschaftlichen Vereins Hamburg*, 30, 255–303.

Sellis, T. K., N. Roussopoulos, and C. Faloutsosj. 1987. The R+-tree: A dynamic index for multidimensional objects. In P. M. Stocker, W. Kent, and P. Hammersley (eds.), *VLDB'87, Proceedings of 13th International Conference on Very Large Data Bases, September 1–4, 1987, Brighton, England*, 507–518. Palo Alto, CA: Morgan Kaufmann.

Shin, Y.-J., and P. Cury. 2001. Exploring fish community dynamics through size-dependent trophic interactions using a spatialized individual-based model. *Aquatic Living Resources*, 14, 65–80.

Shugart, H. H. 1984. *A theory of forest dynamics: The ecological implications of forest succession models*. New York: Springer-Verlag.

Shugart, H. H., T. M. Smith, and W. M. Post. 1992. The potential for application of individual-based simulation models for assessing the effects of global change. *Annual Review of Ecology and Systematics*, 23, 15–38.

Silbernagel, J. 1997. Scale perception—from cartography to ecology. *Bulletin of the Ecological Society of America*, 78, 166–169.

Silvertown, J. 1991. Modularity, reproductive threshold, and plant population dynamics. *Functional Ecology*, 5, 577–582.

Silvertown, J. W., S. Holtier, J. Johnson, and P. Dale. 1992. Cellular automaton models of interspecific competition for space—the effect of pattern on process. *Journal of Ecology*, 80, 527–534.

Silvertown, J. W. 1992. *Introduction to plant population ecology*. Essex, England: Longman Scientific and Technical.

Simberloff, D. 1981. The sick science of ecology: Symptoms, diagnosis, and prescription. *Eidema*, 1, 49–54.

———. 1983. Competition theory, hypothesis-testing, and other community ecological buzzwords. *American Naturalist*, 122, 626–635.

Smith, E. P., and K. A. Rose. 1995. Model goodness-of-fit analysis using regression and related techniques. *Ecological Modelling*, 77, 49–64.

Smith, T. M., and M. Huston. 1989. A theory of the spatial and temporal dynamics of plant communities. *Vegetatio*, 83, 49–69.

Smith, T. M., and D. L. Urban. 1988. Scale and resolution of forest structural pattern. *Vegetatio*, 74, 143–150.

Sommer, U. 1996. Can ecosystem properties be optimized by natural selection? *Senckenbergiana Maritima*, 27, 145–150.

Soulé, M. E. 1986. *Conservation biology: The science of scarcity and diversity*. Sunderland, MA: Sinauer.

Spencer, R.-J. 2002. Experimentally testing nest site selection: Fitness trade-offs and predation in turtles. *Ecology*, 83, 2136–2144.

Stacey, P. B., and W. D. Koenig (eds.). 1990. *Cooperative breeding in birds: Long-term studies of ecology and behavior*. Cambridge: Cambridge University Press.

Starfield, A. M. 1997. A pragmatic approach to modeling for wildlife management. *Journal of Wildlife Management*, 61, 261–270.

Starfield, A. M., and A. L. Bleloch. 1986. *Building models for conservation and wildlife management*. London: Collier Macmillan.

Starfield, A. M., K. A. Smith, and A. L. Bleloch. 1990. *How to model it: Problem solving for the computer age*. New York: McGraw-Hill.

Stelter, C., M. Reich, V. Grimm, and C. Wissel. 1997. Modelling persistence in dynamic landscapes: Lesson from a metapopulation of the grasshopper *Bryodema tuberculata*. *Journal of Animal Ecology*, 66, 508–518.

Stephan, T., and C. Wissel. 1999. The extinction risk of a population exploiting a resource. *Ecological Modelling*, 115, 217–226.

Stephens, P. A., F. Frey-Roos, W. Arnold, and W. J. Sutherland. 2002a. Model complexity and population predictions. The alpine marmot as a case study. *Journal of Animal Ecology*, 71, 343–361.

———. 2002b. Sustainable exploitation of social species: A test and comparison of models. *Journal of Applied Ecology*, 39, 629–642.

Stillman, R. A., A. E. Poole, J. D. Goss-Custard, R.W.G. Caldow, M. G. Yates, and P. Triplet. 2002. Predicting the strength of interference more quickly using behaviour-based models. *Journal of Animal Ecology*, 71, 532–541.

Stillman, R. A., A. D. West, J. D. Goss-Custard, S.E.A. le V. dit Durell, M. G. Yates, P. W. Atkinson, N. A. Clark, M. C. Bell, P. J. Dare, and M. Mander. 2003. A behaviour-based model can predict shorebird mortality rate using routinely collected shellfishery data. *Journal of Applied Ecology*, 40, 1090–1101.

Stoll, P., and J. Weiner. 2000. A neighborhood view of interactions among individual plants. In U. Dieckmann, R. Law, and J.A.J. Metz (eds.), *The geometry of ecological interactions: Simplifying spatial complexity*, 11–27. Cambridge: Cambridge University Press.

Stoll, P., J. Weiner, H. Muller-Landau, E. Müller, and T. Hara. 2002. Size symmetry of competition alters biomass-density relationships. *Proceedings of the Royal Society London B*, 269, 2191–2195.

Storch, I. 2002. On spatial resolution in habitat models: Can small-scale forest structure explain capercaillie numbers? *Conservation Ecology*, 6 (available at www.ecologyandsociety.org).

Strand, E. 2003. Adaptive models of vertical migration in fish. Ph.D. diss., University of Bergen, Norway.

Strand, E., G. Huse, and J. Giske. 2002. Artificial evolution of life history and behavior. *American Naturalist*, 159, 624–644.

Suter, G. W. 1981. Ecosystem theory and NEPA assessment. *Bulletin of the Ecological Society of America*, 62, 186–192.

———. 1996. Abuse of hypothesis testing statistics in ecological risk assessment. *Human and Ecological Risk Assessment*, 2, 331–347.

Sutherland, W. J. 1996. *From individual behaviour to population ecology*. New York: Oxford University Press.

Sutton, T. M., K. A. Rose, and J. J. Ney. 2000. A model analysis of strategies for enhancing stocking success of landlocked striped bass populations. *North American Journal of Fisheries Management*, 20, 841–859.

Symonides, E., J. Silvertown, and V. Andreasen. 1986. Population cycles caused by overcompensating density-dependence in an annual plant. *Oecologia*, 71, 156–158.

Tesfatsion, L. 2002. Agent-based computational economics: Growing economies from the bottom up. *Artificial Life*, 8, 55–82.

Tews, J. 2004. The impact of climate change and land use on woody plants in semiarid savanna: Modeling shrub population dynamics in the Southern Kalahari. Ph.D. diss., University of Potsdam, Germany.

Tews, J., U. Brose, V. Grimm, K. Tielbörger, M. Wichmann, M. Schwager, and F. Jeltsch. 2004. Animal species diversity driven by habitat heterogeneity/diversity: The importance of keystone structures. *Journal of Biogeography*, 31, 79–92.

Thierry, B. 1985. Social development in three species of macaque (*Macaca mulatta*, *M. fascicularis*, *M. tonkeana*): A preliminary report on the first ten weeks of life. *Behavioural Processes*, 11, 89–95.

———. 1990. Feedback loop between kinship and dominance: The macaque model. *Journal of Theoretical Biology*, 145, 511–521.

Thompson, W. A., I. Vertinsky, and J. R. Krebs. 1974. The survival value of flocking in birds: A simulation model. *Journal of Animal Ecology*, 43, 785–820.

Thorpe, J. E., M. Mangel, N. B. Metcalfe, and F. A. Huntingford. 1998. Modelling the proximate basis of salmonid life-history variation, with application to Atlantic salmon, *Salmo salar* L. *Evolutionary Ecology*, 12, 581–599.

Thrall, P. H., S. W. Pacala, and J. A. Silander. 1989. Oscillatory dynamics in populations of an annual weed species *Abutilon theophrasti*. *Ecology*, 77, 1135–1149.

Thulke, H., V. Grimm, M. S. Müller, C. Staubach, L. Tischendorf, C. Wissel, and F. Jeltsch. 1999. From pattern to practice: A scaling-down strategy for spatially explicit modelling illustrated by the spread and control of rabies. *Ecological Modelling*, 117, 179–202.

Tikhonov, D. A., J. Enderlein, H. Malchow, and A. Medvinsky. 2001. Chaos and fractals in fish school motion. *Chaos, Solitons and Fractals*, 12, 277–288.

Tilman, D., and D. Wedin. 1991. Oscillations and chaos in the dynamics of a perennial grass. *Nature*, 353, 653–655.

Topping, C. J., T. S. Hansen, T. S. Jensen, J. U. Jepsen, F. Nikolajsen, and P. Odderskær. 2003. ALMaSS, an agent-based model for animals in temperate European landscapes. *Ecological Modelling*, 167, 65–82.

Topping, C. J., and J. U. Jepsen. 2002. Simulation models of animal behavior are useful tools in landscape and species management. *IALE Bulletin*, 20, 1–2.

Topping, C. J., S. Ostergaard, C. Pertoldi, and L. A. Bach. 2003. Modelling the loss of genetic diversity in vole populations in a spatially and temporally varying environment. *Annales Zoologici Fennici*, 40, 255–267.

Trani, M. K. 2002. The influence of spatial scale on landscape pattern description and wildlife habitat assessment. In J. M. Scott and P. Heglund (eds.), *Predicting species occurrences: Issues of accuracy and scale*, 141–156. Washington, DC: Island Press.

Turchin, P. 1998. *Quantitative analysis of movement: Measuring and modeling population redistribution in animals and plants*. Sunderland, MA: Sinauer Associates.

———. 2003. *Complex population dynamics: A theoretical/empirical synthesis*. Princeton, NJ: Princeton University Press.

Turchin, P., L. Oksanen, P. Ekerholm, T. Oksanen, and H. Henttonen. 2000. Are lemmings prey or predators? *Nature*, 405, 562–565.

Turner, M. G., G. J. Arthaud, R. T. Engstrom, S. Hejl, J. Liu, S. Loeb, and K. McKelvey. 1995. Usefulness of spatially explicit population models in land management. *Ecological Applications*, 5, 12–16.

Turner, M. G., Y. Wu, L. L. Wallace, W. H. Romme, and A. Brenkert. 1994. Simulating winter interactions among ungulates, vegetation, and fire in northern Yellowstone Park. *Ecological Applications*, 4, 472–496.

Tyler, J. A., and K. A. Rose. 1994. Individual variability and spatial heterogeneity in fish population models. *Reviews in Fish Biology and Fisheries*, 4, 91–123.

Tyre, A. J., H. P. Possingham, and D. B. Lindenmayer. 2001. Matching observed pattern with ecological process: Can territory occupancy provide information about life history parameters? *Ecological Applications*, 11, 1722–1737.

Uchmański, J. 1985. Differentiation and frequency distributions of body weights in plants and animals. *Philosophical Transactions of the Royal Society of London B.*, 310, 1–75.

———. 1999. What promotes persistence of a single population: An individual-based model. *Ecological Modelling*, 115, 227–242.

———. 2000a. Individual variability and population regulation: An individual-based model. *Oikos*, 90, 539–548.

———. 2000b. Resource partitioning among competing individuals and population persistence: An individual-based model. *Ecological Modelling*, 131, 21–32.

———. 2003. Ecology of individuals. In R. S. Ambasht and N. K. Ambasht (eds.), *Modern trends in applied terrestrial ecology*, 275–302. New York: Plenum Press.

Uchmański, J., and V. Grimm. 1996. Individual-based modelling in ecology: What makes the difference? *Trends in Ecology and Evolution*, 11, 437–441.

———. 1997. Individual-based modelling: What is the difference? Reply. *Trends in Ecology and Evolution*, 12, 112.

Ulbrich, K., and J. R. Henschel. 1999. Intraspecific competition in a social spider. *Ecological Modelling*, 115, 243–252.

Ulbrich, K., J. R. Henschel, F. Jeltsch, and C. Wissel. 1996. Modelling individual variability in a social spider colony (*Stegodyphus dumicola*: Eresidae) in relation to food abundance and its allocation. *Revue Suisse de Zoologie*, suppl. vol. (vol. hors série), 661–670.

Umeki, K. 1997. Effect of crown asymmetry on size-structure dynamics of plant populations. *Annals of Botany*, 79, 631–641.

van Nes, E. H. 2002. Controlling complexity in individual-based models of aquatic vegetation and fish communities. Ph.D. diss., University of Wageningen, The Netherlands.

van Nes, E. H., E.H.R.R. Lammens, and M. Scheffer. 2002. PISCATOR, an individual-based model to analyze the dynamics of lake fish communities. *Ecological Modelling*, 152, 261–278.

Van Winkle, W., K. A. Rose, and R. C. Chambers. 1993. Individual-based approach to fish population dynamics: An overview. *Transactions of the American Fisheries Society*, 122, 397–403.

Van Winkle, W., H. I. Jager, S. F. Railsback, B. D. Holcomb, T. K. Studley, and J. E. Baldrige. 1998. Individual-based model of sympatric populations of brown and rainbow trout for instream flow assessment: Model description and calibration. *Ecological Modelling*, 110, 175–207.

Vehrencamp, S. L. 1983. A model for the evolution of despotic versus egalitarian societies. *Animal Behaviour*, 31, 667–682.

Verboom, J., K. Lankester, and J.A.J. Metz. 1991. Linking local and regional dynamics in stochastic metapopulation models. *Biological Journal of the Linnean Society*, 42, 39–55.

von Neumann, J., and A. W. Burks. 1966. *Theory of self-reproducing automata*. Urbana: University of Illinois Press.

Vose, D. 2000. *Risk analysis: A quantitative guide*. Chichester: John Wiley and Sons.

Vucetich, J. A., and S. Creel. 1999. Ecological interactions, social organization, and extinction risk in African wild dogs. *Conservation Biology*, 13, 1172–1182.

Waldrop, M. M. 1992. *Complexity: The emerging science at the edge of order and chaos*. New York: Simon and Schuster.

Walker, J., P.J.H. Sharpe, L. K. Penridge, and H. Wu. 1989. Ecological field theory: The concept and field tests. *Vegetatio*, 83, 81–95.

Waller, L. A., D. Smith, J. E. Childs, and L. A. Real. 2003. Monte Carlo assessments of goodness-of-fit for ecological simulation models. *Ecological Modelling*, 164, 49–63.

Walling, L. L. 2000. The myriad plant responses to herbivores. *Journal of Plant Growth and Regulation*, 19, 195–216.

Walter, G. H., and R. Hengeveld. 2000. The structure of the two ecological paradigms. *Acta Biotheoretica*, 48, 15–46.

Watson, J. 1968. *The double helix: A personal account of the discovery of the structure of DNA*. New York: Atheneum.

Watt, A. S. 1947. Pattern and process in the plant community. *Journal of Ecology*, 35, 1–22.

Weiner, J. 1982. A neighborhood model of annual-plant interference. *Ecology*, 63, 1237–1241.

———. 1990. Asymmetric competition in plant populations. *Trends in Ecology and Evolution*, 5, 360–364.

———. 1995. On the practice of ecology. *Journal of Ecology*, 83, 153–158.

Weiner, J., P. Stoll, H. Muller-Landau, and A. Jasentuliyana. 2001. The effect of density, spatial pattern, and competitive symmetry on size variation in simulated plant populations. *American Naturalist*, 158, 438–450.

Weisfeld, M., and B. McCarty. 2000. *The object-oriented thought process*. Indianapolis, IN: SAMS Press.

Werner, E. E., and S. D. Peacor. 2003. A review of trait-mediated indirect interactions in ecological communities. *Ecology*, 84, 1083–1100.

Werner, F. E., J. A. Quinlan, R. G. Lough, and D. R. Lynch. 2001. Spatially explicit individual-based modeling of marine populations: A review of advances in the 1990s. *Sarsia*, 86, 411–421.

West, A. D., J. D. Goss-Custard, R. A. Stillman, R.W.G. Caldow, S.E.A. le V. dit Durell, and S. McGrorty. 2002. Predicting the impacts of disturbance on wintering waders using a behaviour based individuals model. *Biological Conservation*, 106, 319–328.

Westoby, M. 1984. The self-thinning rule. *Advances in Ecological Research*, 14, 167–225.

Wiegand, T., F. Jeltsch, I. Hanski, and V. Grimm. 2003. Using pattern-oriented modeling for revealing hidden information: A key for reconciling ecological theory and application. *Oikos*, 100, 209–222.

Wiegand, T., F. Knauer, P. Kaczensky, and J. Naves. 2004b. Expansion of the brown bear (*Ursus arctos*) into the eastern Alps: A spatially explicit population model. *Biodiversity and Conservation*, 13, 79–114.

Wiegand, T., S. J. Milton, and C. Wissel. 1995. A simulation model for a shrub ecosystem in the semiarid Karoo, South Africa. *Ecology*, 76, 2205–2221.

Wiegand, T., K. A. Moloney, J. Naves, and F. Knauer. 1999. Finding the missing link between landscape structure and population dynamics: A spatially explicit perspective. *American Naturalist*, 154, 605–627.

Wiegand, T., J. Naves, T. Stephan, and A. Fernandez. 1998. Assessing the risk of extinction for the brown bear (*Ursus arctos*) in the Cordillera Cantabrica, Spain. *Ecological Monographs*, 68, 539–570.

Wiegand, T., E. Revilla, and F. Knauer. 2004a. Dealing with uncertainty in spatially explicit population models. *Biodiversity and Conservation*, 13, 53–78.

Wiegand, T., H. A. Snyman, and K. Kellner. 2004. Inverse pattern-oriented modeling for calibration and optimization of simulation models: The example of a grassland model. Environmental Research Centre Leipzig-Halle, Leipzig. Unpublished manuscript.

Wilson, W. G. 1998. Resolving discrepancies between deterministic population models and individual-based simulations. *American Naturalist*, 151, 116–134.

———. 2000. *Simulating ecological and evolutionary systems in C*. Cambridge: Cambridge University Press.

Winkler, E., and J. Stöcklin. 2002. Sexual and vegetative reproduction of *Hierarcium pilosella* L. under competition and disturbance: A grid-based simulation model. *Annals of Botany*, 89, 525–536.

Wissel, C. 1989. *Theoretische Ökologie—Eine Einführung*. New York: Springer.

———. 1992a. Aims and limits of ecological modelling exemplified by island theory. *Ecological Modelling*, 63, 1–12.

———. 1992b. Modelling the mosaic-cycle of a Middle European beech forest. *Ecological Modelling*, 63, 29–43.

———. 2000. Grid-based models as tools for ecological research. In U. Dieckmann, R. Law, and J.A.J. Metz (eds.), *The geometry of ecological interactions: Simplifying spatial complexity*, 94–115. Cambridge: Cambridge University Press.

Wissel, C., T. Stephan, and S.-H. Zaschke. 1994. Modelling extinction and survival of small populations. In H. Remmert (ed.), *Minimum animal populations*, 67–103. Ecological Studies 106. Berlin: Springer.

With, K. A. 1997. The application of neutral landscape models in conservation biology. *Conservation Biology*, 11, 1069–1080.

Wolff, W. F. 1994. An individual-oriented model of a wading bird nesting colony. *Ecological Modelling*, 72, 75–114.

Wölfl, M., L. Bufka, J. Cerveny, P. Koubek, M. Heurich, H. Habel, T. Huber, and W. Poost. 2001. Distribution and status of lynx in the border region between Czech Republic, Germany and Austria. *Acta Theriologica*, 46, 181–194.

Wolfram, S. 2002. *A new kind of science*. Champaign, IL: Wolfram Media.

Wood, S. N. 1994. Obtaining birth and death rate patterns from structured population trajectories. *Ecological Monographs*, 64, 23–44.

Wu, H., P.J.H. Sharpe, J. Walker, and L. K. Penridge. 1985. Ecological field theory: A spatial analysis of resource interference among plants. *Ecological Modelling*, 29, 215–243.

Wyszomirski, T. 1983. A simulation model of the growth of competing individuals of a plant population. *Ekologia Polska*, 31, 73–92.

———. 1986. Growth, competition and skewness in a population of one-dimensional individuals. *Ekologia Polska*, 34, 615–641.

Wyszomirski, T., I. Wyszomirska, and I. Jarzyna. 1999. Simple mechanisms of size distribution dynamics in crowded and uncrowded virtual monocultures. *Ecological Modelling*, 115, 253–273.

Yoda, K., T. Kira, H. Ogawa, and K. Hozumi. 1963. Self-thinning in overcrowded pure stands under cultivated and natural conditions. *Journal of Biology (Osaka City University)*, 14, 107–129.

Yodzis, P. 1989. *Theoretical ecology*. New York: Harper and Row.

Zeeman, E. C. 1977. *Catastrophe theory: Selected papers, 1972–1977*. Boston: Addison-Wesley.

Zeide, B. 1987. Analysis of the 3/2 power law of self-thinning. *Forest Science*, 33, 517–537.

———. 1989. Accuracy of equations describing diameter growth. *Canadian Journal of Forest Research*, 19, 1283–1286.

———. 1991. Quality as a characteristic of ecological models. *Ecological Modelling*, 55, 161–174.

———. 2001. Natural thinning and environmental change: An ecological process model. *Forest Ecology and Management*, 154, 165–177.

Zeigler, B. P. 1976. *Theory of modelling and simulation*. Malabar, FL: Krieger.

Zeigler, B. P., H. Praehofer, and T. G. Kim. 2000. *Theory of modeling and simulation: Integrating discrete event and continuous complex dynamic systems* (2nd edition). Boston: Academic Press.

Zhivotovsky, L. A., A. Bergman, and M. W. Feldman. 1996. A model of individual adaptive behavior in a fluctuating environment. In R. K. Belew and M. Mitchell (eds.), *Adaptive Individuals in Evolving Populations*, 131–153. Santa Fe Institute Studies in the Sciences of Complexity, vol. 26. Reading, MA: Addison-Wesley.

Index

actions: concurrent, 113; continuous time vs. discrete time, 111–12; definition of, 109, 391; designing, 112–13; examples of, 110; hierarchy of, 110; in object-oriented software, 263; observer, 115. *See also* scheduling

adaptive behavior, 391

adaptive traits, 79–84; advantages of for modeling behavior, 81–82; definition of, 74, 79–81, 391; direct vs. indirect, 82–83, 391; effects at community and ecosystem level, 219, 233; in example models: 141–42, 159, 177, 211, 215–16; formulation of, 255–60; in plants, 199, 211, 215, 217; to produce emergence, 83

agent-based models, 18, 19, 72, 114, 126, 271–72, 350; definition of, 391

allometric theory of self-thinning, 207

analytical approximation of individual-based models, 369–72

analytical models, 365–79; benefits of, 368–69; critics of, 369; monographs and textbooks on, 366; to understand and analyze individual-based models, 372–75. *See also* classical models; classification of models

analyzing individual-based models, 34–35, 62–63, 312–48; vs. analyzing real systems, 312; currencies for, 319–23; from the bottom up, 314, 317–18; via goodness-of-fit to census data, 321; by simplification, 324, 373–75; strategies and techniques for, 315–27; using system-level and analytical concepts, 373–78; tools for in software platforms, 281; by using unrealistic scenarios, 325–27. *See also* independent predictions; parameterization; statistics; sensitivity analysis; robustness of model results

applied ecology and theory in individual-based ecology, 58, 172, 386

artificial economies, 234

artificial ecosystems, 234

artificial evolution, 93, 234–42; to formulate traits, 258–59

artificial neural networks. *See* artificial evolution

assumptions: ad hoc, 17, 158; hidden, 384

automation of software testing, 292–93, 298

Axelrod's model of social interaction, 388

Bayesian statistics, 255, 342

bear model of Wiegand et al., 332, 334

beech forest model (BEFORE) of Neuert et al., 5–8, 226–34, 250, 253, 346, 373

behavior: emergent, 391; of individuals and systems, 391. *See also* group and social behavior

behavioral ecology, 139, 170; integration with population ecology, 158

biodiversity, modeling links to traits of individuals, 233

Boids model of Reynolds, 83, 125–28, 139–45, 388

Boolean algebra in modeling adaptive traits, 254

bottom-up models, 49, 367

buffer mechanisms in demographics of social animals, 150–53, 156–57

calibration, 47, 222, 315. *See also* parameterization

canid model of Pitt et al., 154–63, 263

CAS. *See* complex adaptive systems

categories of models. *See* classification of models

causality, 279: in object-oriented programming, 276; and scheduling, 115